BREAKING ICE

BREAKING ICE:

Renewable Resource and Ocean Management in the Canadian North

Edited by Fikret Berkes
(University of Manitoba),
Rob Huebert
(University of Calgary),
Helen Fast
(Department of Fisheries and Oceans),
Micheline Manseau
(Parks Canada and University of Manitoba),
and Alan Diduck
(University of Winnipeg)

UNIVERSITY OF
CALGARY
PRESS

Published by the
University of Calgary Press
2500 University Drive NW
Calgary, Alberta, Canada T2N 1N4
www.uofcpress.com

We acknowledge the financial support of the Government of Canada through the Book Publishing Industry Development Program (BPIDP), the Alberta Foundation for the Arts and the Alberta Lottery Fund – Community Initiatives Program for our publishing activities. We acknowledge the support of the Canada Council for the Arts for our publishing program.

Canada Council   Conseil des Arts
for the Arts   du Canada

Canada

ALBERTA
LOTTERY FUND

Cover design, Mieka West.
Internal design & typesetting,
    Jason Dewinetz.

∞ This book is printed on Eco book 100; which is a 100% post-consumer recycled, ancient-forest-friendly paper. Printed and bound in Canada by HOUGHTON BOSTON.

LIBRARY AND ARCHIVES OF CANADA CATALOGUING IN PUBLICATION:

Breaking ice : renewable resource and ocean management in the Canadian north / edited by Fikret Berkes … [et al.].

(Northern lights series ; 7)
Accompanied by a DVD.

 Co-published by Arctic Institute of North America.

Includes bibliographical references and index.

ISBN   1-55238-159-5

1 Renewable natural resources–Canada, Northern–Management.
2 Marine resources–Canada, Northern–Management.
3 Sustainable development–Canada, Northern.
4 Canada, Northern–Environmental conditions.
5 Canada, Northern–Social conditions.
6 Canada, Northern–Economic conditions.
I Berkes, Fikret
II Arctic Institute of North America.
III Series.

GC1023.15B74 2005   333.7'09719
C2005-900601-3

CONTENTS

LIST OF FIGURES

LIST OF TABLES

LIST OF BOXES

LIST OF ACRONYMS

ACIA	Arctic Climate Impact Assessment
AEPS	Arctic Environmental Protection Strategy
AFL	Arctic Foods Ltd.
AMAP	Arctic Monitoring and Assessment Programme
AMCS	Arctic Marine Conservation Strategy
ARI	Arctic Research Institute
BSBMP	Beaufort Sea Beluga Management Plan
BSIMPI	Beaufort Sea Integrated Management Planning Initiative
CACAR	Canadian Arctic Contaminants Assessment Report
CAIPAP	Canadian Arctic Indigenous People's Against POPS
CAPP	Canadian Association of Petroleum Producers
CAS	Complex adaptive systems
CATS	Complex-adaptive-tourism-system
CBM	Community-based management
CFFS	Conservation of Flora and Fauna Secretariat
CFCS	Chloroflurocarbons
CFIA	Canadian Food Inspection Agency
CHCDC	Coral Harbour Community Development Corporation
CINE	Centre for Indigenous People's Nutrition and Environment
CITES	Convention on International Trade in Endangered Species
CLCS	Commission on the Limits of the Continental Shelf
COS	Canada's Oceans Strategy
CWMA	Churchill Wildlife Management Area
CWS	Canadian Wildlife Service
CYFN	Council of Yukon First Nations
DFO	Department of Fisheries and Oceans
DIAND	Department of Indian and Northern Affairs
DOE	Department of Environment
DSD	Department of Sustainable Development, Nunavut
EEZ	Exclusive Economic Zone
EMAN-North	Northern Ecological Monitoring and Assessment Network
EU	European Union
FAO	Food and Agriculture Organization (United Nations)
FedNor	Federal Economic Development Agency in Northern Ontario
FGRS	Fisheries (General) Regulations
FJMC	Fisheries Joint Management Committee
FN	First Nations
FWS	Fish and Wildlife Service (US)

GN	Government of Nunavut
GNT	Government of Northwest Territories
GNWT	Government of Northwest Territories
GPA	Global Programme of Action
GPS	Global Positioning System
HTC	Hunters and Trappers Committee
HTO	Hunters and Trappers Organization
ICC	Inuit Circumpolar Conference
IFA	Inuvialuit Final Agreement
IFAW	International Federation for Animal Welfare
IFCS	Intergovernmental Forum on Chemical Safety
IFMP	Integrated Fisheries Management Plans
IGC	Inuvialuit Game Council
IIBA	Inuit Impacts Benefit Agreement
IISD	International Institute for Sustainable Development
IM	Integrated Management
INAC	Indian and Northern Affairs Canada
INC	Intergovernmental Negotiating Committee
IPEN	International POPs Elimination Network
IQ	Inuit Qaujimajatuqangit
IRC	Inuvialuit Regional Corporation
IRRCMP	Inuvialuit Renewable Resource Conservation Management Plan
ISR	Inuvialuit Settlement Region
ITC	Inuit Tapirisat of Canada
ITK	Inuit Tapirisat Katami
IWCO	Independent World Commission on the Ocean
JPMC	Joint Park Management Committee
KAF	Kivalliq Arctic Foods
KWB	Kivalliq Wildlife Board
LOMA	Large Ocean Management Area
LRTAP	Long Range Transboundary Air Pollution
MEA	Millennium Ecosystem Assessment
MEQ	Marine Environmental Quality
MMPA	Marine Mammal Protection Act (US)
MMR	Marine Mammal Regulations
MPA	Marine Protected Area
NAFO	North Atlantic Fisheries Organization
NCP	Northern Contaminants Program
NFA	Nunavut Final Agreement
NGO	Non-government organization
NLCA	Nunavut Land claims agreement

NPA	National Programme of Action
NPC	Nunavut Planning Commission
NRI	Nunavut Research Institute
NRTEE	National Round Table on the Environment and the Economy
NTCL	Northern Transportation Company Limited
NTI	Nunavut Tunngavik Incorporated
NTRE	Non-tourism related externalities
NWMB	Nunavut Wildlife Management Board
NWTFRS	Northwest Territories Fisheries Regulations
OECD	Organization for Economic Co-operation and Development
OMRN	Oceans Management Research Network
PAME	Protection of the Arctic Marine Environment
PBAC	Polar Bear Administrative Committee
PBSC	Polar Bear Specialist Group (IUCN)
PBTC	Polar Bear Technical Committee
PC	Parks Canada
PCBS	Polychlorinated biphenyls (a class of chemical pollutants)
POPS	Persistent organic pollutants
RCAP	Royal Commission on Aboriginal Peoples
RCC	Regional Contaminants Coordinator
RCMP	Royal Canadian Mounted Police
RWED	Resources, Wildlife, and Economic Development (Northwest Territories)
RWO	Regional Wildlife Organization
SARS	Severe acute respiratory syndrome
SMC	Senior Management Committee
TAH	Total Allowable Harvest
TDI	Tolerable daily intake
TEK	Traditional Ecological Knowledge
TEKMS	Traditional Ecological Knowledge and Management System
UNECE	United Nations Economic Commission for Europe
UNEP	United Nations Environment Programme
WG	Working Group
WHO	World Health Organization
WMAC (NWT)	Wildlife Management Advisory Council (Northwest Territories)
WNP	Wapusk National Park
WSWG	West Side Working Group

PREFACE & ACKNOWLEDGMENTS

A large number of people and groups are responsible for the production of this book and DVD. The project, "Integrated management, complexity and diversity of use: responding and adapting to change," which resulted in this volume, was supported by a Strategic Grant from the Social Sciences and Humanities Research Council of Canada (SSHRC) and the Department of Fisheries and Oceans (DFO). The project was carried out by the Integrated Management Node, and benefited from the cooperation of the Ocean Management Research Network (OMRN), the National Secretariat of the OMRN (headed by Tony Charles at Saint Mary's University) and the two other nodes of the OMRN, Linking Science and Local Knowledge node (headed by Alison Gill at Simon Fraser University) and the Sustainability Node (headed by Peter Sinclair at Memorial University). We thank them all for ongoing support and stimulation.

The original project had co-applicants and collaborators from nine universities, and project partners from three government agencies (DFO Central and Arctic Region; Environment Canada; and Parks Canada) and six other agencies: Aurora College (Inuvik), Canadian Arctic Resources Committee (Ottawa and Yellowknife), Fisheries Joint Management Committee (Inuvik), Kivalliq Inuit Association (Rankin Inlet), Nunavut Arctic College (Iqaluit), and the Tuktu and Nogak Project (Iqaluktuuttiaq). Many of these universities and agencies are represented among the contributors to this book. Some of the other key people who played leadership roles in the Node and/or contributed to the development of ideas in this book include Ranjana Bird Prasad, David Fennell, Joan Eamer, Gary Kofinas, Jack Mathias, and Magdalena Muir.

The following people reviewed chapters of the volume (in alphabetical order): Rick Armstrong, Derek Armitage, Burton Ayles, Nigel Bankes, Fikret Berkes, Jason Boire, Doug Chiperzak, Doug Clark, Don Cobb, Johan Colding, Ann Dale, Iain Davidson-Hunt, Alan Diduck, Graham Dodds, Nancy Doubleday, Helen Fast, Alan Fehr, David Fennell, Chris Furgal, Patricia Gallaugher, Alison Gill, Sandra

Grant, Emdad Haque, Masood Hassan, Cliff Hickey, Rob Huebert, Dyanna Jolly, Anne Kendrick, Mina Kislalioglu Berkes, Gary Kofinas, Igor Krupnik, Steve Light, Jack Mathias, Barbara Neis, Claudia Notzke, Brenda Parlee, Alan Penn, Robert Siron, Scott Slocombe, Peter Timonin, Shirley Thompson, Dan Topolniski, Chris Trott, and David VanderZwaag. As well, two anonymous referees for the University of Calgary Press reviewed the book as a whole.

The DVD, "Watching, Listening and Understanding Changes in the Environment," was prepared to highlight community-based monitoring of the environment. It is a co-production of the University of Manitoba, Natural Resources Institute (Anne Kendrick, Micheline Manseau, and Brenda Parlee, producers) and the Rainbow Bridge Communications Co., *www.rbcc.ca* (Barb Allard and Sheryle Carlson, producers; Derek Sharplin, editor). The narration is by Lucy Tulugarjuk, the script and editing by Laren Bill, Anne Kendrick, Micheline Manseau, and Brenda Parlee.

Footage from Baffin Island: the interviewer is Nancy Anilniliak and the interviewee Davidee Kooneeliusie; video by Paula Hughson, photos by Davidee Kooneeliusie and Micheline Manseau. Footage from Old Crow: the interviewer is Laren Bill and the interviewee Randall Tetlichi; DVD by Laren Bill; additional footage from the Gwich'in Settlement Region by Brenda Parlee and the Inuvialuit Settlement Region by the Inuvialuit Communications Society. Footage from Lutsel K'e: the interviewer is Brenda Parlee and the interviewees Henry Basil, Shawn Catholique, Stan Desjarlais, Steve Ellis, and August Enzoe; video by Tsats'i Catholique, photos by Anne Kendrick and Phil Lyver. Additional Lutsel K'e footage: Lutsel K'e Wildlife, Lands and Environment Committee. Music: Throat singing by Lucy and Kayla Tulugarjuk; Inuvialuit Drummers, Courtesy of CBC-North, Yellowknife. Funding for the video was provided by the SSHRC/DFO, the Nunavut Field Unit and Western Canada Service Centre of Parks Canada, and the Walter and Duncan Gordon Foundation.

In the production of this book, Jacqueline Rittberg acted as the IM Node managing editor, Julie Dekens as the editorial assistant handling chapter reviews, Eleanor (elly) Ayr Bonny as the editorial assistant handling copy-edited chapters, and Apurba Deb as proofreader. Mark Ouellette prepared the figures. We gratefully acknowledge the excellent work of the University of Calgary Press under Director Walter Hildebrandt.

Fikret Berkes, Rob Huebert, Helen Fast,
Micheline Manseau, and Alan Diduck
The Editors

CHAPTER 1
INTRODUCTION

Fikret Berkes (University of Manitoba)
and *Helen Fast* (Department of Fisheries and Oceans)

INTEGRATED MANAGEMENT, COMPLEXITY, DIVERSITY, CHANGE

This volume is the output of the project, "Integrated management, complexity and diversity of use: responding and adapting to change." The project focused on the Canadian North, defined to include the Beaufort Sea, the Arctic Ocean coast and islands, Hudson Bay, and Hudson Strait. The use of the idea of integrated management (Cicin-Sain and Knecht 1998), complexity, and the diversity of resource use was intended to capture the issue of competing demands on the environment and the interrelationships among them. There was also a practical reason relevant to policy: integrated management is one of the three principles on which the 1997 *Oceans Act* is based (s. 30). Canada's Oceans Strategy (2002) is constructed as a framework for integrated management.

The subtitle of the project, responding and adapting to change, captures the second major theme in the project and in this volume. We defined change broadly to include social and cultural change, economic and technological change, development pressures, globalization, and larger-scale issues such as climate change and Arctic ecosystem contamination. One notable fact about the Canadian North is the rate at which change has been occurring. We refer not only to visible technological change, such as TV, internet, and GPS units now used by indigenous hunters in daily life, but also to fundamental changes in both the social and the biophysical environment in the North.

The pace of change is striking, for example, in the area of contaminants, as evident from the contents of the Canadian Arctic Contaminants Assessment reports of 1997 and 2003 (Jensen *et al.* 1997; INAC 2003). The urgency of the contaminants issue can be deduced from the negotiations leading up to the international agreement designed to address the problem, the 2001 *Stockholm Convention on Persistent Organic Pollutants* (Downie and Fenge 2003). Also fundamentally important is the issue of climate change, judging by the work

1

of the multinational Arctic Climate Impact Assessment carried out under the Arctic Council (ACIA 2004) and the urgency of local and indigenous observations (Krupnik and Jolly 2002).

The title of the Krupnik and Jolly (2002) book is telling: "The earth is faster now" is a quotation from an Alaska elder, Mabel Toolie, referring to rapid change and the declining ability of her people to read the weather. British polar scientists Clarke and Harris (2003, 1) echo the same sentiment in scientific language: "The capacity of marine ecosystems to withstand the cumulative impact of a number of pressures, including climate change, pollution and overexploitation, acting synergistically is of greatest concern." Change and the unpredictability and vulnerability created by change have been common themes in public meetings as well, such as the Beaufort Sea 2000 Conference (Ayles *et al.* 2002).

However, our understanding of change and its impact on social and environmental systems has been rather incomplete. Particularly in Canada, both marine research and northern research have suffered in recent decades. According to the findings of the Task Force on Northern Research (NSERC and SSHRC 2000, 1), the North has been facing "unprecedented social, physical and environmental challenges." Yet, the Canadian research capability and the level of research activity in the North have declined over the years, requiring a rebuilding of research programs and training opportunities for a new generation of northern researchers. Taken together, these are some of the main considerations that shaped the objectives of the project.

This volume carries out the overall goal of the project and the parent body, the Ocean Management Research Network (OMRN 2004) through three specific objectives: (1) to research and learn from the experience in the area of integrated management, complexity and diversity of resource use; (2) to apply critical thinking to the phenomena of change and the way in which societies respond and adapt to new challenges; and (3) to understand the dynamics of change and explore policy options to build capacity to adapt to new challenges.

CONCEPTS AND DEFINITIONS
Sustainability, the Ecosystem Approach and Integrated Management

This book is interdisciplinary, using various concepts that cut across social and natural sciences; some definitions are therefore needed to establish a common vocabulary. The book contributes to the overall goal of *sustainability* in the Canadian North, whereby "the needs of the present and local population can be met without compromising the ability of future generations or populations in other locations to meet their needs" (MEA 2003, 215). Sustainability, as used here, is a process (and not an end point) that includes ecological, social, cultural, and economic dimensions. The question of "what is to be sustained" is not self-evident and has to be addressed on a case-by-case and area-by-area basis. The term *ecological system* (ecosystem) is used in the conventional ecological sense to refer to "a dynamic complex of plant, animal and microorganism communities and their nonliving environment, interacting as a functional unit" (MEA 2003, 210). Ecosystems are not biological constructs untouched by human influences but have humans as integral component. To emphasize that social and ecological systems are in fact linked, and that the delineation between social and natural systems is artificial and arbitrary, we use the terms *social-ecological system* and *social-ecological linkages* (Berkes and Folke 1998, 4).

Ecosystem approach is a "strategy for the integrated management of land, water and living resources that promotes conservation and sustainable use in an equitable way" (MEA 2003, 52) or "an approach to management that recognizes the complexity of ecosystems and the interconnections among component parts" (Canada's Oceans Strategy 2002, 36). The related concept of *ecosystem-based management* is "the management of human activities so that ecosystems, their structure, function and composition are maintained at appropriate temporal and spatial scales" (Canada's Oceans Strategy 2002, 36). The ecosystem-based management idea overlaps with the notion of *integrated management*, defined by Canada's Oceans Strategy (2002, 36) as

> a continuous process through which decisions are made for the sustainable use, development and protection of areas and resources. IM acknowledges the interrelationships that exist among different uses and the environments they potentially affect. It is designed to overcome the fragmentation inherent in a sectoral management approach, analyzes the implications of development, conflicting uses and promotes linkages and harmonization among various activities.

We use the term *cross-scale interactions* to refer to two kinds of linkages: *horizontal* (across geographic space or across sectors) and *vertical* (across levels of organization) (Young 2002). Harmonization refers to the horizontal linkages that are necessary to coordinate activities (*e.g.*, protected areas, transportation, oil and gas exploration, fishery management) to overcome fragmentation of decision making by sector. We make a distinction between *level* ("the discrete levels of social organization, such as individuals, households and communities and nations") and *scale* ("the physical dimensions, in either space or time, of phenomena or observations") (MEA 2003, 212, 214).

Many of the linkages in the Canadian North involve *Aboriginal peoples* (defined to include First Nations people, the Inuit and the Metis) and working with their knowledge of the environment. The term *indigenous knowledge* (IK) is used to mean local knowledge held by indigenous peoples, or local knowledge unique to a given culture or society. We use *traditional ecological knowledge* (TEK) more specifically to refer to "a cumulative body of knowledge and beliefs, evolving by adaptive processes and handed down through generations by cultural transmission" (Berkes 1999, 8). The word traditional is used to refer to historical and cultural continuity, recognizing that societies are constantly redefining what is considered "traditional." Inuit Qaujimajatuqangit (IQ) is the preferred term for TEK as used by some Inuit groups. We use the terms Western resource management science, scientific resource management, and conventional resource management interchangeably. We recognize that all societies have their own knowledge systems, but we identify Western science and scientific method as representing a particular kind of knowledge which is used as the basis of resource management by centralized bureaucracies everywhere in the world.

Governance, Institutions, and Co-Management

We use *governance* in the broader sense of coordination of social systems, and *institutions* in the broader sense of working rules and norms (Ostrom 1990). Processes of governance may or may not include the state; hence governance is possible even without the government, and institutions can exist without the presence of government agencies. The conventional governance involves top-down or state-centric governance in which *command-and-control* management is the norm, a "the policy framework in which environmental and resource management rules are prescribed by the regulator, leaving little flexibility for actors in the implementation" (MEA 2003, 209). But evolving notions of governance also includes an alternative: "In this second approach, which is more society-centered, the focus is on coordination and self-governance as such, manifested in different types of networks and partnerships" (Pierre and Peters 2000, 3).

There have been lively debates over governance, involving mixtures of self-governing, co-governing and hierarchical forms of governing, and incorporating state, private and civil society actors. In many countries, changes in patterns of governance have involved shifts in the balance and relation between government and society, and between public and private sectors. There is a trend toward more complementary patterns of interaction between formal governance and civil society, and a sharing of responsibility and accountability by public and private actors (Kooiman 1993).

Some authors have been using the term "*good* governance" to refer to these changes and trends (Rhodes 1997). Elements of good governance include participation (the involvement of resource users in the decisions that affect their livelihoods), accountability (the ability of the parties affected by a decision to demand and receive an explanation), transparency (openness of decision making), and legitimacy (the acceptance by users of the authority of rule-makers) (Jentoft 1999). In the area of resource management, participation is often seen as the basic element of good governance (McCay and Jentoft 1996; Wiber *et al.* 2004). The *subsidiarity principle* articulates the objective that decisions affecting peoples' lives should be made by the lowest capable social organization, and emphasizes the importance of local-level institutions in governance (McCay and Jentoft 1996).

Institutions are "the rules that guide how people within societies live, work, and interact with each other" (MEA 2003, 21). North (1994) defines institutions as "humanly devised constraints that structure human interaction ... made up of formal constraints (rules, laws, constitutions), informal constraints (norms of

Rankin Inlet youth at the Meliadine Territorial Park, Rankin Inlet, Nunavut. The two young girls are Beatrice Pissuk (on the left) and Roseanne Shimout (on the right). Photo by Steve Newton, 2003.

behaviour, conventions and self-imposed codes of conduct), and their enforcement characteristics." Institutions are not merely rule sets but have normative and cognitive dimensions as well (Jentoft *et al.* 1998), a consideration that is important in view of the differences between northern indigenous peoples and the dominant Canadian society in terms of social values, perceptions and knowledge.

Co-management systems in the Canadian North may be understood in this broader sense of governance and institutions. Co-management involves the sharing of responsibilities among the expanded set of players in governance; it involves networks and partnerships of a diversity of actors and their institutions. *Co-management* may be defined as "the sharing of power and responsibility between the government and local resource users" (Berkes *et al.* 1991, 12) or "a system that enables a sharing of decision-making power, responsibility and risk between government and stakeholders" (NRTEE 1998, 14). Co-management, as defined by the NRTEE, implies a formal agreement between at least one government and another group, and specifies government as a partner. Community-based management is sometimes considered an aspect of co-management, "but when community-based management does not include government as a partner in the decision making process, it is not co-management" (NRTEE 1998, p.13).

Even though co-management is often examined in terms of the formal arrangement under northern land claims and other agreements, the functional side of co-management is about joint problem-solving. Working partnerships develop over a period of time through learning and building of mutual respect and trust, as in the co-management of the Beverly-Qamanirjuaq caribou herd (Kendrick 2000). Thus, the actual power-sharing is the result, rather than the starting point, of the process of co-management.

Learning through networks and partnerships is important, and co-management oriented to problem-solving shows two characteristics. The first is the dynamic learning characteristic of *adaptive management*, or learning-by-doing in an iterative way (Holling 1978; Lee 1993). The second is the linkage characteristic of participatory or co-operative management (Pinkerton 1989; NRTEE 1998). Folke *et al.* (2002, 20) have used the term *adaptive co-management* to refer to this "process by which institutional arrangements and ecological knowledge are tested and revised in a dynamic, ongoing, self-organized process of trial and error." It is an inclusive and collaborative process in which stakeholders share management power and responsibility, as in Pierre and Peters' (2000) "second approach" to governance.

Adaptive management is a key concept because it was designed to address issues of *uncertainty*, "an expression of the degree to which a future condition (*e.g.*, of an ecosystem) is unknown" (MEA 2003, 215). Resource management operates in an environment of uncertainty in northern ecosystems, perhaps more so than elsewhere. This is because scientific information is sketchy, the normal range of environmental variation is large, and the rate of change is fast and getting faster, due to such major perturbations such as environmental contamination and climate change. Hence, it is difficult to address many of the issues of northern Canada with an approach that starts with the assumptions that research can provide the necessary data and that the future can be predicted and controlled. The lessons of a half-century of fisheries management suggest that such abilities to predict and control are an illusion (Charles 2001).

If the future is inherently unpredictable, what are the management options? Adaptive management, with its emphasis on feedback learning from an intervention and the use of that information in the design and implementation of the next intervention (Lee 1993), is one tool to deal with uncertainty. A second tool is the *precautionary principle*, "the management concept stating that in cases where there are threats of serious or irreversible damage, lack of full scientific certainty shall not be used as a reason for postponing cost-effective measures to prevent environmental degradation" (MEA 2003, 213). There are a number of alternative statements of the precautionary principle; the *Oceans Act* (s. 30) defines it simply as "erring on the side of caution."

The study of change is one of the objectives of this volume. The approach used in the volume does not deal either with social change (Kulchyski *et al.* 1999) or with environmental change alone (Clarke and Harris 2003); it deals with social-ecological system change. The concept of resilience is a promising tool for dealing with change because it provides a way of analyzing the dynamics of how systems persist, transform themselves, or collapse. *Resilience* is the capacity of a system to tolerate disturbance without collapsing into a qualitatively different state. Hence resilience thinking pays special attention to *thresholds*, the points where systems flip from one equilibrium state to another; and *surprises* that are said to occur when perceived reality departs qualitatively from expectation (Holling 1986). According to the Resilience Alliance (2004):

> A resilient ecosystem can withstand shocks and rebuild itself when necessary. Resilience in social systems has the added capacity of humans to anticipate and plan for the future.... Resilience as applied to ecosystems, or to integrated systems of people and the natural environment, has three defining characteristics:
>
> - The amount of change the system can undergo and still retain the same controls on function and structure;
> - The degree to which the system is capable of self-organization; and
> - The ability to build and increase the capacity for learning and adaptation.

In brief, resilience is "the capacity of a system to tolerate impacts of drivers without irreversible change in its outputs and structure" (MEA 2003, 214). A *driver* is "any natural or human-induced factor that directly or indirectly causes a change in an ecosystem" (MEA 2003, 210). More generally, drivers or external drivers, such as climate change, are those key factors that cause change. The identification of such drivers in a complex system is an important step in resilience thinking. It can then lead to an exploration of policy options to build capacity to adapt to change (Folke *et al.* 2002; Berkes *et al.* 2003).

Uncertainty, resilience, and a number of other concepts used in this volume have something in common: they are all related to complex systems. A *complex system* is one that has a number of attributes not observed in simple systems, including non-linearity, uncertainty, scale, and self-organization (Levin 1999). In complex systems, an *emergent property* is a "phenomenon that is not evident in the constituent parts of the system but that appears when they interact in the system as a whole" (MEA 2003, 211). It is a property that cannot be predicted or understood simply by examining the system's parts. For example, consciousness cannot be understood by examining neurons and their connections, but emerges as a property of the whole organism. Similarly, resilience is an emergent property of integrated social-ecological systems; it cannot be understood by examining the parts of the system but emerges out of the consideration of how a system tolerates the impacts of drivers, shocks, and change (Gunderson and Holling 2002).

BACKGROUND AND CONTEXT
International Context

Historically, marine and coastal waters have supported two major activities: harvesting of marine products and transportation. But as the use of coastal areas has intensified, a number of additional activities have become important, including aquaculture, oil and gas exploration, and tourism and recreation. Through stock depletion, habitat degradation and pollution, coastal resources have declined in Canada and throughout the world, while potentially competing uses have intensified, producing a major crisis in oceans management. By the late 1980s, it had become clear that ocean and coastal resources were not sustainable under the conventional approaches of managing single activities and species. New management approaches, embracing environmental considerations, were urgently needed to replace this sectoral approach.

National and international organizations have been looking for alternatives to conventional management. One of the alternative approaches is embodied in the Code of Conduct for Responsible Fisheries, initiated by the United Nations Food and Agriculture Organization (FAO). Although focusing on only one sector, the Code provides a comprehensive set of guidelines that includes the Precautionary Principle, ecosystem stewardship, dispute resolution, international law, and international trade in fish products. Taking another cut at the issue, the Lisbon Principles for sustainable ocean governance use a multi-sector, multi-species governance approach (Table 1.1). They provide a more comprehensive but a smaller set of guidelines than the Code, and include the principles of responsibility, scale-matching, precaution, adaptive management, full-cost allocation, and participatory decision making (Costanza *et al.* 1999).

The international trends are toward ecosystem approaches and integrated management, and the consideration of uses and impacts from a variety of activities at the ecosystem level. These trends are not confined to the area of oceans management. Similar trends are also apparent in international efforts for the sustainable management of forest ecosystems and agro-ecosystems. Several

Table 1.1

**THE LISBON PRINCIPLES FOR THE SUSTAINABLE GOVERNANCE OF
THE OCEANS AND COASTAL AREAS**

Responsibility principle	The responsibility of individuals or corporations to use environmental resources in an ecologically sustainable, economically efficient and socially just manner.
Scale-matching principle	The importance of assigning decision making to the scale of governance which has the most relevant ecological information, which considers ownership and actors, and which internalizes costs and benefits.
Precautionary principle	The need to take uncertainty about potentially irreversible environmental impacts into account.
Adaptive management principle	The requirement to continuously monitor social, economic, and ecological systems because they are dynamic and have some level of uncertainty.
Full cost allocation principle	The need to identify and allocate all internal and external costs and benefits (social and ecological) of alternative uses of environmental resources.
Participation principle	The importance of full stakeholder participation in the formulation and implementation of decisions about environmental resources.

international initiatives have been examining alternative management options for ecosystems, including the Millennium Ecosystem Assessment (MEA 2003), the projects carried out under the Resilience Alliance (Gunderson and Holling 2002; Berkes *et al.* 2003; Resilience Alliance 2004), and the projects under the banner of Sustainability Science (Kates *et al.* 2001).

Many of these international efforts are participatory in nature, engaging the knowledge of resource users, their livelihoods and well-being, adaptive learning, and institutions of self-governance. This is consistent with the shift from a narrow concern with "government" to a broader concern with "governance" that involves state, private, and civil society actors. Participatory approaches have become increasingly important in resource management in part because of our understanding of ecosystem complexity and uncertainty. The complex nature of larger environmental problems means that the "objective, disinterested technical expert" no longer has a central role. The age of expert management is over, replaced by participatory problem solving in which the risks of decision making are shared among users and managers (Ludwig 2001).

These international trends are consistent with those in Canada. A report of the Royal Society of Canada examined aquatic research in the Canadian North and suggested a pluralistic strategy incorporating all critical perspectives. To meet the broader range of societal expectations, research should not only address the perspectives of scientists, policy makers, and managers, but also the perspectives of civil society actors, including northern Aboriginal groups (RSC 1995). Such pluralistic science extends the range of scientific inquiry beyond the conventional positivist, expert-knows-best science (Ludwig 2001) and creates space for the incorporation of locally generated knowledge. There has been local input into the research questions being asked, and the use of local knowledge, for example, in the areas of Arctic contaminants and climate change, in the Canadian North, and specifically in areas under Aboriginal land claims.

Context of Aboriginal Land Claims

A number of treaties were signed in the late nineteenth through the early twentieth century, covering parts of the Canadian North but excluding much of the Northwest Territories, Yukon, Quebec, Labrador, and British Columbia. When the Supreme Court of Canada recognized the existence of Aboriginal title in the early 1970s, Canada's ownership of nearly half of the country's land mass came under question. A new era of Aboriginal claims through comprehensive land claims agreements was born, removing the uncertainty and opening the way for development.

The *James Bay and Northern Quebec Agreement* of 1975 was the first one to be signed. A number of agreements have been signed since, including the *Inuvialuit Final Agreement* of 1984, the *Gwich'in Comprehensive Land Claim Agreement* of 1992, the *Sahtu Dene and Metis Comprehensive Land Claim Agreement* of 1993, the *Yukon First Nations Umbrella Final Agreement* of 1993, and the *Nunavut Land Claims Agreement* of 1993. The last-mentioned agreement is the largest in terms of geographical area, and it resulted in the creation of the new Nunavut Territory amid much fanfare.

The new generation of treaties are referred to as comprehensive agreements because they spell out the nature of the arrangement between the Government of Canada and Aboriginal groups, under a large number of headings, including self-government powers, control over social services such as education and health, compensation payments, environmental assessment, land use regulations, and the management of land and resources.

For the purposes of this volume, the provisions for land and resource management are crucially important. Aboriginal control over the environment in the land claims areas is mostly in the form of joint jurisdiction, legally specifying Aboriginal rights and responsibilities. The formalization of power-sharing between the central and local/regional governments is deemed important, for example, by the Report of the Royal Commission on Aboriginal Peoples (RCAP 1996) because it means that indigenous hunting and other resource use rights are recognized by law and are (at least in theory) enforceable. Each of the comprehensive claims agreements has a section or sections that specify the sharing of jurisdiction for fisheries and wildlife management, creating co-management boards as the main instruments of resource management.

The various agreements establish institutional structures in the form of management boards and joint committees. For example, Article 5, Part II of the *Nunavut Land Claims Agreement* specifies the membership of the Nunavut Wildlife Management Board (a co-management body), the board's bylaws, powers, duties, and responsibilities. Section 14 of the *Inuvialuit Final Agreement* establishes four co-management institutions, including one for fisheries and marine mammal management, the Fisheries Joint Management Committee (FJMC) (Fast *et al.* 2001). Some of the agreements specify the use of Aboriginal knowledge in the process of co-management. For example, the *Nunavut Land Claims Agreement* provides for an Inuit traditional knowledge study of the bowhead whale that has since been carried out (Hay *et al.* 2000).

Strong provisions for resource and environmental co-management, and some thirty years of experience in joint problem solving, set the North apart from other coastal regions of Canada. In striving toward ocean and coastal co-management under the *Oceans Act*, many of the key lessons are not from the east and west coasts but from the Arctic (NRTEE 1998). Formal co-management is probably not essential for successful joint problem solving (Kendrick 2000). However, the experience in the North shows that joint management strongly parallels the emergence of Aboriginal land claims (Berkes *et al.* 2001). Participatory management, not only in fisheries and wildlife, but in a range of areas including integrated management, protected areas, ecosystem and human health, contaminants research, environmental assessment, climate change, has followed increasing political power in the North.

Context of the Oceans Act and Canada's Oceans Strategy

The context of the *Oceans Act* is of central significance for this volume, as it addresses the problems of the conventional sectoral management of coastal and ocean resources, and it directs the Department of Fisheries and Oceans to build partnerships with Aboriginal land claims agencies, coastal communities, and other stakeholders in the marine environment.

Canada is a maritime nation, with a coastline of 244,000 kilometres, and a continental shelf covering 3.7 million square kilometres. Canada's Exclusive Economic Zone extends 200 nautical miles from shore and is equivalent to over 30 per cent of Canada's total land mass (Canada's Oceans Strategy 2002). About seven million Canadians live in coastal communities, and many coastal communities depend on the ocean and its resources for their livelihoods. Canada's ocean-based industries generate over $22 billion annually in direct economic activity and contribute over $83 billion to international trade. The growth of coastal and oceans-related activities has resulted in imbalances and degradation of the marine environment. These changes are increasingly evident in biodiversity loss, water quality issues, habitat loss, and the introduction of invasive species.

The *Oceans Act* (Canada 1997) came into force in 1997. With this Act, Canada became the first country in the world with comprehensive oceans management legislation. The Act is an important model, not only to meet Canada's needs but also as an example for other countries (Hanson 1998). It describes oceans management as a collective responsibility that requires collaboration among all levels of government and stakeholders. The list of stakeholders identified includes the private sector, Aboriginal organizations, and local communities. The strategy for implementing this responsibility was subsequently described in Canada's Oceans Strategy (2002). The three policy objectives of this strategy are: a) understanding and protecting the marine environment; b) supporting sustainable economic opportunities; and c) international leadership (Canada's Oceans Strategy 2002).

Understanding the marine environment is predicated on scientific understanding to delineate ecosystem boundaries, identify key ecosystem functions, develop risk assessment techniques, develop performance indicators, and assess the health of the ecosystem. With this information, governments and other stakeholders will be able to make informed decisions on steps that need to be taken to protect the marine environment.

Supporting sustainable economic opportunities also depends on a sound understanding of the marine environment. Some of the major industrial activities in the marine environment include fisheries, offshore energy and mineral resource development, and shipping. Sustainable management of Canada's oceans requires that the benefits of development and economic activity be balanced with the costs of lost economic opportunities and continued environmental degradation.

Finally, Canada has a responsibility to influence international priorities, decisions and processes, particularly as they pertain to sovereignty and security, and the provision of support for sustainable ocean resources. Canada's Oceans Strategy will continue to evolve as new knowledge is gathered, and as DFO, other agencies and stakeholders gain experience in oceans management (Canada's Oceans Strategy 2002). Over a period of time, the process of integrated management is expected to facilitate sound decision making at the level of large-scale ecosystems, multiple users and issues of marine environmental quality.

THE RESEARCH AGENDA LEADING TO THIS VOLUME

The subject of integrated management, complexity, and diversity of use is relevant to all three coastal areas of Canada, and so is the theme of responding and adapting to change. The North is a particularly suitable setting for studying integrated management with respect to change and complexity of coastal areas. Resource use activities in the North are less intensive than those in the Atlantic and Pacific, and resource use rights are relatively more clearly defined, due to Aboriginal land claims agreements. The coastal areas in the North, from the Inuvialuit Region to northern Quebec and more recently to Labrador, are covered by comprehensive land claims agreements. Wide-ranging participatory decision-making processes, spearheaded by the co-management provisions of these agreements evolving since the 1970s, offer lessons in the solution of Aboriginal rights and resource conflicts elsewhere.

To meet the objectives of learning from experience, analyzing change, and exploring policy options to build capacity to adapt to change, the Integrated Management Node of the Oceans Management Research Network organized

Lake Hazen, Quttinirpaaq National Park. Photo by Micheline Manseau, 2002.

itself into five working groups: Ecotourism and Development, Security and Sovereignty, Community-Based Monitoring, Community and Marine Ecosystem Health, and Resilience and Adaptation.

The Ecotourism and Development Working Group examined a diversity of activities in the coastal zone. For example, in the Mackenzie River estuary and the Beaufort Sea, activities include tourism (Notzke 1999), Inuvialuit beluga hunting (Dressler *et al.* 2001), and protected area planning (Mathias and Fast 1998). These potentially conflicting activities are occurring while the area is facing increasing pressures for oil and gas development, marine food web contamination (O'Neil *et al.* 1997; Jensen *et al.* 1997), and climate change (Krupnik and Jolly 2002). The Working Group examined potentially sustainable and non-extractive industries, such as ecotourism (Fennell 1998), and studied selected geographic areas, such as West Hudson Bay and the Beaufort Sea, building on previous work.

The Security and Sovereignty Working Group started with an agenda that included the exploration of the nature of security in the post-Soviet circumpolar north. A new understanding of security is concerned with international co-operation through such new institutions as the Arctic Council (Huebert 1999). An expanded notion of security may extend to environmental security and change. For example, climate change is related to the thinning of ice, reduction in the extent of ice cover, and the extension of navigation season in the Arctic. This, in turn, is related to resource development, environmental security and, given possible change of the international status of the Northwest Passage, sovereignty (CARC 2002).

The Community-Based Monitoring Working Group started with the premise that involving local and regional organizations to monitor the environment would help implement Canada's Oceans Strategy and foster good environmental stewardship in general. Such monitoring would be sensitive to local concerns; it would be based on local and traditional knowledge of the environment and local priorities in defining what is to be monitored and how (see the DVD). Indigenous approaches to monitoring (O'Neil *et al.* 1997) could enrich the set of tools used by Western science, and traditional environmental knowledge could complement scientific knowledge, as in the example of climate change (Riedlinger and Berkes 2001).

The Community and Marine Ecosystem Health Working Group brought together researchers and practitioners around issues of community health, vulnerability and food security, nutrition and local economies, contaminants, and marine environmental quality (MEQ). It explored the connection between environmental health and community health, the issue of marine environmental quality from an indigenous point of view, and food security in the Arctic in the face of environmental change. Persistent organic pollutants (POPs) in the food chain have had a major impact on the harvest of wild foods, especially Arctic marine mammals (Downie and Fenge 2003). What is at stake is Arctic self-reliance and a way of life based on the potentially sustainable use of local renewable resources (Doubleday 1996).

The Resilience and Adaptation Working Group started with a research agenda using the notion of complexity, drawing attention to the importance of uncertainty, scale, self-organization, and resilience (Levin 1999; Gunderson and Holling 2002). It discussed the basic issue of how to cope with and recover from the shocks and stresses of rapid change. Folke *et al.* (2002) have suggested that policies for resilience may involve building adaptive capacity through the creation of flexible multi-level governance systems that can learn from experience and generate knowledge to cope with change. As applied to the Arctic, this may mean strengthening local institutions, fostering international institutions, such as the Arctic Council, and building cross-scale linkages from the local level to the international.

Working groups included university academics, graduate students, government researchers and resource managers, and practitioners. A major strength of the project was its integration with the work of the Department of Fisheries and Oceans (DFO) Central and Arctic Region, and the development of a rich array of interlinking partnerships through the Oceans Management Research Network. Northern partners included Aurora College, the Canadian Arctic Resources Committee, Fisheries Joint Management Committee of the Inuvialuit Region, the Land Administration of the Kivalliq Association, the Nunavut Research Institute, and the Tuktu and Nogak Project. As the project progressed, there were additional linkages with the Gwich'in Renewable Resources Board, Lutsel K'e Dene First Nation, eastern Hudson Bay Inuit communities, and a number of additional groups that contributed to the DVD.

The plan follows the three objectives of the volume: (1) to research and learn from the experience in the area of integrated management, complexity, and diversity of resource use, (2) to apply critical thinking to the phenomena of change and the way in which societies respond and adapt to new challenges, and (3) to understand the dynamics of change and to explore policy options to build capacity to adapt to new challenges.

The first section, which contains Chapters 2 through 6, is about learning from the continuous process of sustainable use, development, and environmental protection. The chapters challenge the fragmentation inherent in the sectoral management approach and offer alternative ways of understanding issues. The first three chapters are broadly on the themes of food, human health, and environmental health. Chapters 2 (Myers *et al.*) and 3 (Thompson) focus on a key issue for northern Aboriginal people: the land-based economy, wild foods, and contamination. Chapter 2 discusses the importance of subsistence activities in the context of changing social, economic, and environmental conditions. Chapter 3 applies vulnerability analysis to northern communities that are at risk from rapid and sweeping changes. Chapter 4 (Cobb *et al.*) examines marine environmental quality and the way science and traditional knowledge looks at environmental quality indicators. The next two chapters turn to the integrated management experience in Canada's western Arctic. Chapter 5 (Fast *et al.*) examines one detailed case of public participation of integrated management in the Inuvialuit Region. Chapter 6 (Schlag and Fast) is about communication and education regarding marine stewardship, a key concept in Canada's Oceans Strategy. It examines stewardship and sustainability issues through the lens of a younger person.

Chapters 7 to 10, which make up the second section, seek to apply critical thinking to respond and adapt to new challenges. The first two chapters of the section examine the links between traditional ecological knowledge (or Inuit knowledge, IQ), scientific knowledge, indicators, and monitoring. Chapter 7 (Manseau *et al.*) examines the role and place of traditional knowledge in resource management. Chapter 8 (Parlee *et al.*) illustrates this theme by explaining indigenous concepts of ecosystem health indicators among the *Denesoline* of the Northwest Territories. Chapters 9 and 10 explore new ways of doing things in the area of northern economic development, focusing on eastern Hudson Bay. Chapter 9 (Lemelin) is about ecotourism and its development in the Churchill area. It focuses on wildlife tourism relating to polar bears and uses a chaos and complexity approach. Chapter 10 (Junkin) examines the experience with local economic development based on local wildlife resources, with a focus on caribou harvesting. It is a discussion of commercial (rather than subsistence) use of wildlife and its links to development.

The third section, encompassing Chapters 11 to 14, addresses the questions of strengthening local institutions and building linkages from the local level to the international, and the creation of flexible multi-level governance systems that can learn from experience. Chapter 11 (Berkes *et al.*) sets the scene, and illustrates the idea, of building institutional linkages across levels of social and political organization, using examples from the Inuvialuit Region's Fisheries Joint Management Committee, narwhal and polar management, and persistent organic pollutants. The next three chapters deal with renewable resource and ocean management in innovative ways, building conceptual frameworks based, respectively, on ideas of adaptive co-management (Chapter 12, Kristofferson and Berkes), social learning (Chapter 13, Diduck *et al.*) and law, hierarchy, and resilience (Chapter 14, Bankes). Chapter 12 considers an Arctic char example from the Central Arctic, treating it as a case of adaptive management (learning-by-doing) and co-management under the *Nunavut Land Claims Agreement*. Chapter 13 turns to the experience of polar bear and narwhal management in the Nunavut Territory. The case is analyzed in terms of social learning from management successes and failures, getting at the dynamics of change. The notion of resilience is examined further in Chapter 14 and applied to law and the regulation of resource use in the Nunavut Territory.

The fourth and last section explores governance, policy, and future directions. Chapter 15 (Huebert) uses a current case of international dispute (the dispute over a small island between Canada and Greenland in the High Arctic) to explore new challenges in Arctic governance. Chapter 16 (Armitage and Clark) surveys a wide variety of material from the North and the South regarding northern research priorities. It synthesizes a great deal of thought regarding issues, priorities, and research directions for renewable resources and oceans management, and looks for common themes and areas of convergence. The chapter should be examined in concert with the DVD in the back pocket of the book, the DVD that provides northern perspectives from the people themselves. Finally, Chapter 17 (Huebert *et al.*) provides a conclusion for the volume.

The book provides a unique approach that links the health of people, communities, and ecosystems; it draws heavily on theories related to adaptive management and resilience; and it explores the recent experience of new institutions established under northern land claims agreements. The overall message is that Canada and the people of the North have crafted new and potentially workable approaches to resource management and sharing, co-management institutions, and ways of sharing knowledge and learning from local and traditional knowledge. Some of these lessons (for example, regarding co-management) are relevant for addressing renewable resource and ocean management issues elsewhere in Canada and internationally. As one of the referees pointed out, however, the message of the book is tempered by the apparent shortcomings of the overall governance system in dealing with the tremendous changes of the last five decades, but also with emerging problems such as climate change and youth who lack the connection to the land of elders.

Throughout the book, we integrate knowledge from social and management-oriented research that supports the health and sustainability of Canada's oceans and coastal communities. Although our theme is the North, we use the Oceans

Management Research Network as our wider forum to bring critical thinking and new perspectives from all three coasts and the international experience. Linking research to management and policy applications breaks the ice, connecting academics, government managers, policy-makers, Aboriginal groups, and industry – groups that have been operating as solitudes for a long time. The new policy environment under the *Oceans Act* has the potential to facilitate a sharing of knowledge and understanding among the solitudes, leading to a shared vision to address new challenges.

REFERENCES

ACIA. 2004. Arctic Climate Impact Assessment. Arctic Council. *http://www.acia.uaf.edu/*

Ayles, B.G., R. Bell and H. Fast. 2002. The Beaufort Sea Conference 2000 on renewable marine resources of the Canadian Beaufort Sea. *Arctic* 55 (suppl.): iii–v.

Berkes, F. 1999. *Sacred Ecology: Traditional Ecological Knowledge and Resource Management*. Philadelphia and London: Taylor & Francis.

Berkes, F., J. Colding and C. Folke, eds. 2003. *Navigating social-ecological systems: Building resilience for complexity and change*. Cambridge: Cambridge University Press.

Berkes, F., and C. Folke, eds. 1998. *Linking social and ecological systems. Management practices and social mechanisms for building resilience*. Cambridge: Cambridge University Press.

Berkes, F., P.J. George, and R.J. Preston. 1991. "Co-management." *Alternatives* 18(2): 12–18.

Berkes, F., J. Mathias, M. Kislalioglu, and H. Fast. 2001. "The Canadian Arctic and the *Oceans Act*: The development of participatory environmental research and management." *Ocean & Coastal Management* 44: 451–69.

CARC. 2002. "On thinning ice." *Northern Perspectives* 27(2). *http://www.carc.org/*

Canada. 1997. *Oceans Act*. S.C. 1996, c. 31. In force 31 January 1997.

Canada's Oceans Strategy. 2002. Our oceans, our future. Policy and operational framework for integrated management of estuarine, coastal and marine environments of Canada. Ottawa: Fisheries and Oceans Canada.

Charles, A. 2001. *Sustainable fishery systems*. Fish and Aquatic Resources Series 5. Oxford: Blackwell Science.

Cicin-Sain, B., and R.W. Knecht. 1998. *Integrated coastal and ocean management*. Washington, DC: Island Press.

Clarke, A. and C.M. Harris. 2003. "Polar marine ecosystems: Major threats and future change." *Environmental Conservation* 30: 1–25.

Costanza, R., F. Andrade, P. Antunes *et al.* 1999. "Ecological economics and sustainable governance of the oceans." *Ecological Economics* 31(2): 171–87.

Doubleday, N.C. 1996. "'Commons' concerns in search of uncommon solutions: Arctic contaminants, catalyst of change?" *The Science of the Total Environment* 186: 169–79.

Downie, D., and T. Fenge, eds. 2003. *Northern lights against POPs: Combatting toxic threats in the Arctic*. Montreal: McGill-Queen's University Press.

Dressler, W., F. Berkes, and J. Mathias. 2001. "Beluga hunters in a mixed economy: Managing the impacts of nature-based tourism in the Canadian Western Arctic." *Polar Record* 37: 35–48.

Eddy, S., H. Fast, and T. Henley. 2002. "Integrated management planning in Canada's northern marine environment: Engaging coastal communities." *Arctic* 55: 291–301.

Fast, H., and F. Berkes. 1998. "Climate change, northern subsistence and land based economies." In *Canada country study: National cross-cutting issues*, vol. 8, edited by N. Mayer and W. Avis. Downsview: Environment Canada, 205–26.

Fast, H., J. Mathias, and O. Banias. 2001. "Directions toward marine conservation in Canada's Western Arctic." *Ocean & Coastal Management* 44: 183–205.

Fennell, D.A. 1998. "Ecotourism in Canada." *Annals of Tourism Research* 25: 231–235.

Folke, C., S. Carpenter, T. Elmqvist *et al.* 2002. *Resilience for sustainable development: Building adaptive capacity in a world of transformations.* Paris: ICSU, Rainbow Series No. 3. *http://www.sou.gov.se/mvb/pdf/resiliens.pdf*

Gunderson, L.H., and C.S. Holling, eds. 2002. *Panarchy. Understanding transformations in human and natural systems.* Washington, DC: Island Press.

Hanson, A.J. 1998. "Sustainable development and the oceans." *Ocean & Coastal Management* 39: 167–77.

Hay, K., D. Aglukark, D. Igutsaq, J. Ikkidluaq, and M. Mike. 2000. Final Report of the Inuit Bowhead Knowledge Study. Iqaluit: Nunavut Wildlife Management Board.

Holling, C.S. 1986. "The resilience of terrestrial ecosystems: local surprise and global change." In *Sustainable development of the biosphere*, edited by W. C. Clark and R. E. Munn. London: Cambridge University Press, 292–317.

——— , ed. 1978. *Adaptive environmental assessment and management.* New York: Wiley.

Huebert, R. 1999. "Canadian Arctic security issues." *International Journal*, Spring: 203–29.

INAC 2003. *Canadian Arctic contaminants assessment report II.* Ottawa: Indian and Northern Affairs Canada.

Jensen, J., K. Adare, and R. Shearer, eds. 1997. *Canadian Arctic contaminants assessment report.* Ottawa: Indian and Northern Affairs Canada.

Jentoft, S. 1999. "Legitimacy and disappointment in fisheries management." *Marine Policy* 24: 141–148.

Jentoft, S., B.J. McCay, and D.C. Wilson. 1998. "Social theory and fisheries co-management." *Marine Policy* 22: 423–436.

Kates, R.W., W.C. Clark, R. Corell *et al.* 2001. "Sustainability science." *Science* 292: 641–42.

Kendrick, A. 2000. "Community perceptions of the Beverly-Qamanirjuaq Caribou Management Board." *Canadian Journal of Native Studies* 20: 1–33.

Kooiman, J., ed. 1993. *Modern governance: New government – society interactions.* London: Sage.

Krupnik, I., and D. Jolly, eds. 2002. *The earth is faster now: Indigenous observations of Arctic environmental change.* Fairbanks AK: Arctic Research Consortium of the United States.

Kulchyski, P., D. McCaskill, and D. Newhouse, eds. 1999. *In the words of the elders: Aboriginal cultures in transition.* Toronto: University of Toronto Press.

Lee, K.N. 1993. *Compass and gyroscope: Integrating science and politics for the environment.* Washington, DC: Island Press.

Levin, S.A. 1999. *Fragile dominion: Complexity and the commons*. Reading, MA: Perseus Books.

Ludwig, D. 2001. "The era of management is over." *Ecosystems* 4: 758–64.

Mathias, J., and H. Fast. 1998. Options for a marine protected area in the Inuvialuit Settlement Region: Focus on beluga habitat. Inuvik: Report for the Inuvialuit Game Council on Behalf of the Fisheries Joint Management Committee.

McCay, B.J., and S. Jentoft. 1996. "From the bottom up: Issues in fisheries management." *Society & Natural Resources* 9: 237–50.

MEA. 2003. *Ecosystems and human well-being: A framework for assessment*. Millennium Ecosystem Assessment. Washington, DC: World Resources Institute/Island Press.

NSERC and SSHRC. 2000. *From crisis to opportunity: Rebuilding Canada's role in northern research*. Final Report to NSERC and SSHRC from the Task Force on Northern Research. Ottawa: Minister of Public Works and Government Services.

NRTEE. 1998. *Sustainable strategies for oceans: A co-management guide*. Ottawa: National Round Table on the Environment and the Economy.

Newton, S.T., H. Fast, and T. Henley. 2002. "Sustainable development for Canada's arctic and subarctic communities: A backcasting approach to Churchill, Manitoba." *Arctic* 55: 281–90.

North, D. C. 1994. "Economic performance through time." *American Economic Review* 84: 359–68.

Notzke, C. 1999. "Indigenous tourism development in the Arctic." *Annals of Tourism Research* 26: 55–76

OMRN. 2004. Ocean management research network. *http://www.omrn.ca/*

O'Neil, J., B. Elias, and A. Yassi. 1997. "Poisoned food: Cultural resistance to the contaminants discourse." *Arctic Anthropology* 34: 29–40.

Ostrom, E. 1990. *Governing the commons. The evolution of institutions for collective action*. New York: Cambridge University Press.

Pierre, J.B., and G. Peters, eds. 2000. *Governance, politics and the state*. Basingstoke: Macmillan.

Pinkerton, E., ed. 1989. *Co-operative management of local fisheries*. Vancouver: UBC Press.

RCAP. 1996. *Report of the Royal Commission on aboriginal peoples*. Vol. 2, Part 2. Ottawa: Supply and Services Canada.

RSC. 1995. *Aquatic science in Canada. A case study of research in the Mackenzie Basin*. Vancouver: The Aquatic Science Committee of the Royal Society of Canada.

Resilience Alliance. 2004. "What is resilience?" *http://www.resalliance.org/ev_en.php*

Rhodes, R.A.W. 1997. *Understanding governance: policy networks, governance, reflexivity and accountability*. Buckingham, UK: Open University Press.

Riedlinger, D., and F. Berkes. 2001. "Contributions of traditional knowledge to understanding climate change in the Canadian Arctic." *Polar Record* 37: 315–28.

Wenzel, G.G. 1991. *Animal rights, human rights: Ecology, economy and ideology in the Canadian Arctic*. Toronto: University of Toronto Press.

Wiber, M., F. Berkes, A. Charles, and J. Kearney. 2004. "Participatory research supporting community-based fishery management." *Marine Policy* 28:459–68.

Young, O. 2002. *The institutional dimensions of environmental change: Fit, interplay and scale*. Cambridge, MA: MIT Press.

UNDERSTANDING THE ISSUES:
LEARNING FROM EXPERIENCE

CHAPTER 2

FEEDING THE FAMILY IN TIMES OF CHANGE

Heather Myers (University of Northern BC)
Helen Fast (Department of Fisheries and Oceans)
Mina Kislalioglu Berkes (University of Manitoba)
Fikret Berkes (University of Manitoba)

INTRODUCTION

Northern peoples have traditionally depended on harvests of land and marine wildlife for their food, as well as for cultural definition and social connection. This tradition remains strong in some communities, while in others there appears to be a decline in harvest participation and production, raising questions about food security, nutrition, and health. Country foods have also been considered an important component of future economic development in the region, and could provide affordable, nutritious food for the region's residents. But changes are now occurring in Arctic communities, instigated by, among other influences, integration into the southern Canadian economy and incipient cultural colonialism regarding lifestyle choices. Added to these influences, Arctic residents are now receiving information about contaminant loads in Arctic wildlife and peoples, and about environmental change, usually generated by activities outside the Arctic. In addition, they are faced with economic development and conservation decisions that may affect how they can conduct traditional land-based pursuits. All of these new influences loop back to some fundamental questions of food security – a concept comprising access to acceptable, affordable, nutritious food: how will Arctic residents make choices regarding food sources in the future?

We will discuss a number of components that affect the future of food production and consumption in Arctic communities:

- a measure of domestic food production, consumption and sharing activities will provide a baseline for understanding the scope of traditional food use;

- an evaluation of the impacts of environmental change on the sustainability and utilization of marine resources will help to determine the capacity/limitations for future harvest;
- an analysis of community comprehension of contaminants in wildlife and humans, given a decade of information programs, will help to evaluate any consequent changes in behaviour and attitude;
- an assessment of economic development options and preferences regarding country foods will also measure changes in attitudes/opportunities.

An understanding of these components should contribute to policy development regarding Canada's Arctic oceans, as well as a research agenda to ensure ocean quality, resource management, use, and conservation.

SUBSISTENCE FOOD PRODUCTION AND CONSUMPTION

The basis of understanding many issues about the Arctic lies in the importance of northern peoples' traditional food harvesting, consumption, and sharing. Hunting, fishing and gathering, and processing of food and other products of the harvest, are important components of social, cultural, and economic life in the North. Traditional self-reliance in food production and the ethics of sharing and reciprocity have been maintained over millennia, despite the introduction of imported foods in recent years. The terms domestic food, traditional food, and country food are used synonymously, in this chapter, to refer to foods that are available from local natural resources and which are culturally accepted. Subsistence refers to the practices of producing such foods and related by-products for use within the household or for exchange with other households.

Even though there have been many socio-economic changes, these foods have remained important and desirable for most northern peoples. Northern Aboriginal identity is partly defined in terms of living off the land and producing food from lands and waters. Having access to and consuming wild food is important for core cultural values, such as sharing. The distribution of subsistence harvests to relatives and neighbours remains a widespread practice. For example, among the Cree people of western James Bay, about 50 per cent of all respondents reported sharing their food with three or more households. Even in the relatively large communities of Moose Factory and Attawapiskat, sharing with three to six other households was common (Berkes *et al.* 1994).

Both land use and a traditional economy persist in the northern parts of Canada's provinces and territories. This land-based economy has remained a cornerstone of the mixed economies of many northern communities, and despite predictions to the contrary, it has not been replaced by the modern wage economy (George and Preston 1987). But much of the value of the traditional economy is "invisible" to conventional economic analysis. Hunting brings food to the table but little cash transaction to the economy. Since the products of

Table 2.1

HARVEST ESTIMATES

Year	Region	Harvest	Per capita production
1978–9	E. James Bay, PQ	809,181 kg	115 kg/yr (7,022 people)
1984–5	Keewatin, NWT	895,298 kg	224 kg/yr (3,999 people)
1980	Northern Quebec	1,100,179 kg	285 kg/yr (3,857 people)
1990	W. James Bay, ON	686,500 kg	106 kg/yr (6,475 people)

hunting seldom pass through the market, government statistics do not place any value on subsistence, and hunters are technically defined as "unemployed." Yet, subsistence hunting is obviously important for the economy of small northern communities, many of them Aboriginal.

In the Northwest Territories, Usher (1989) estimated that subsistence production and processing added about 10 per cent to total labour income, and an estimated 80 per cent of native households participated in the domestic economy. Even though harvesting was done on a part-time basis, the average Arctic hunter took 1,000 to 1,500 kg of meat and fish annually with a replacement value of $10,000 to $15,000 (Usher 1989). Later estimates maintain this level of participation: 73–79 per cent of Nunavut Inuit males in 1999; 60 per cent of NWT Inuit and Inuvialuit in 1998 (Conference Board of Canada 2002, 34–35).

Berkes and Fast (1996) compiled and standardized all available harvest estimates, mostly from the 1980s, by community and by region (see Table 2.1 below). The general finding is that many Inuit communities in the Arctic obtained in the order of 200 kg per person per year of meat, mostly from wildlife (including waterfowl and marine mammals) and from fish. Various indigenous groups living in the Subarctic harvested in the order of 100 kg per person per year (Berkes and Fast 1996).

The replacement value or income-in-kind from country food may be substantial in these small, semi-isolated indigenous communities. On a per-household basis, the yearly value of the subsistence harvest for the Western James Bay Cree (1,116 households) was about $7,030. If other subsistence products, such as fuelwood, berries, and medicinal plants were taken into account, the in-kind income increased to $8,400 per year per household or $9.4 million for the region (Berkes *et al.* 1994). Adjusted to constant 1991 dollars, the average value of subsistence harvests (not including plant products) were in the order of $15–17,000 per household per year in the Arctic, and $6–9,000 in the Subarctic (Berkes and Fast 1996). On a territorial basis, in Nunavut this harvest carries real significance, where the replacement cost of country food harvested is (conservatively) estimated at $30–50M per year (Conference Board of Canada 2002, 31).

Studies that have included data on both the traditional and the non-traditional economy, including wage income and transfer payments, have made it possible to calculate the relative value of traditional harvests in the overall economy (Berkes and Fast 1996, Usher 1989, Treseder *et al.* 1999). For example, in western James Bay, the traditional economy comprised 25 per cent of the total economy. Across Canada, the range was from a high of 58 per cent in Sanikiluaq in Hudson Bay to a low of 11 per cent, excluding land-based commodities, in northern Manitoba (Berkes *et al.* 1994). Overall, the traditional harvest value was one-third that of the entire cash economy, easily exceeding the income from any other single source.

The above numbers may be considered to be indicative of the quantitative significance of the subsistence economy, but they should be treated with caution. They are based largely on questionnaire studies and are subject to the limitations of such studies (Usher and Wenzel 1987). Despite this, the major conclusion is that the traditional economy of many northern indigenous groups has remained alive and quantitatively significant. However, the amounts of country food harvested have declined perhaps in most areas.

A detailed comparison by Usher (2002) of Inuvialuit (western Arctic Inuit) harvests in the 1960s versus the 1990s provides insights into the nature of such changes. Using various data sources, Usher (2002) found that the total country food harvest declined from about 677,000 kg/yr in the

Rankin Inlet youth models elegant sealskin wedding gown: new livelihood opportunities may be developed from traditional resources. Photo by Steve Newton, 2004.

1960s to 333,000 kg/yr in the 1990s. The Inuvialuit population in the region nearly doubled during this time, while the total country food harvest declined by about one-half, so that per capita harvests in the 1990s were about one-third of those in the 1960s. Usher (2002) attributes most of this change to the decline of the dog team and the fur trade. Changes in the composition of the harvest support Usher's analysis. The harvest of the land animals is about the same, and caribou (a human food) has actually increased. Regarding the groups of species used partly for dog food in the 1960s, the harvest of marine fish is about one-quarter of what it was, and that of marine mammals is about one-half. Many of these shifts are accompanied by other changes.

There have been a number of transformations in the modes of production in the Arctic which have contributed to the changes in harvest and consumption of traditional foods since the 1960s, including such drastic changes as the collapse of fur markets in the 1980s (Wenzel 1991), the erosion of land-based knowledge

(Ohmagari and Berkes 1997), and gradual and perhaps more positive changes such as the use of the living resources of the Arctic to underpin sustainable, culturally appropriate economic development options (Treseder *et al.* 1999). These changes are being compounded by new and continuing influences, some of which we will discuss below. Many of these changes have impacted not only the harvest and the local diet, but also the relationship of indigenous peoples to their environment.

COUNTRY FOOD CONSUMPTION AND NUTRITION

The study of country food consumption has developed separately from the study of harvests. There is no direct way to convert harvest values into actual human food intakes, as the fraction of the harvest that becomes table food varies by area, by season, and by the proportion of desirable foods in the mix. A large number of studies since the 1970s has documented that traditional foods are important quantitatively – and even more important qualitatively. For example, among the Dene and Inuit of the Northwest Territories, Yukon, and Nunavut, only about one-third of the total food energy in the contemporary diet comes from traditional foods. But more than one-half of the protein comes from traditional foods (Kuhnlein *et al.* 2001). Thus, the question of the quality of the food is crucial.

Country foods are the most nutritious food available to northerners, providing protein, omega-3 fatty acids, key vitamins, and minerals. Imported foods are not only expensive, because they have to be transported into the northern communities, but also often a poor source of quality food. In northern indigenous communities characterized by high unemployment and low income, "affordable" imported foods, and the ones most commonly chosen, tend to be the less nutritious ones – high in refined carbohydrates, fats, sugars, and salt, but low in vitamins, fibre, and protein (Kuhnlein *et al.* 2001). Thus, it is easy for them to constitute a large proportion of the energy (or calories) consumed when imported foods form part of the diet. Indeed, Kuhnlein and Receveur (2003) have recently reported that, on the days that country foods are eaten, northerners' diets are high in various vitamins and minerals, as well as protein. On days that imported foods are eaten, their diets are high in sucrose, carbohydrates, and fats.

Work done at the Centre for Indigenous Peoples' Nutrition and Environment (CINE) has demonstrated that nutrient densities per 1,000 kcal of the traditional food portion of the diet are higher than those in the market food portion of the diet in seven of eight nutrient categories. As consumed by Inuit women of a Baffin Island community, there was more protein, iron, zinc, magnesium, vitamin A, and copper in the traditional portion of the diet and more calcium in the market food portion (Kuhnlein *et al.* 2001).

Thus, any decline in the consumption of these country foods is of concern from the point of view of dietary health. Worse, the switch from these foods to a diet of refined carbohydrates, high sugar intake and fats low in omega-3 fatty acids is a cause for concern with respect to increasing rates of heart disease, diabetes, obesity, and other diet-related problems among northern indigenous peoples. Given this background of the importance of traditional foods in Arctic communities, we now move on to discuss the impacts of environmental and social change in the Arctic, how these changes are perceived and understood by northern residents, and the implications of change for future economic and development decisions.

IMPACTS OF ENVIRONMENTAL CHANGE

The various factors of environmental change, and in particular the impact of contaminants, in raising concerns about traditional foods, should be considered in a historical context. The changes that indigenous peoples have had to deal with in the last two hundred or so years have been enormous. As summarized in Table 2.2, even in the last forty or so years, there have been many changes in the Arctic affecting the harvest and consumption of traditional foods and the relationship of indigenous peoples to the land. Other factors which could be added to the list of influences include the impacts of development, such as hydroelectric development in James Bay, oil and gas development in the western Arctic, and mining in Yukon and northern Manitoba (Berkes and Fast 1996), involvement in the wage economy (Kruse 1991; Myers 1982), climate change (Krupnik and Jolly 2002), and concerns over the health of the land due to contaminants in country foods (Jensen *et al.* 1997).

Concerns about pollution and contamination in the Arctic are not new. Food-chain accumulation of radioactive fallout from nuclear tests became an issue as early as the 1950s and the 1960s, and challenged the notion of the Arctic as a pristine environment. In the late 1960s and the 1970s, the contamination of northern aquatic ecosystems by methyl mercury became a major concern. This toxic form of mercury was found in fish harvested for commercial purposes and subsistence food, as a result of release of inorganic mercury into the environment from chlor-alkali plants in northern Ontario and Quebec, and as a consequence of the creation of hydroelectric reservoirs in northern Manitoba and Quebec (Berkes 1980).

By 1980, evidence began to accumulate of higher-than-expected values of persistent organic pollutants (POPs), such as pesticides and PCBs, in the Arctic. Although the initial reaction was to look for local sources, such as PCBs from DEW-line stations and pollution by the flow of Siberian rivers and the Mackenzie, attention soon shifted to the role of aerial transport as the main source. Cool temperatures in the Arctic favour the deposition of POPs from the atmosphere to land and water.

POPs are widely used in the South and can be transported long distances in the atmosphere. They bioaccumulate in the fatty tissues of organisms and they resist degradation. As in the radioactive fallout and mercury cases, species higher in the food chain tend to accumulate more POPs, resulting in widespread Arctic marine ecosystem contamination (Muir *et al.* 1992). The general conclusion is that colder temperatures result in degradation-resistant organic pollutants being

Table 2.2

CHANGES IN MODES OF PRODUCTION RELATED TO HARVEST AND CONSUMPTION OF
TRADITIONAL FOODS IN THE CANADIAN ARCTIC AND THE SUBARCTIC SINCE THE 1960S

Centralized settlements instead of seasonal mobility	Reduction of family involvement in harvesting, impacting the transmission of traditional environmental knowledge by reducing on-the-land learning
Adoption of mechanized transportation in place of dog teams	Decreased need for fish and marine mammals for dog food; increased need for cash income for purchase of equipment and fuel
Individualized hunting in place of group cooperation	Gradual abandonment of communal caribou drives, beluga hunts, and fish-trap fishing; loss of some social norms for appropriate behaviour on the land
Commercialization or de-commercialization of resources	Development of commercial hunts for some species; development of cultural tourism and ecotourism; loss of fur markets due to the animal rights lobby
Involvement in formal economy jobs	Responding to availability of non-traditional jobs (*e.g.*, mining; service industry employment) that provide cash incomes necessary for mechanized transportation and fuel
Harvesting as part of a mixed economy	Specialization of hunters; development of reciprocal arrangements between hunters and non-hunters within family groups; balancing of harvest costs/needs with job expectations and income; adjustment of traditional food sharing practices within the community
Availability and accessibility of market foods	Changes in tastes and values; acceptability of non-traditional foods as a result of media and schooling; adjustments to traditional food sharing practices
New knowledge about contaminant levels in wildlife	Adjusting modes of production to target foods with low contamination levels; minimizing those with high levels
New knowledge about the continued desirability of country food consumption	Reaffirming traditional wisdom about the desirability of country foods and the healthy influence of going on the land
The need for sustainable income-generating economic options	Searching for ways and incentives to develop options that can generate income, as well as being supportive of land-based activities

deposited in northern regions, continuing to accumulate long after their use has been reduced or stopped elsewhere (Muir *et al.*1992; Jensen *et al.* 1997).

Although the phenomenon of food chain accumulation was well known to ecologists, no data were available until the mid-1980s on contaminant intake through the diet of northern peoples and on levels of accumulation. A report by Dewailly *et al.* (1989) regarding high levels of PCBs in the breast milk of Inuit women from the Hudson Bay area came as a shock to many. In response to accumulating evidence on contaminants in country food, a project was carried out between 1985 and 1987 in the Inuit community of Broughton Island, Nunavut, known to have relatively high levels of traditional food harvests. The results showed that blood PCBs in many individuals, including two-thirds of those under fifteen years of age, exceeded the tolerable levels set by Health and Welfare Canada (Kinloch *et al.* 1992).

Accumulating evidence of Arctic ecosystem contamination prompted government research and response, coordinated by an intergovernmental Technical Committee on Contaminants in the Northern Ecosystems and Native Diets. The work of this committee and the follow-up studies have generated a large database on organochlorines, radionuclides and heavy metals. The work included the Broughton Island study, but was nevertheless criticized for failure to address human health concerns and for lack of indigenous representation. As a result, in 1989 the technical committee was expanded to include five Aboriginal parties: the Council of Yukon First Nations, the Dene Nation, the Metis Nation-NWT, the Inuit Tapirisat of Canada, and the Inuit Circumpolar Conference.

The task of addressing Aboriginal health concerns involved, among other things, the participation of Aboriginal people to help identify research priorities. For example, the Dene people of the Mackenzie Valley region had been reporting changes in fish quality at least since the early 1980s. Liver of burbot (*Lota lota*) is traditionally consumed by the Dene, but Dene fishers reported that burbot livers were becoming small and dark in colour, and were unfit for consumption. Initially oil contamination from Norman Wells was suspected. Laboratory studies revealed that the change in appearance was associated with a low fat content. Only low levels of low-boiling aromatic hydrocarbons were found in the abnormal livers, similar to control fish. Thus, no clear connection could be established between the liver condition and petroleum contamination. Later, the livers were found to contain unexpectedly high concentrations of toxaphene (up to 5,000 ng/g wet weight) and other organochlorines (Lockhart *et al.* 1987), being deposited by global transport processes.

The revamped technical committee, now including indigenous representatives, developed a five-year (1991–97) Northern Contaminants Program (NCP). In 1997, NCP published the summary of the findings of the research programs it had funded, the Canadian Arctic Contaminants Assessment Report or CACAR I (Jensen *et al.* 1997); these focused on determining levels, geographic extent, and sources of contaminants in the Arctic environment and peoples. The program was renewed with a five-year mandate and funded a further five years of research, focusing on impacts and risks to human health, as well as temporal trends of contaminants of concern in key Arctic species. Benefit-risk communication is now undertaken by the Aboriginal partners and territorial health departments, and considers the amount of country food consumption as well as the benefits of such consumption. The NCP has put considerable time into careful information (Indian and Northern Affairs Canada 2003, x–xi), including development of school curricula; Regional Contaminants Coordinators (RCCs); training courses for front-line workers; community tours by RCCs, health experts, scientists and an Aboriginal partner; Elder-scientist retreats; one-on-one and small-group communications.

The history of the impacts of environmental change in the Arctic, especially those related to contaminants, has been one of rapid change, surprise, discovery, and re-discovery. The role of the indigenous peoples has changed from one of passive recipient to one of active participant. This change in role may be related to the emergence of Aboriginal land claims, the assertion of increasingly greater authority of indigenous peoples over their land, and the acquisition of a greater voice in resource management and other issues.

However, the communication of research results and advisories to the community has remained a problem. In earlier studies, including the one in Broughton Island, results were communicated by "experts" in ways that caused alarm and confusion in communities. Many people ceased to eat country foods altogether, which brought a set of more immediate health problems and undermined the very real health benefits still to be derived from a country food diet. The increasing involvement of Aboriginal representatives through the NCP has improved communication, both in the design of projects and in providing dietary advice to communities based on the results. Consumers of traditional foods are indeed being exposed to contaminant levels that are of concern, but in view of the importance of traditional foods, most health advisories in the 1990s sought to balance risks versus benefits, rather than aiming to ban consumption (Kuhnlein *et al.* 2001). The benefits include the nutrition, taste, social and cultural values, health effects, educational and physical benefits of being on the land, and economic considerations. The risks include the contaminants and their uncertain health effects.

The risk-benefit analysis also needs to consider the risk of *not* harvesting and consuming traditional food (Kuhnlein 2002). Given the various social and economic changes impacting northern peoples (*e.g.,* Table 2.2), the contaminants problem is only one factor among many. The connection between people and the land is under multiple stresses. However, the risk of losing the connection to the land would, in turn, have additional consequences for the well-being of both the people and the land. In the final analysis, it may be that the message about contaminants is still either confused, not getting through to northern peoples, or being ignored.

COMMUNITY PERCEPTIONS OF CONTAMINANTS

The risks from contaminants in country foods are complicated to understand, even more so when one factors in language differences, conceptual differences, and cultural differences. Early lessons from the NCP showed the importance of communicating more carefully – listening to local observations of changes in fish and wildlife; understanding the likely behavioural responses to negative information and risk; balancing potential risks versus known benefits; and informing human subjects of study results and implications.

Northern residents' perceptions about contaminants have been influenced by early alarms, then assuaging messages, and more recently, a sort of confusing double message reflected in scientists, governments and Aboriginal organizations arguing before international audiences that the Arctic and its peoples are

Box 2.1
Synopsis of new knowledge from the NCP, 2003 regarding northern food supply.
Synopsis of some key findings from NCP II (from CACAR II Highlights, Indian and Northern Affairs Canada, 2003).

New findings have refined the understanding of pathways, regional differences in levels, and human contaminant intakes. Of interest are findings that levels of many POPs are decreasing (except for dieldrin and endosulfan), but there is uncertainty about what patterns characterize mercury levels in various environments. Some new contaminants are now being found. Mercury, heavy metals, and POPs levels in fish, wildlife, birds, and marine mammals show different patterns – between and within species and regions. For example, PCBs and dieldrins are dropping in beluga whales, but DDT and toxaphene are staying the same, and chlordane and endosulfan are increasing (Indian and Northern Affairs Canada, 2003, vi); mercury and cadmium levels in seal kidneys and livers are higher in some ringed seal populations than others (many are higher than guidelines set for fish consumption), but POPs in seal blubber are similar across the north and declining in some locations. Contaminants levels in marine invertebrates and fish and land mammals are relatively low, but in freshwater fish, mercury levels are rising – in some cases, exceeding subsistence consumption levels, but not commercial sale levels (Indian and Northern Affairs Canada, 2003, vi–vii).

Human health messages in NCPII have confirmed that nutritional, economic, social, and cultural benefits of eating country foods outweigh the currently known risks. There are many reasons that support the use of country foods: physical fitness, nutrition and disease prevention, social/cultural/spiritual benefits, economic benefits. Yet, in Kivalliq and Baffin communities, more than 25 per cent of the population is consuming more mercury than allowed for in the Total Daily Intake (TDI). Ten per cent of mothers in Baffin, and 16 per cent of those in Nunavik, have mercury blood levels in "increasing risk" levels, but blood levels of Inuit in Kivalliq and Kitikmeot, and those of Aboriginal peoples in other parts of the territories, are much lower, within the "acceptable" range (Indian and Northern Affairs Canada 2003, ix). pops intakes are generally below TDIS, but in the Inuvialuit, Kitikmeot, Kivalliq, and Baffin regions, many people (25–50 per cent) are taking in more than the TDI for chlordane and toxaphene. In the Baffin region, pcb intakes are also higher than the TDI. Interestingly, there have been few studies of the health effects of contaminants on northern peoples, though one is underway in Nunavik on vulnerable groups – women of child-bearing age, pregnant women, fetuses, and children. Contradictory results exist about the effects of mercury on children's development, and early evidence suggests that some level of protection is offered by vitamin E, selenium, certain fatty acids, and fish protein. In Nunavik, some relationship has been suggested between mothers' PCB blood levels and birth size, infection rate/immune abilities, and between DDT and infection rates in children and infants.

being contaminated, then arguing at home that eating traditional foods is "still the best." People are definitely wondering if food is safe to eat, but the messages remain mixed and the nuances difficult to grasp. Animals in one region may have higher-than-tolerable levels of one contaminant, yet that species in another region will not; one species in an area may have higher levels of contaminants, while another species in the same region will not (see inset Box 2.1). Food guide-lines suggest that certain types of food should be avoided by certain groups of people (such as *muktuk* by women of child-bearing age, pregnant women, or young children). Having alarmed northern residents in the 1980s, and calmed (or nuanced!) them in the 1990s with information that country food is "still the best," the NCP reports released in March 2003 suggested that there may now be identifiable impacts of existing human contaminant levels (Dewailly *et al.* 2003). It is understandable that northern people may be confused or uncertain about contaminants.

Indeed, a recent survey in Nunavut and Labrador,[1] focusing on contaminants comprehension among three target population groups, showed that only about 30 per cent in Nunavut understood the concept of contaminants, as defined and communicated by the NCP, while slightly more people in Labrador knew about and understood the issue (Myers and Furgal, under review). Primarily, hunters tended to be the most informed, and elders to some extent, but women with children were relatively unaware of this issue. People tended to relate the word/concept to concrete items they could see (garbage on the land, rusted met-als, development on the land), though some related the issue to environmental processes, pollution, viruses, or food poisoning. Many were certain that if an animal had contaminants, they would be able to see it, and either avoid it or cook it well enough to reduce the risk. This level of comprehension obviously has critical implications for food choices by Inuit, particularly for women with children who are making key diet choices for their families, affecting the growth and development of their children and the next generation.

Forty per cent of respondents in Nunavut said they had some concern about eating country food, but often defined the source of the concern as "skinny animals," sick animals, radio stories, bears at the dump, stomach ache, old meat. Very few specifically mentioned PCBs. Despite any concerns by respondents, all reported eating caribou and seal, with varying levels of consumption of other foods. Asked if there were country foods they did not eat, respondents mentioned all types of food, without pattern – reasons included availability, seasonality, taste preferences, "skinniness," and quota limitations – not contaminants. Regard-ing the nuanced information about different levels of contaminants in different species and foods, and the different susceptibilities of consumers (primarily women, children and fetuses), respondents did not reflect this understanding, and felt instead that they would *see* different levels of contamination and that everyone should eat country food because it is good for one (Myers and Furgal, under review).

Despite extensive public information efforts, comprising print and radio media, videos, school curriculum packages, public meetings, and community-based health committees, it is clear that information is not getting through to key parts of the population. A number of things may be contributing to this state of knowledge. Primarily, the obvious uncertainty about the science, especially reflected in the mixed messages, may have "turned off" the general listener. As well, general knowledge of Western science, not to mention understandable Inuktitut terminology for scientific concepts, is limited. This may be exacerbated by the tension between Western science and traditional ecological knowledge.

Part of this difficulty may be explained by the basic differences between the Aboriginal worldview and the Western one. Conventional contaminant research is, by definition, analytical and reductionist in approach, and therefore a high degree of specificity is a desired end point. By contrast, Aboriginal knowledge of the environment is holistic, inclusive, and fluid. Toxicologists have often noted that there is no precise or accurate term for contaminants in Inuktitut. One word used has been *ulurianaqtuq* (Baker Lake), which may be translated as, "the product of something dangerous." This non-specific "dangerous something" could mean an oil spill, toxic chemicals, garbage, viruses, or other (Jensen *et al.* 1997).

Such translations of scientific terminology may be frustrating to a toxicologist who is trying to communicate specific information on the relationship between toxicant levels (or intakes) and undesirable effects. But it may well be considered an extension of a worldview that is less specific than the Western one, but broader and more inclusive. Omura (2002) points out that systematic generalizations are in general regarded negatively by the Inuit. Such statements are thought to be oversimplifying and generalizing complex phenomena, and therefore "childish," without much sense (*ihuma*). A similar attitude is adopted by the Yupik of Alaska (Morrow 1990). Such considerations put the translation and communication problem in a different light. If difficulties in contaminant research are partly a result of differences between the two worldviews in observing and understanding the environment, then more innovative methods of communication may be necessary.

On the other hand, a profound, culturally based commitment to country food, and the lack of affordable alternatives, may create a state of denial among some northerners. Finally, the evaluation of risk has been extremely difficult to communicate with any accuracy, coloured as it is by scientific uncertainty, trade-offs with culturally-held values, personal preferences/tolerances, and language difficulties. In Arctic communities, faced with a number of pressing social and economic issues, people may also simply be suffering from information overload.

Complicating these communications/comprehension problems is the fact that communications habits in northern communities are different than those assumed to exist by southern-based scientists and officials – for instance, making authoritative statements or directing peoples' behaviour is an uncomfortable task for young or middle-aged Inuit – acceptable communications styles tend to be more passive. This is further complicated by the very high degree of personnel turnover, so that, for instance, health committee representatives, or health professionals, or teachers may become somewhat knowledgeable, or attend a meeting or two, then change jobs or leave town – and all that training is gone, but assumed by the larger system to still exist.

Adding to the difficulties of adapting to a changing future are other intergenerational communication issues. It is apparent to many northerners that continuing to enjoy the benefits of the renewable resource economy depends on transferring to succeeding generations the extensive traditional environmental knowledge (*Inuit Qaujimajatuqangit* [IQ]) held by elders and others. This transfer of knowledge is not occurring at the level or rate needed to ensure it remains viable. In the Kivalliq Region of Nunavut, for example, David Alagalak, President of the Kivalliq Wildlife Board has stated that there remain only fifteen years to access much of the IQ of the Kivalliq Region. Beyond that, elders will have passed on and the knowledge will have been lost (Hudson Bay Ocean Working Group 2003). In the Inuvialuit Settlement Region in the western Arctic there is general agreement among community leaders, parents, and elders that Inuvialuit youth will not be ready to assume leadership responsibilities as the present generation retires. Youth share this sense of failure. They attribute their lack of capacity not to lack of interest, but rather to an experience of being abandoned and forgotten. Youth want to acquire the skills and knowledge by experiencing the land first-hand and learning from family and elders (Schlag and Fast, this volume).

The perceptions of the contaminants issue may reflect a tension between traditional knowledge ("country food is best") and Western science ("invisible contaminants are in the meat"). The invisibility and long-term nature of the effects make it hard to prove the existence of contaminants. Government officials regularly receive anecdotal information about fish with no eyes, seals with no fur or very thin, caribou with oddly shaped hooves, caribou with low fat content or poor quality coats (Fast *et al.* 2001; McDonald *et al.* 1997; Fisheries and Oceans Trip Report 2001). Respondents to the Contaminants Comprehension survey often said that if animals were contaminated they would "see it" (Myers and Furgal, under review), and indicators cited reflected those that have been used/referred to since before the contaminants issue was raised – "skinny animals," unusual behaviour, white spots in the meat. Faith in visible, concrete knowledge about the Arctic environment and wildlife, based on traditional knowledge, remains firm, in contrast to Western science transmitted by southern scientists in sometimes incomprehensible language and concepts.

Despite the messages they are receiving about contaminants, the Comprehension survey discovered that about 30% of Nunavut respondents "would change their food choices if they were found to be contaminated," but few respondents (about 10%) thought they had actually been exposed–and that they could avoid contaminants by food selection or cooking, for example (Myers and Furgal, under review).

There are a number of implications of both the contaminants issue and the state of knowledge among northern peoples. The existence of contamination in wildlife threatens northerners' accepted view of wildlife as a valued source of food, or as a never-ending/unquestioned component of the Arctic environment. Questions about the survival and health of certain species, such as polar bears and whales, strike at the very essence of traditional ecological identity and integrity, and threaten the spiritual and physical health of Arctic peoples.

In addition, on a practical level, hunters' roles in northern communities may be undermined – from a valued role in providing for the community, they may be cast in a questioned role as producers of "contaminated" food. Elders' dependence on country food produced by their families and the community may be undermined if hunters feel less confident or committed to such production. The sharing of food within extended families and communities, which reinforces important social and cultural values, may be undermined if country food is less available or desirable, though sharing of imported foods may replace it. Country food stores operated by some community-based organizations may be caught between less demand and less production, thus undermining a potentially important economic opportunity.

Income is potentially threatened by these food changes, since producers may be less able to earn cash income from their harvest and consumers will find it costly to purchase (nutritious) imported foods. Less measurable but equally important is the loss of confidence, pride, and identity associated with traditional production, processing and consumption of food.

ASSESSMENT OF ECONOMIC OPTIONS AND PREFERENCES
REGARDING COUNTRY FOOD

The uncertainty and lifestyle changes affecting northern peoples have serious impacts on options for both formal and informal incomes. Most residents of small remote Nunavut communities currently get, and expect to continue to get, a lot of their food from subsistence harvesting. Wage-paying work and cash-earning options remain limited, encouraging people to produce at least some of their own food (depending upon affordable costs for harvesting). This is encouraged by the Government of Nunavut and Nunavut Tunngavik Incorporated.

Subsistence provides a type of security which cannot be had from cash incomes or from wage employment. The Government of Nunavut is not expected to grow much more, meaning few new job opportunities in that sector. The Bathurst Mandate (Nunavut Visioning Exercise) encourages *Nunavummiut* to "build on their strengths": the arts and crafts economy, tourism, and harvesting of natural resources. Harvesting includes subsistence harvest; market harvest for local sale; organized community hunts where hunters are reimbursed for bringing country food into the community for distribution; and commercial harvest for large-scale resale in both domestic and export markets (Junkin 2002, 11).

The Nunavut Land Claims Agreement (1999) gives Inuit "security of tenure" to the wildlife and participation in its management with the territorial and federal governments through co-management boards (Junkin 2002, 9). This commitment is expected at the federal level as well. In community meetings in 2001–3, DFO was told that Canada's *Oceans Strategy* must also support Inuit efforts to maintain their culture and lifestyle. This includes rights to maintain harvesting activities and the need for a "coordinated, transparent and inclusive approach to the management of marine and ocean resources" (Terriplan 2003, 4–5).

Spring fishing is an important family event, an opportunity for subsistence harvesting and also for intergenerational training. Photo by Heather Myers.

The role of land and marine resources in the future of the Nunavut economy requires more detailed planning. Currently, the marine sector accounts for 5–10 per cent of the total NWT economy and more than 10 per cent of the Nunavut economy (G.S. Gislason & Associates Ltd. 2002). The critical questions are: what will people focus on as food and economic sources in the future? What problems, considerations and opportunities might affect these choices? Questions arise regarding regional understanding and decision-making about future domestic and commercial food production, and whether – or how – social expectations will need to be adjusted.

There is some concern that if all young Nunavummiut carry on the subsistence values of their parents, their increasing numbers might put a strain on carrying capacity of the marine resources. The future importance of domestic harvesting cannot be predicted: many point to changing lifestyles and increased use of store-bought food by young people, while others note that employment does not necessarily reduce harvest participation, and that reduced participation rates may only be related to youthfulness and may increase as youth mature and take on family/social obligations (Condon *et al.* 1995; Kruse 1991). Food Security surveys in Nunavut in 2000–2001 showed young respondents actively participating in harvesting and consumption of country food (Myers *et al.* in press). If low employment remains a condition in Nunavut communities, interest in self-reliant domestic food production is likely to continue. As an economic development opportunity, as well, there could be interest in commercial production of country food.

Commercial fisheries have been conducted for decades in Arctic communities, often on a small family- or community-based scale. More recently, some of these have been organized, with collector boats gathering the catch from dispersed fishers and transferring it to processing plants. Some larger freezer-packer boats and experimental fisheries have also been tried out, often encountering difficulties with equipment and transportation (Myers 2000). The experiences in fisheries suggest that small-scale ventures may be more sustainable in northern communities, whereas large-scale enterprises tend to bring with them serious problems in terms of transportation, cost, infrastructure, capacity in work skills and management, and marketing (Myers 2000). While serving local markets can help to diversify and develop the local economy, expanding into other markets, particularly beyond the territory, may be difficult for northern entrepreneurs in

Table 2.3

VALUE OF THE MARINE ENVIRONMENT IN TERMS OF COMMERCIAL FISHERIES, ARTS & CRAFTS, AND SPORT USE

		GDP	Employment	Income
Commercial Fisheries	Nunavut	$8M	140 PYS	$3.5M
Arts and Crafts	NWT	$1.6M	30 PYS	$1.2M
	Nunavut	$12M	200 PYS	$10M
Sports hunts/fishing	NWT	$1M	10 PYS	$0.6M
	Nunavut	$3.4M	55 PYS	$2.0M

terms of knowing and serving that "other" market. In Sanikiluaq, for example, a small scallop fishery has been initiated, which generates a break-even source of income for a few families, selling to the local market. Should they wish to expand into other markets, however, they will encounter barriers of high transportation costs, the high cost of building a commercial processing plant that meets federal standards, and federal testing (M. Fleming, pers. comm., 11 July 2003).

There is a clear desire to develop more commercial fisheries in Nunavut. Currently, there are no commercial marine fisheries in NWT, but in Nunavut there are commercial fisheries for char, clams, shrimp, and turbot. Char is the oldest commercial fishery in Nunavut, often taking place in small, dispersed fisheries by family-based or community-based groups. An estimated 90 tonnes was landed in 2001, with almost two hundred operators and hired hands working in this fishery. Of the turbot quota, 5,000 tonnes goes to Nunavut fishers, 2,500 tonnes to company quotas, and 1,500 tonnes to a competitive fishery involving southern Canadian companies. Some of the catch is processed in Pangnirtung (365 tonnes), but much of it goes south. There are seventeen offshore shrimp licences, most of which are leased to, and fished by, non-Nunavut companies in return for royalty payments plus employment and training opportunities. A commercial clam fishery is experimental, with less than 10 per cent of the 55,000-kg annual quota having been caught in recent years. It is sold to local people for $8.50 per kg, but cannot be exported because it cannot be tested without access to a bivalve testing facility. Overall, harvesting and processing of finfish and shellfish, as well as royalties, which are a significant component, provide an important input to the Nunavut GDP, employment, and income (see Table 2.3) (G.S. Gislason & Associates Ltd. 2002, 12–14). Between 1992 and 1995, $4.3 million was invested by various levels of government to support the Baffin Region fisheries, in order to locate commercial stocks, support pilot projects, explore new technologies, and improve marketing and lobbying on behalf of the industry (GNWT 1996).

Further value is derived from the marine-based economy in the form of non-food-related economic activities such as arts and crafts production and sports hunts/sport fishing (Table 2.3). Arts and crafts can provide a significant component of community economies, from products including carvings, sealskin products, prints, wall hangings, and tapestries. Relatively little information is available on the economic value of this sector since most artists work from home and StatsCan does not collect information. Marine tourism includes polar bear hunts, cruises, angling and sea kayaking, boat tours, wildlife viewing, and so on. Polar bear hunts are the largest of these activities in NWT and Nunavut, with an average hunt generating about $30,000 in northern revenues. Of this, $25,000 goes to the outfitter and the rest to related expenses. There are about forty polar bear hunts in the NWT and seventy-five in Nunavut, generating expenditures of $1.2 million and $2.25 million respectively. Sport fishing in Nunavut focuses on Arctic char, pike, grayling, cisco, and other species (G.S. Gislason & Associates Ltd. 2002).

The optimism about future development of marine resources, whether for subsistence or commercial purposes, must be tempered by consideration of several knowledge and management needs. Generally, marine fishery development in the Arctic faces obstacles of remoteness, ice conditions, expensive technology and related training needs, expensive transportation, quota regulations, and limited knowledge of Arctic fishery potentials. There is a need for clear procedures for issuing fishing licences, deciding quotas, and giving TEK its rightful place in establishing quotas and complementing scientific stock assessment numbers. These and other concerns have come out in community consultations with Nunavut communities and hunters and trappers organizations (Terriplan 2003):

Management knowledge and capacity: *The need for better research and information on harvesting data and inventories, and thus greater certainty in quotas and regulations; the use of Inuit Qaujimajatuqangit or TEK; the need for researchers to report back findings.* The lack of scientific information results in uncertainty with regard to setting quotas for a variety of inshore and offshore marine species and complicates developing community- or commercially-based experimental fisheries; there are only two technicians and one biologist for the region (J. Maurice, pers. comm., July 14, 2003). Communities on the other hand, are eager to explore the commercial potential of other species. Traditional knowledge and scientific knowledge of stocks are sometimes not in agreement (whether for fish or marine mammals), and this makes quota setting, harvest regulation, and the distribution of fishing licenses very difficult (Terriplan 2003, 15–16). In Arctic Bay, residents argued that quota systems encourage misuse of resources – traditional rules are better to maintain resources; they also wished for greater consultation before regulations are developed. On the other hand, employment participation in both inshore and offshore fisheries require skilled workers, which are lacking in Nunavut.

Infrastructure and licensing: *Adequate ports, harbours and processing plants are needed; more exploratory fisheries are needed for local economic development; fairer distribution of fishing licences; more visible DFO staff.*

There are currently no deepwater port facilities to support significant commercial harvesting activities in Nunavut, and only four registered processing plants. Several communities commented on confusing procedures for applying for/distributing commercial and exploratory fishing licences, and a lack of clarity regarding how narwhal quotas were established (Terriplan 2003, 17). Others were concerned with equity regarding non-Nunavut-based trawlers. Not one Nunavut fishing company owns its own offshore fishing vessel, and Nunavut fishers only have the options of selling or transferring their quota, leasing a southern vessel, or going into contracts with foreign vessels. Furthermore, Nunavut fishers have limited access to adjacent turbot and shrimp quotas, contributing to the challenges for expansion (J. Maurice, pers. comm., July 14, 2003). DFO presence was desired for research and more expeditious information transfer.

Environmental quality and change: *Evaluation of climate change implications on conditions such as ice break-up; protection of wildlife health, including waste and sewage management, water quality and health implications for some stocks (e.g., clams) harvested by local people.* Sewage treatment is a growing issue in Nunavut communities and becomes very pointed where local clam populations are affected. Many communities also want to see more research regarding contaminants and wildlife health. In Resolute and Arctic Bay, mine tailings are a local concern, and in Sanikiluaq, impacts of hydro development are a concern regarding marine ecosystem health. In many communities, people voiced concerns about climate change and the impacts of ship traffic or tourism on marine life.

Impacts of industrial fishery development: *Assessment of the impact of trawlers on the turbot fishery in Cumberland Sound.* There is concern about the impacts of large trawlers on the Cumberland Sound turbot fishery and implications for small-scale fishermen and species that feed on turbot, which include seal, beluga, and narwhal. The communities are unclear as to the roles of the HTOs, the NWMB and DFO in setting quotas (Terriplan 2003, 23).

RESEARCH AND POLICY ISSUES REGARDING FOOD SECURITY

In summary, it is apparent that the influences bearing on the country food production systems of northerners are not only historical – they continue today. Despite the abiding value of country foods to northern peoples and economies, their relationship and dependence on those foods is being pressured to change. These changes affecting modes of production as well as consumption of traditional foods, both historical and more current, are summarized in Table 2.2.

Yet, despite these changes, and given Arctic peoples' location, environment, resources, culture, and economic circumstances, country foods remain an important subsistence and commercial resource. It is clear that these peoples must still find ways to feed their families. To the important subsistence harvest of food from the Arctic lands and oceans, there is added the potential for some degree of commercial development of these resources. While there may be unknowns regarding the future extent of fishery use and country food production, it does seem clear that country foods will continue to be very important for reasons of health, culture, and economy.

It is also clear that the changing Arctic environment may affect this sector of activity, though the evidence from Nunavut and Labrador suggests that peoples' valuation of country food is not (yet) much affected by the information they have received. Contaminant levels will likely have an effect on the potential for commercial development, however.

Finally, in terms of human management of resources and their use, it is clear that capacity lags behind need, for fishery knowledge and management, and for commercial development, particularly at larger scales, whether for population data, infrastructure, or skills.

Considering these changes leads the authors to some recommendations regarding both research and policy development, in the realms of northern oceans, resources, and economies. There are a number of questions and information needs arising from our survey of these issues, which affect the ability of northerners to feed themselves in either subsistence or commercial ways:[2]

> *Research is needed to understand the nature of changes in country food consumption at both local and regional levels.* It is clear that traditional foods continue to be important quantitatively – but even more so qualitatively, in terms of nutrition and culture. Though total country food harvest has declined over the years, much of this change may be related to the decline of the dog team and the fur ban, rather than a decline of country food used for direct human consumption.

> *Research and policy discussion are needed on the question of sustainable, culturally appropriate (formal and informal) economic options based on the land and country foods.* This should take into account community and local-level characteristics and needs. The message has not changed for decades: country food is important to northerners; its pursuit must be supported in ways that are appropriate to northerners.

> *More public discussion and policy development are needed regarding the elements that should be included in the "new" risk-benefit approach to contaminants in country food, and the currency or currencies to be used.* For example, is mortality and morbidity the only relevant currency, or should there be an accounting of human satisfaction and cultural values of living off the land? Is the apparent "lack of awareness" by northern residents of the contaminants information essentially this – a re-weighting of risk factors? What factors do influence northern peoples' behaviour

in the face of new information like that about contaminants? And of course, there must be ongoing research into the levels and impacts of contaminants in the Arctic environment, wildlife, and peoples.

Research and decisions are needed regarding what kinds of indicators (social sustainability indicators; human satisfaction indicators) should be used to track change. These will need to draw on *Inuit Qaujimajatuqangit* as well as modern/southern values – those valued by young and old in Arctic communities. They will also need to blend values regarding environmental quality, domestic and commercial economies, social continuation and connection.

More commitment and support is needed from DFO *to develop commercial/ economic opportunities for northern peoples, based on northern resources and values.* For example, policy, supporting legislation, and action are needed to assess Arctic commercial fishery quotas. This should assess the viability of commercial fisheries and the development of appropriate-scaled infrastructure. This would comprise the population dynamics of target fisheries and their ecosystems and the level of fishing pressure they can sustain; then an assessment of the size of fishery that can be sustained biologically and economically, the training of a workforce, and development of appropriate infrastructure for fishing and transportation.

Given the size and capacities of northern communities and resources, measures of "success" need to be adapted to a scale which is appropriate, sufficient and sustainable in the north. This will probably mean that smaller but longer-term activities are more acceptable, more locally meaningful, and more economically viable in the Arctic setting.

Given the sweeping and rapid nature of changes, even in the last forty years (Table 2.2), new ways are needed with which to anticipate and deal with change and surprise. The experience of "feeding the family," as examined in this chapter, indicates a changing world in which surprise is likely. The conventional policy emphasis has been on managing the environment and resources. This has proven to be difficult to do in rapidly changing environments, such as the Arctic, where the future is unpredictable. Instead, perhaps the policy emphasis should be on building adaptive capacity, so that local institutions, and local people feeding their family, are better able to respond to surprises by learning to deal with change. This is the area of resilience thinking and the subject of the next section of the book.

REFERENCES

Berkes, F. 1980. "The mercury problem: an examination of the scientific basis for policy-making." In *Resources and the environment*, edited by O.P. Dwivedi. Toronto: McClelland and Stewart, 269–87.

Berkes, F., P.J. George, R.J. Preston, A. Hughes, J. Turner, and B.D. Cummins. 1994. "Wildlife harvesting and sustainable regional native economy in the Hudson and James Bay Lowland, Ontario." *Arctic* 47: 350–60.

Berkes, F., and H. Fast. 1996. "Aboriginal peoples: The basis for policy-making towards sustainable development." In *Achieving Sustainable Development*, edited by A. Dale and J.B. Robinson. Vancouver: UBC Press, 204–64.

Condon R., P. Collings, and G. Wenzel. 1995. "The Best Part of Life: Subsistence Hunting, Ethnicity, and Economic Adaptation Among Young Adult Inuit Males." *Arctic* 48(1): 31–46.

Conference Board of Canada. 2002. *Innovation and traditional resource-based economies study, May 2002*. Ottawa: Conference Board of Canada.

Dewailly, E., A. Nantel, J.P. Weber, and F. Meyer. 1989. "High levels of PCB in breast milk of Inuit women from Arctic Quebec." *Bulletin of Environmental Contamination and Toxicology* 43: 641–46.

Dewailly, E., G. Muckle, C. Furgal, and P. Ayotte. 2003. "Regional, national and international risk management: how to deal with cohort study results." In *Canadian Arctic Contaminants Assessment Symposium, Ottawa, March 4–7, 2003, Abstracts*: O-H11.

Fast, H.B., S. Eddy and O. Banias, eds. 2001. *Charting a Coordinated Approach to Management of the Western Hudson Bay Region. Proceedings of the Western Hudson Bay Workshop, Winnipeg, MB, October 23–25, 2000*. Ottawa: Canadian Manuscript Report of Fisheries and Aquatic Science 2590.

Fisheries and Oceans Canada. Trip Report: Churchill, Kivalliq and Iqaluit Community Tour, March 7–19, 2001. Unpublished.

George, P. and R.J. Preston. 1987. "'Going in between': The impact of European technology on the work patterns of the West Main Cree of northern Ontario." *Journal of Economic History* 47: 447–60.

GNWT. 1996. Towards a Baffin Fisheries Strategy, Part I – Background Report. Department of Resources, Wildlife and Economic Development. Unpublished.

Government of Nunavut, Bathurst Mandate. *http://www.gov.nu.ca/Nunavut/English/departments/bathurst*

G.S. Gislason & Associates Ltd. 2002. The Marine-Related Economy of NWT and Nunavut. Draft report prepared for Fisheries and Oceans Canada. Unpublished.

Hudson Bay Ocean Working Group. 2003. Notes from Traditional Ecological Knowledge Focus Group, Winnipeg, June 25–26, 2003. *http://www.umanitoba.ca/academic/institutes/natural_resources/im-node/hudson_bay/index.shtml*

Indian and Northern Affairs Canada. 2003. *Canadian Arctic Contaminants Assessment Report II: Highlights*. Ottawa.

Jensen, J.K. Adare and R. Shearer, eds. 1997. *Canadian Arctic contaminants assessment report*. Ottawa: Indian and Northern Affairs Canada

Junkin, B. 2002. Economic Development Activities & their Orientation. Prepared for the Economic Development Committee of the Hudson Bay Oceans Integrated Management Group. Unpublished.

Kinloch, D., H.V. Kuhnlein and D.C.G. Muir. 1992. "Inuit foods and diet: A preliminary assessment of benefits and risks." *The Science of the Total Environment* 122: 247–78.

Krupnik, I. and D. Jolly, eds. 2002. *The Earth Is Faster Now: Indigenous Observations of Arctic Environmental Change*. Fairbanks: Arctic Research Consortium of the United States.

Kruse, J. 1991. "Alaska Inupiat Subsistence and Wage Employment Patterns: Understanding Individual Choice." *Human Organization* 50(4): 317–26.

Kuhnlein, H.V., O. Receveur, and H.M. Chan. 2001. "Traditional food systems research with Canadian indigenous peoples." *International Journal of Circumpolar Health* 60: 112–22.

Kuhnlein, H.V. 2002. Indigenous peoples' food systems: Perspectives on contaminant risks and nutrient benefits. Paper Presented at the Monbukagakusho International Symposium, National Museum of Ethnology, Osaka. Unpublished.

Kuhnlein, H.V., and O. Receveur. 2003. CINE Dietary exposure assessments of arctic indigenous peoples for risks and benefits, *Canadian Arctic Contaminants Assessment Symposium, Ottawa, March 4–7, 2003; Abstracts*. Ottawa: Indian and Northern Affairs Canada.

Lockhart, W.L., D.A. Metner, D.A.J. and D.C.G. Muir. 1987. "Hydrocarbons and complaints about fish quality in the Mackenzie River, Northwest Territories, Canada." *Water Pollution Research Journal of Canada* 22: 616–62.

McDonald, M.A., L. Arragutainaq and Z. Novalinga. 1997. *Voices from the Bay: Traditional Ecological Knowledge of Inuit and Cree in the Hudson Bay Bioregion*. Ottawa: Canadian Arctic Resources Committee.

Morrow, P. 1990. "Symbolic actions, indirect expressions: Limits to interpretations of Yupik society." *Etudes/Inuit/Studies* 14: 141–58.

Muir, D.C.G., R. Wagemann, B.T. Hargrave, D.J. Thomas, D.B. Peakall and R.J. Norstrom. 1992. "Arctic marine ecosystem contamination." *The Science of the Total Environment* 122: 75–134.

Myers, H. 1982. "Traditional and modern sources of income in the Lancaster Sound Region." *Polar Record* 21(130): 11–22.

——— . 2000. "Options for appropriate development in Nunavut communities." *Etudes/Inuit/Studies* 24(1): 25–40.

——— , and C. Furgal. Under review. Communicating about contaminants in Nunavut and Labrador.

——— , S. Powell, and G. Duhaime. In press. "Setting the table for food security: Policy impacts in Nunavut." *Canadian Journal of Native Studies*.

Nunavut Land Claims Agreement. 1999. Government of Canada. Ottawa: Queens Printer.

Ohmagari, K., and F. Berkes. 1997. "Transmission of indigenous knowledge and bush skills among the Western James Bay Cree women of subarctic Canada." *Human Ecology* 25: 197–22.

Omura, K. 2002. Problems on use of traditional ecological knowledge in resource management: A case from the Canadian Arctic. Paper Presented at the Monbukagakusho International Symposium, National Museum of Ethnology, Osaka. Unpublished.

Schlag, M., and H. Fast. This volume. Marine Stewardship and Canada's Oceans Agenda in the Western Canadian Arctic: A Role for Youth.

Treseder, L., J. Honda-McNeil, M. Berkes, F. Berkes, J. Dragon, C. Notzke, T. Schramm, and R.J. Hudson. 1999. *Northern Eden: Community-Based Wildlife Management in Canada*. London: International Institute for Environment and Development.

Terriplan Consultants Ltd. 2003. IER Planning, Research and Management and updated by J. Maurice. Toward Integrated Ocean Resource Management in the Baffin Island Region of Nunavut: Results of the Baffin Marine Issues Scan. Report prepared for Fisheries and Oceans Canada.

Usher, P.J. 1989. *Towards a strategy for supporting the domestic economy of the Northwest Territories*. Yellowknife: Background study prepared for the NWT Legislative Assembly's Special Committee on the Northern Economy.

——. 2002. "Inuvialuit use of the Beaufort Sea and its resources, 1960–2000." *Arctic* 55 (suppl. 1): 18–28.

——, and G. Wenzel. 1987. "Native harvest surveys and statistics: A critique of their construction and use." *Arctic* 40: 145–60.

Wenzel, G.W. 1991. *Animal Rights, Human Rights: Ecology, Economy and Ideology in the Canadian Arctic*. Toronto: University of Toronto Press.

NOTES

1 Awareness, comprehension and perception of contaminants issues in two regions of the Canadian north, funded by Northern Contaminants Program II.

2 In all of these examples, it is assumed that information will draw on both southern scientific, and northern traditional knowledge.

CHAPTER 3

SUSTAINABILITY AND VULNERABILITY:

ABORIGINAL ARCTIC FOOD SECURITY IN A TOXIC WORLD

Shirley Thompson (University of Manitoba)

It is not so much that humanity is trying to sustain the natural world, but rather that humanity is trying to sustain itself. The precariousness of nature is our peril, our fragility.

— Amartya Sen, Nobel Laureate Economist

The central and most distinguishing feature of the modern Arctic indigenous economy continues to be its dependence on wildlife and the habitat that supports it.

— Arctic Indigenous Peoples' Secretariat, 2002

Environmental change impacts food security in Aboriginal communities in Canada's Arctic. Northern Aboriginal communities are widely recognized as having mixed, subsistence-based economies in which the harvesting of country food for primarily domestic consumption plays a significant role in their food security and culture (Usher *et al.* 2003). Since time immemorial, Canada's Aboriginal peoples were self-sufficient, through subsistence-based activities, in the harsh Arctic climate without causing degradation to their environment (Usher *et al.* 2003). However, colonization and modernization makes their food supply and subsistence activities vulnerable.

The undermining of sustainable societies, which preserve ecological capital, by unsustainable societies, which draw down ecological capital through trade (*e.g.,* overfishing, fur trading, mining) and formulating government policies outside of ecological and cultural context, is discussed by a variety of authors (Churchill 1999; Parajuli 1997). This chapter focuses on the factors compromising sustainability and food security of Inuit in Nunavut. "Toxic" impacts in the Arctic that undermine sustainable livelihoods go beyond merely mercury and persistent organic pollutants (POPs) contamination and include climate change and Eurocentric autocratic government policy that has included relocating Aboriginal peoples. Hofrichter (2000, 1) applies the term "toxic culture" to show how social arrangements encourage and excuse the deterioration of the environment, culture, and human health. Atmospheric pollution, over which northerners have little or no control, debases the sustainable lifestyles of Aboriginal peoples, as does depleting resources (*e.g.,* Peary Caribou are endangered in the high Arctic and in Banks Island and threatened in low Arctic), restricting hunting and gathering (*e.g.,* harvesting limits exist for polar bears, Beluga whales, etc.) and government policies that disregard place-based knowledge (*e.g.,* mandatory schooling focused on Euro-Canadian culture, and relocation) (Marcus 1995).

Inuit "traditional way of life," modernized to include new technology such as snowmobiles, has a small ecological footprint. An ecological footprint is the number of hectares required to sustain individuals, communities, and nations based on population size, average consumption per person, infrastructure for housing, transportation, industrial production, and the resource intensity of the technology being used (Redefining Progress 2004). As most food in the global economy travels more than 5,000 kilometres, any subsistence activity reduces transportation and environmental impact. Today's global food system is dependent on mechanization, large inputs of fertilizer and pesticides, mono-cropping, green revolution- and bio-technologies, processing or refrigeration, as well as vast transportation, marketing, and supermarket networks (Gottlieb 2002). In contrast, Inuit hunting and gathering derives food from local sources in the natural, unmanipulated environment.

Although environmental change is nothing new, the number of assaults and their magnitude are increasing with toxic contamination, ozone depletion, resource scarcity, and mass extinction of species, of which a number of Arctic species are at risk. Nature's constant state of flux has accelerated, impacted by settlement, mining, industry, infrastructure, and military activity so that environmental change is described by Inuit elder Mabel Toolie as "the Earth is faster now" (Caleb Pungowiyi quoted in Krupnik and Jolly 2002, 7). Such changes alter the quality and quantity of environmental resources, thereby diminishing sustainability of Aboriginal Arctic communities. Dependence on a compromised local physical environment, without wealth to import resources from elsewhere, limits options for survival. Impacted by the toxic aspects of modernity, Inuit communities in Nunavut are vulnerable, as their coping capacity is severely limited by poverty and minimal infrastructure. This chapter first looks at the importance and vulnerability of food security and subsistence activities before looking at the impact of environmental change on food security in Nunavut's Inuit communities due to: (1) poverty undermining food security; (2) contamination causing toxic impacts; and (3) government policies restricting access to land and resources.

FOOD SECURITY THROUGH SUSTAINABLE LIVELIHOODS

Chambers and Conway (1992) define sustainable livelihood as comprised of the capabilities, assets, and activities required to obtain a means of living. Sustainable livelihoods require people to have an intimate knowledge of the land to engage in resource use. Knowledge and values of 'ecosystem people' (Parajuli 1997) who have a reciprocal relationship with their respective ecosystems is very different from those of 'biosphere people' who draw on resources from afar and often transform those resources through industrial processes. For Aboriginal peoples, this knowledge comes from cultural teachings, as well as generations of resource harvesting from traditional areas. Indigenous communities have unique lifestyles intimately adapted to local climate, vegetation, and wildlife and may be particularly threatened by environmental change (Parajuli 1997). Inuit peoples in Nunavut, as ecosystem peoples, have evolved knowledge about plants and animals, their habitat preferences, local distribution, life histories, and their seasonal behaviour, with an interest in ensuring the long-term availability of the natural resources of their own localities. The transmission of this knowledge from one generation to the next, along with a philosophy of sharing, has shaped customs that have promoted sustainability among nomadic peoples (Environment Canada 2003). Much of this knowledge is used to obtain food, medicine, and other necessities, and in avoiding crises caused by natural calamities (Berkes, Folke and Gadgil 1995).

Sustainable livelihoods are especially important in environments that are fragile, marginal, and vulnerable, like the Arctic (Chambers and Conway 1992). Inuit people in the Arctic live in a harsh climate, with wildlife and other natural resources forming the basis of Inuit society, culture, and economy (Environment Canada 2003). They describe their ancestors as "curious and inventive. They constantly searched for new ways of doing things to make their life more secure. Thus, they became very knowledgeable about the land and ways of hunting" (Northwest Territories Education 1991, 1). One Inuit hunter explains how good trappers practice sustainable trapping:

> If you're a good trapper, you know which animals to trap at certain times of the year, and you know which animals not to trap in a given year, because they're at the bottom of their cycle. Most animals are in a seven-year cycle ... so if you know that, as a trapper, you can sustain your living. That's what sustainable means. (Elmer Ghostkeeper quoted in Wuttunee 1995, 207)

Rather than serving the purpose of accumulation at the level of enterprise, harvesting country foods tends to be inseparable from the social system (Usher *et al.* 2003). Economic and social relations among households in a community are guided by kinship principles, which are the primary determinant of access to resources and the organization of labour for productive activities as well as of the distribution of goods and services for consumption, ensuring food security. Successful households do not accumulate wealth for their own private use; they share their produce with other households, distributing their excess production widely through the kinship system (Usher *et al.* 2003).

Country Foods Role in Food Security

Country foods play a vital role in food security, as well as the nutritional, social, cultural, economic, and spiritual well-being of Inuit communities in Nunavut. Egede explains that country foods form an essential basis of personal and community well-being for Inuit:

> Inuit foods give us health, well-being and identity. Inuit foods are our way of life.... Total health includes spiritual well-being. For us to be fully healthy, we must have our foods, recognizing the benefits they bring. Contaminants do not affect our souls. Avoiding our food from fear does. (Egede 1995, 2)

These foods are the product of a social system and spiritual relations connected with being on the land and hunting, representing far more than a meal, but rather a healthy culture (Usher *et al.* 1995). Among Aboriginal peoples of the Canadian North, the integration of the body (*i.e.,* physical actuality and functionality of the human body), and the soul (*i.e.,* spirit, mind, immediate emotional state, or even the expression of consciousness) (Borré 1994) is accomplished through capturing, sharing, and consuming country food. The following quotes by two Inuit people below speak to the necessity of eating country foods:

> Whales and sea mammals are considered to be the best food to feed the [Inuit] body.... Without these types of foods, we the Inuit would have been gone a long time ago. Therefore, in order to live a full and healthy life and to keep the generations going, we, the Inuit, need the food that has brought us to where we are today. (Angela Gibbons, Salliq (Coral Harbour), March 1995 quoted in Freeman *et al.* 1998, 48)

> When one eats meat, it warms your body very quickly. But when one eats fruit or other imported food, it doesn't help keep you very warm. With imported food ... you're warm just a short time period. But [our] meat is different; it keeps you warm. It doesn't matter if it is raw meat or frozen meat ... it has the same effect. (Ussarqak Quajaukitsoz, July 1995 quoted in Freeman *et al.* 1998, 46)

Country foods are key to physical health and well-being, according to several studies with Inuit communities. More than 80 per cent of respondents in five Inuit areas (n = 1, 721 individual interviews) agreed that harvesting and eating country foods provided a wealth of nutritional and cultural benefits (Kuhnlein *et al.* 2000). Country foods are important sources of lipids, vitamins, minerals, and protein and other important nutrients (CACAR 1997). In dietary studies, days eating country food provided less saturated fat, sucrose, and total carbohydrate and provided more vitamin E, iron, and zinc than days without country food (Kuhnlein *et al.* 2001; Van Oostdam *et al.* 2003). Reduced country food consumption in northern

Aboriginal populations, coupled with decreasing physical activity, is associated with obesity, dental caries, anemia, lowered resistance to infection, and diabetes (Szathmary *et al.* 1987; Thouez *et al.* 1989).

Country food use, as a percentage of total dietary energy, varied from a low of 6 per cent in communities close to urban centres to a high of 40 per cent in more remote areas (CACAR 2003), according to twenty-four-hour dietary recall among Yukon Dene, Metis, and Inuit. The term "country food," or "traditional food," refers to the mammals, fish, plants, berries, and waterfowl/seabirds harvested from local stocks. More than 250 different species of wildlife, plants and animals were identified as making up the diet of Arctic peoples in workshops with 10,121 Dene and Metis, Yukon First Nation and 1,875 Inuit residents (Receveur *et al.* 1996; Kuhnlein *et al.* 2000; Van Oostdam *et al.* 2003), which provide a diverse and healthy diet. The most frequently mentioned country food items are caribou, moose, salmon, whitefish, grayling, trout, coney, scoter duck, ciso, walleye, spruce hen, pike, ptarmigan, Arctic char, Canada goose, muskox, eider duck, crowberry, beluga muktuk, ringed seal, narwhal muktuk, partridge, and cloudberry.

Hunting, fishing, trapping, and gathering in order to obtain country food is a nutritional necessity for most Inuit communities in Nunavut. In most northern communities, fresh, nutritious store-bought food is expensive and rare and must be imported from great distances.

Subsistence Activities in a Market Economy

Although wage employment and the market system are now quite familiar to northern communities, the "traditional economy" has not disappeared. In the traditional economy the household is a basic unit of both production and consumption, in contrast to industrial economies in which, typically, firms produce and households consume (Usher *et al.* 2003). Although subsistence hunting is still substantial, the rates are declining. This drop in hunting and fishing reduces the pressure on local carrying capacity, allowing the same resource base to sustain a larger population without stress (Usher *et al.* 2003). Unless a major and persistent harvest disruption has occurred, generally subsistence activities have evolved and survived, although lessened as people integrate market activities into their daily lives (Usher *et al.* 2003).

A balance between traditional and wage economies has yet to be achieved in many Inuit communities, as individuals and communities struggle to adapt to the demands of industrial employment in a boom-and-bust economy, while retaining their connection to the land and to their traditional way of life. The abrupt decline in oil and gas exploration in the Mackenzie Delta and Beaufort Sea during the 1980s and the recent closure of gold mines in response to falling world prices confirm that Aboriginal communities remain vulnerable to the boom-and-bust cycles of resource industries (NRTEE 2001).

As well as containing some of the richest fisheries, having extensive continental shelves over which shoaling fish congregate, the Arctic is rich in non-renewable

Caribou harvesting facilities in Coral Harbour, Nunavut. Photo by Steve Newton, 2001.

resources. It holds some of the world's largest deposits of coal, iron, copper, lead, and uranium, as well as oil, gas, and gold (Freeman *et al.* 1998). For almost three-quarters of the twentieth century, Aboriginal communities in the Canadian North had no control over non-renewable resource development (NRTEE 2001).

A legacy of social and cultural dislocation within Aboriginal communities is linked, in part, to non-renewable resource development. A recent study conducted by Pricewater Coopers in the Fort Liard area, Northwest Territories, found higher rates of alcohol consumption and alcohol-related crime to be associated with a recent increase in economic activity (NRTEE 2001). Clearly, a shift to a market economy creates a social and cultural disruption, without adequate supports in place to ensure cultural survival.

VULNERABILITY LIMITING FOOD SECURITY

Vulnerability represents the interface between threats to human well-being and the capacity of people and communities to cope with those threats (UNEP 2002). Both natural phenomena (storms, fires, etc) and human activities (using CFCs depletes the ozone, testing nuclear weapons releases radioactivity, employing cars and industries introduces toxic chemicals, etc.) pose threats. Although everyone is vulnerable to environmental risks, human exposure to environmental threats varies, as does the adaptive ability of societies and individuals. Often vulnerability is discussed only in the positive terms of resilience, capacity, and adaptive management. This management approach places the responsibility on the Inuit in Nunavut to absorb and counteract negative environmental impacts caused by the industrial economy in the South, rather than focusing on the source of the problems and demanding change. In coping with environmental change, the World Health Organization includes socio-economic conditions and the social and physical environments as key determinants of people's health and well-being in its definition of health. Health is:

> The extent to which an individual or group is able, on the one hand, to develop and satisfy needs; and, on the other hand, to change or cope with the environment. Health is therefore seen as a resource for everyday life, not the objective of living; it is seen as a positive concept emphasizing social and personal resources, as well as physical capacities. (World Health Organization 1984, 2)

The availability of resources (*e.g.*, wealth, technology, skills, infrastructure, education, management capabilities, demographic makeup, etc.) plays a role in people's adaptation and coping response (UNEP 2002). Indigenous and poor communities in isolated rural environments in the North are more vulnerable to environmental change, due to economic insecurity, inadequate drinking water, wastewater and transportation infrastructure, and reduced health and education services. The economic development strategy for Nunavut considers the many barriers to a better quality of life:

> Physical infrastructure is limited, the workforce is under-skilled, essential services are under-developed. Government is by far the largest component of all economic activity. Although Inuit family and community ties are very strong in Nunavut, some of the most important supports underlying successful modern economies are under great strain, like the health and school systems. An economic development strategy for Nunavut must recognize that in developmental terms the Nunavut economy is far behind other jurisdictions in Canada. (Canada-Nunavut Business Service Centre 2003)

Archibald and Grey (2000) point to the underlying shortages of housing, infrastructure, and employment as the cause of the health and suicide 'crisis' in Nunavut: "Provide people with proper housing, water, sewage, jobs and the means to provide adequate food and health statistics would improve." (Quoted in Kinnon 2002, 12). Inadequate housing and homelessness are growing problems: one in six residents of Iqaluit lacks proper housing (Inuit Tapariit Kanatami 2002). Poor quality and overcrowded housing lead to family tensions and violence, poor health conditions, including high levels of respiratory ailments and communicable diseases (Assembly of First Nations 2001).

The inadequate facilities for water and sewage result in the Inuit in Nunavut, particularly children, having significantly higher incidence of water-borne diseases compared to other Canadians. Although the Canada food/water-borne illness rate is 97.8/100,000, in Nunavut it is triple that at 291/100,000 and in the

Kivalliq region of Nunavut it is quadruple that at 408/100,000 (Archibald and Grey 2000; Kinnon 2002). Contaminants in drinking water include organisms such as giardia, salmonella and *E. coli* bacteria, dissolved metals, pesticides, and industrial chemicals. Many water systems and sanitation systems are substandard (Kinnon 2002).

It is widely recognized that more needs to be done (*e.g.*, infrastructure development, upgrading of water treatment facilities, and increased training in many communities, as well as efforts in local communities to improve sanitation and decrease contamination) to broaden safety efforts to protect Inuit communities from water-borne health hazards (Kinnon 2002). Given its small population and large infrastructure needs, the North's share of overall infrastructure funding is inadequate. The government of Canada uses a per capita allocation formula as the basis for infrastructure funding, but this does not address the urgent needs of the North and provide a minimum threshold to allow communities to provide safe water supplies, health services and secondary education.

More than any other Aboriginal group in Canada, Inuit must travel far to access many health services, especially specialized services. Removal from home communities and family support is emotionally stressful (Hanarahan 2002). Suicide rates are five to seven times the national averages in Aboriginal communities, but particularly high in Nunavut. Compared to 13 deaths per 100,000 Canadians, overall Nunavut has 77 deaths per 100,000 and 94 deaths per 100,000 people in the Qikiqtani region of Nunavut (Kinnon 2002). While Aboriginal communities are gaining greater control over social and economic development and health programs, these programs are often based on non-Indigenous models.

Overall education levels of the northern population are lower than for the overall Canadian population. Of the adult population of Nunavut, only 1.4 per cent of Inuit have a university degree and 2.9 per cent of Inuit have high school graduate certificates (Statistics Canada 2001). It should be noted that many people attend trade school without completing high school, because schools in Nunavut frequently do not offer grade twelve. Also, within the formal education system there is little or no education provided on traditional knowledge, which isolates youth from the land and their elders. Traditional knowledge remains strong among the Inuit elders in Nunavut; however, its transmission to future generations faces many barriers because of the rift caused by Western education and colonialization:

> The Western education system continues to fail to teach the values, beliefs and principles which underlie Traditional Knowledge. In addition, time spent in residential schools or day schools has limited the opportunity for Traditional Knowledge to be passed on to younger generations.... In the changing world where Euro-Canadian power and control appeared insurmountable, many elders questioned the

value of their knowledge to younger generations in the modern world. At the same time young people became less receptive to the language, the information and the style of traditional teachings which contradicted everything they were taught and learned in school. Young students have less time for year-round exposure to Traditional Knowledge on the land and much more exposure through the media to the dominant society. (Brockman 1997)

Traditional Ecological Knowledge is a body of knowledge built up by a group of people through generations of living in close contact with nature (Battiste and Henderson 2000). Traditional or indigenous knowledge is cumulative and dynamic, building on the historic experiences of a people (Battiste and Henderson 2000). While those concerned about biological diversity will be most interested in knowledge about the environment, this information must be understood in a manner that encompasses knowledge about the cultural, economic, political, and spiritual relationships with the land. UNESCO cites the importance of traditional ecological knowledge and the need for government to provide active support for its transmission – not just in isolated communities but in universities and other educational and international organizations.

Governmental and non-governmental organizations are encouraged to sustain traditional knowledge systems through active support to the societies that are keepers and developers of this knowledge, their ways of life, their languages, their social organization and the environments in which they live, and fully recognize the contribution of women as repositories of a large part of traditional knowledge.... Governments, in co-operation with universities and higher education institutions, and with the help of relevant United Nations organizations, should extend and improve education, training and facilities for human resources development in environment-related sciences, also utilizing traditional and local knowledge. (UNESCO 1999)

In the past, Indigenous coping mechanisms included adaptive behaviour, such as regular seasonal migration and changes in practices for hunting and gathering, to ensure food security. Although Inuit communities in Nunavut were once nomadic to ensure better hunting and fishing, enclosure in government housing and mandatory schooling requires that families settle down. Today, food security is limited by poverty, food contamination, and Eurocentric Canadian government policy.

Food Insecurity and Poverty
Food insecurity is a problem for Aboriginal people in Canada, particularly in isolated northern communities in the Arctic. Food security means being able to obtain a nutritionally adequate, culturally acceptable diet at all times through

local non-emergency sources. To ensure an active healthy life, both adequate food production or imports and economic access to food at the household level are required at all times (Canadian Dietetic Association 1991).

According to the Aboriginal People's Survey, half of all respondents over fifteen years of age (N = 388,900) reported lack of food availability once or twice per month during the previous twelve months (Statistics Canada 1993). Further, almost half of eight hundred women with children in eight isolated northern communities across Canada were extremely concerned about not having enough money for purchasing food from the store (Lawn 1994). In all of the communities more than half of the women reported running out of money to buy food between two and four times a month. This percentage rose to 80 per cent in four Inuit communities (Lawn 1994; 1995).

Amartya Sen (1992) writes that food supply is not the primary cause for famine, starvation, or food insecurity; rather the lack of entitlement (*e.g.*, lack of employment and income, lack of participation in decision-making, inaccessible and unresponsive government, etc) is the problem. Factors contributing to food insecurity include: low incomes, high food costs, unemployment, inadequate social assistance, and reduced access to country food related to concerns over food safety (Lawn 1994). Poverty curtails options to relocate, import food, or buy the necessary materials (*e.g.*, gas, guns, snowmobiles, boats) to hunt, trap, fish, or gather country foods. Trade is not available to poor people, who engage in subsistence activities, as most wild game has no value in the market economy that devalues subsistence economies and Indigenous cultures.

In many northern communities, country food eases hunger: country food is an economic necessity for most Aboriginal people in Nunavut, as poverty is widespread and nutritious store-bought food is very expensive. Up to 78 per cent of Inuit peoples state they could not afford to feed their families with only store-bought food (Fisk *et al.* 2003). The cost of a standard basket of imported food to provide a nutritionally adequate diet is prohibitively expensive in Arctic communities – with many healthy food choices being economically inaccessible. The cost of living in the North is 50 to 70 per cent higher than in urban centres in the South (Kinnon 2002). Relying on nutritious store-bought food to feed a family of four for one year would cost approximately $12,000 (Fisk *et al.* 2003), as food prices are higher in the North due to transportation costs. As well as northern goods prices being higher (*e.g.*, white bread loaf was priced at $2.60 and two litres of milk costs $5.75 in 1999 in Iqualuit, Nunavut), there are more mouths to feed per family, as 41 per cent of the population is under fifteen years of age in Nunavut (Statistics Canada 2001). As the average earnings in Nunavut are $20,011, less than two-thirds the average earnings for Canadians ($31,757), it is difficult to put food on the table without subsistence activities (Statistics Canada 2001).

Eighty per cent of Nunavut's population relies on subsistence activities (hunting, fishing, gathering), to varying degrees, for food security (NTI 1999). Country food production in the Northwest Territories (NWT) is estimated to have a value of $55 million, or well over $10,000 per Aboriginal household per year if replaced with non-country food (Usher and Wenzel 1989). Note this estimate is in 1989 dollars and applicable to Nunavut, which at that time was part of the NWT.

The North is impacted on by industrial activities in the South, causing contamination of country food from long-range transportation of toxic pollutants in the Arctic. The impact of contaminants on health is threefold: first is the direct impact of toxic contaminants on human health; second is the indirect impact of toxic chemicals diminishing wildlife populations, making country foods scarce (*e.g.,* eider ducks) (Robertson and Gilchrist 1998); and last is people's fear of toxic exposures reducing country food consumption.

Contamination

Toxic contamination by mercury and persistent organic pollutants (POPs) pose special risks for food security and the health of Nunavut and other Arctic communities due to the "cold effect" and biomagnification. Many contaminants travel long distances on prevailing winds from sources beyond the direct control of northerners, where they condense due to the cold weather in the North (CACAR 2003). In the cold climate of the North, persistent organic pollutants (POPs) disappear more slowly and persist longer than in southern regions. A chemical's appearance in the northern landscape, far away from local sources, indicates its persistence (as well as its volatility). A good example of the accumulation of chemicals in the North is the fact that the flame retardant, polybrominated diphenyl ethers (PBDEs), is found at higher levels in the Arctic than over Chicago or the Great Lakes, where the chemical is extensively used (Fisk *et al.* 2003). Another example is chlorofluorocarbons (CFCs). Ozone holes are most pronounced in circumpolar regions as the interaction of CFCs, spring sunshine, and ice-crystals is very effective in destroying ozone. The ozone hole was first discovered in Antarctica in 1985 and then the Arctic.

Although local sources (*e.g.,* harbours, mines, and military sites) are not the primary reason for the widespread presence of contaminants in the Canadian North, they dominate in certain locations. Sites that exist in the North include two thousand military sites, abandoned mines, and exploration sites, former construction sites, and small industrial sites, which release PCBs, mercury, arsenic, and radionuclides (Fisk *et al.* 2003). For example, releases of arsenic from the Giant Gold Mine in the Northwest Territories need to be reduced to under 2,000 kg/yr to ensure there is no risk to fish, wildlife, or human health (Fisk *et al.* 2003). These sites should be a priority for cleanup.

Mercury and PCBs are the major concerns in the North for human exposure. Inuit are more at risk as they eat four times more fish per capita than other Canadians, and they eat the organs and fats of marine mammals, where lipophilic contaminants concentrate (Fisk *et al.* 2003). People who consume more fish or marine mammals are at higher risk because the toxic response is dose-dependent (Fisk *et al.* 2003; Kinloch *et al.* 1992). Many of the traditionally harvested fish, as well as land and marine animals (*e.g.,* ringed seals, beluga whales, narwhal, walrus) are both long-lived, allowing toxics to bioaccumulate over time, as well as being from the higher trophic levels of the food chain, allowing toxics to biomagnify up the food chain.

These factors help explain why 68 to 70 per cent of Inuit mothers from the Nunavik and Baffin (Nunavut) regions exceed the new mercury guideline for maternal blood guideline compared to only a few Dene, Metis or non-Aboriginal women in the North (Fisk *et al.* 2003). At 5.8 µg/litre this level is set to protect the fetus and breastfeeding infant from contaminants. Most Kivaliiq and Baffin communities have more than 25 per cent of their population consuming levels of mercury higher than the level considered safe, called the tolerable daily intake (TDI). This is much higher than in Labrador, Kitikmeot, and Inuvialuit, where 5 per cent of people are affected. The levels of mercury in maternal blood, hair, and umbilical cords share a similar geographic pattern to wildlife levels: 10 per cent of Baffin mothers and 16 per cent of Nunavik mothers have mercury blood levels falling in the "increasing risk" range. This stands in contrast with all Dene, Metis, and non-Aboriginal mothers, who have blood levels in the acceptable range of below 20 µg/litre (Fisk *et al.* 2003).

Grandjean *et al.* (1997) found that higher umbilical cord blood mercury concentrations corresponded with lower performance ratings for children on neurobehavioural tests, particularly in the domains of fine motor function, attention, language, visual-spatial abilities, and verbal memory. Since 1997, Nunavik mothers and infants have participated in a not-yet-published prospective longitudinal study regarding the neurobehavioural effects of perinatal exposure to methyl mercury, PCBs, and organochlorines.

Certain POPs, including PCBs, toxaphene, and chlordane are of concern as well: "levels of some POPs in Canadian Inuit populations are among the highest observed in the world, 5 to 8 times higher than women in southern Canada" (Furgal *et al.* 2003, 5). Those consuming the most country foods are exceeding the TDI levels by many times for toxaphene, chlordane, and PCBs. Inuit mothers have higher levels of PCBs, measured as Aroclor 1260, than Caucasians, Dene, Metis, and other mothers, with Baffin Inuit having the highest levels. Inuit populations that had the greatest levels of the TDIs also had greatly exceeded the PCB maternal blood guideline. Higher cord serum PCB concentrations were associated at birth with lower weight, smaller head circumference, and shorter gestation in a Michigan study (Jacobson *et al.*, 1990). Levels of PCB exposure two to three times higher than in Michigan are being studied in northern Quebec (Muckle *et al.* 2002): the negative effects of prenatal PCB exposure on birth weight and duration of pregnancy remained significant despite the protective effects of omega-3 fatty acids (Muckle *et al.* 2002). Exposure to PCBs was associated with less optimal newborn behavioural function (*e.g.,* reflexes, tonicity, and activity levels): adverse neurological effects of exposure to PCBs were found in infants up to eighteen months of age in a Netherlands study (Huisman *et al.* 1995).

There is solid evidence that mercury contamination has increased in Canadian Arctic animals from pre-industrial times to the present (CACAR 2003). Mercury levels in fish and wildlife, previously at high latitudes, are increasing in many species and locations important to the Inuit. Inuit peoples have high levels of contamination, close to consumption guidelines. While there are no guidelines or standards for human intake of contaminants in seals or other marine mammals, the consumption guidelines developed for fish for subsistence (0.2 µg/g) and for commercial sale (0.5 µg/g) are often surpassed in whale and seal organs. Mercury levels in beluga organs are high enough to damage the whales, if whales respond as other species (CACAR 2003). However, there may be a detoxification mechanism whereby whale organs store mercury in the less toxic form of mercury selenide. Mercury has increased four-fold in the past ten years in belugas from the Beaufort Sea and 2.5-fold western Hudson Bay, but not significantly at any other locations. For seals, the highest levels are from Qausuittuq (Resolute, Nunavut) at 30 µg/g (Fisk *et al.* 2003) and have risen in many locations over the last twenty-five years, particularly three-fold in Mittimatalik (Pond Inlet, Nunavut) from 1976 to 2000. As walruses feed on ringed seals, as well as other foods high on the food chain, high concentrations of mercury are expected, although research on walruses is limited.

Populations of eider ducks have plummeted in the last few decades, and there are concerns about whether harvests are sustainable (Robertson and Gilchrist 1998). Though the rapid decline of eiders is attributed primarily to parasites, contaminants may be partly responsible, according to Robertson and Gilchrist (1998). High levels of cadmium and mercury are found in the liver and kidneys of eider ducks in Arctic and Subarctic areas (Wayland *et al.* 2001), possibly impacting the ducks' immune system.

Across the Arctic, fish that eat other fish (*i.e.,* predatory fish) have higher levels of mercury than non-predatory fish, due to mercury biomagnifying up the food chain. Levels vary among different lakes and rivers, but burbot, trout, inconnu, lake trout, northern pike, walleye, and Artic char were found above subsistence guidelines in at least one lake (Lockhart *et al.* 2003).

In the past, northerners were also exposed to radiocesium from the fallout of atmospheric testing of nuclear weapons, with higher exposures due to caribou meat consumption in the past. Even though doses have declined to an insignificant level now, past exposures warrant concern, as radionuclides cause intergenerational effects, as well as cancer. Doses to northern residents consuming a traditional diet of caribou and other country foods were high at 1,500 (bg/kg) in the 1960s, reducing to 110 in the 1980s before reaching insignificant levels (Tracy and Kramer 2000). The risk from continued consumption of caribou meat is very small (est. 3–4 mSv/year).

Another issue that Canadian Arctic peoples face is climate change and its effect on country foods, food security, and contamination. Over the past thirty-five years, temperatures increased by about 1°C per decade in the North, which has resulted in permafrost thawing in many places (Cohen 1997). Ocean temperatures are predicted to change up to five degrees, which could lead to more precipitation, global erosion, and flooding that would result in greater amounts of contaminants making their way to the Arctic (Cohen 1997). Weather is both a key driver in the ecological dynamics of subsistence resources and in accessibility to harvesting areas.

All of these toxic impacts are made worse by Inuit having little control over development that affects them and little input into government policy, until recently.

Government Policies

The culture and subsistence activities of Inuit peoples have been challenged and influenced by interaction with European explorers, missionaries, whalers, traders, and government bureaucracy. The motive of outsiders, as described by Hugh Brody, was to enact change among the Inuit:

> In the language of the missionary, the Eskimo must be 'saved'; in that of the administrator, he must be 'helped'; in that of most Whites, he must be 'civilized.' Each White justifies his own work by referring to the benefits, medical, moral, intellectual or material, that southern culture can give. (Brody 1975, 101)

For the most part, Inuit peoples were not consulted about the decisions made by outsiders that directly affected their lives (Brody 1975). Brody found that in 1975, Inuit peoples were expressing their desire to be consulted on issues that concerned them. Similarly, Tester and Kulchyski echoed that people should be allowed to "express their needs, define their problems, plan and institute action towards solutions" (1994, 301).

Perhaps the most profound effect of the newcomers was the decision of government administrators to relocate families to "securely establish Canada's title, occupation and administration" (Marcus 1995: 51) from the 1920s to the 1950s. For example, people were relocated to assert Canada's sovereignty to areas in the Queen Elizabeth Islands in Nunavut and Northwest Territories. Relocations were neither voluntary nor requested by Inuit peoples but were forced on community members. While police officers reported the Inuit to be happy and settled in their new homes, various Inuit peoples' stories are of hardship, struggling to find enough food, shelter, and warmth, relying on discarded materials and food at the local dump to support their needs (Marcus 1995). Relocation resulted in the loss of familiarity with the land and its resources. Relocated Inuit peoples had to make adaptations in diet and resource harvesting in order to survive. Sometimes, at the new location, the weather was too cold and wildlife too scarce, leading many people to request a return to more suitable areas. Some did leave the relocated settlements, while others never made it out of their place of exile.

Policy makers assumed that hunting and its associated ways of life would disappear as commercial production and wage employment became prevalent in northern and isolated areas (Murphy and Steward 1956). Migration from the traditional economy to the new economy was seen as the key vehicle for modernization and acculturation. As the notion of "development" came to dominate the purpose and objectives of government administration in the North, the measures of success, and of personal well-being, were those of the southern industrial model (NRTEE 2001). Early labour force surveys and environmental impact statements on northern resource developments used indicators entirely derived from an industrial economy that disregarded subsistence production or cultural survival entirely (NRTEE 2001).

The post-World War II period was one of rapid change in the Canadian North. Cold war defence activity and major resource developments, aided by both public and private investment in transport, energy, and town infrastructure, led to profound macroeconomic change (Usher *et al.* 2003). The wage economy is based largely on mining, oil, construction, and transport. The new resource and administrative towns are centres for economic growth, while the fur trade, which had been the economic mainstay of the small communities, is declining, leaving small communities in economic peril.

The development approach of industry and governments in Canada in the Arctic encourages large scale development projects (*e.g.*, exploitation of hydrocarbon reserves, mineral deposits), which have considerable negative environmental consequences and deplete non-renewable resources. For example, the Queen Elizabeth islands are underlain by oil-bearing rock and have been the subject of extensive drilling since the early 1960s. The government has provided large economic stimulus packages that offer incentives to large investors (NRTEE 2001). However, a focus on sustainability requires the government to shift from non-renewable resource development, jobs, and consumption to sustainable livelihoods. To ensure sustainable Aboriginal communities, a coordinated policy framework that addresses cumulative effects from local and long-range pollution and that prioritizes community life and subsistence activities is required. The most significant risks from non-renewable resource development in the future are likely to arise from the cumulative environmental, social, and cultural impacts of multiple exploration programs, mines, oil and gas facilities, and pipelines, along with the roads and other infrastructure required to support these projects (NRTEE 2001). Cumulative effects management, requiring assessment of the impacts of multiple projects and activities, including long-range pollution that may occur within a large geographical area over an extended period, was not previously carried out (NRTEE 2001).

Northern Arctic communities continue to be vulnerable to imposed Canadian government policies, which support the view of the Arctic as a source of potential wealth for outside development while ignoring the Arctic as the homeland of Aboriginal peoples. The resource-rich North represents to outsiders a potential source of jobs and economic wealth generated from development (Bone 1992). Lands

Ice-fishing through one-metre-thick ice in March. Lake near the Hudson Bay coast, north of Chisasibi, Quebec. Photo by Fikret Berkes.

and resources were developed for the "common good" without due consideration to local Inuit communities, who were regarded as "special interest groups" (Bone 1992). The frontier mentality of Canadian developers considers that northern lands and resources are not needed or will remain undeveloped or underdeveloped by Aboriginal peoples, in order to rationalize ignoring the environmental and health impacts on northern Aboriginal communities (Bone 1992). For the most part, developmental impacts that leave communities with physical and social illness are either regarded as part of the cost of development or as impacting a numerically and politically non-significant population segment (Bone 1992).

Many outside forces are weakening the fur trade and subsistence, including the anti-whaling, anti-sealing, and anti-trapping movements, as well as the government of Canada's wildlife management (Freeman *et al.* 1998). These movements embody cultural imperialism and have led to policies and consumer movements that diminish the food supply and incomes of Inuit and other Indigenous peoples (Freeman *et al.* 1998). Establishment of protected areas and wildlife sanctuaries separate indigenous peoples from lands and resources they rely on for survival. Government policies and educational programming dismiss the cultural and nutritional significance of country foods in the urbanization and homogenization of cultures brought on by globalization (Freeman *et al.* 1998). Government management of resources imposes Western concepts and knowledge about animals as "stocks" and "wildlife" that are to be "managed" for "harvesting" (Freeman *et al.* 1998, 163) with hunting quotas, obliterating traditional Inuit concepts about the land and its resources. This policy undermines "dynamic, evolving, and effective systems of local management and the local knowledge that informs those systems" (Freeman *et al.* 1998, 164). For example, the federal

Department of Fisheries and Oceans failed to respect the rights of Inuit and the role of the Nunavut Wildlife Management Board with respect to the allocation of commercial turbot stocks off Baffin Island, forcing NTI to resort to litigation to obtain relief (NTI 1999).

To ensure meaningful Aboriginal involvement in the decisions that affect them, a devolution of authority over non-renewable resources from the federal government to territorial and Aboriginal governments has been recommended (NRTEE 2001). This should include Aboriginal peoples having direct decision-making authority concerning the availability of land for staking, rather than the current free entry system for mining that is enshrined in the *Canada Mining Regulations* (NRTEE 2001).

In government-Aboriginal relations, there is a growing trend of Aboriginal self-determination and involvement in decisions that affect them. The United Nations recognizes that Aboriginal peoples have an integral role to play in addressing any issue that affects their lands and peoples (UNEP 2002). Northern Aboriginal organizations struggled for participation in various inquiries and environmental impact assessments during the 1970s. They also lodged legal proceedings in respect of adverse effects of industrial development on their lands (NRTEE 2001). Since recognition of Aboriginal rights to land in the 1973 Calder case, Inuit peoples have successfully negotiated landmark agreements, signing three comprehensive land claim settlements (Nunavik, Inuvialuit, and Nunavut). The agreement for the largest land mass, the Nunavut Final Agreement and Articles, is a modern-day treaty, creating the territory of Nunavut on 1 April 1999.

The Inuit of Nunavut exchanged Aboriginal rights to all lands and waters for defined rights for 355,842 square kilometres of Nunavut, of which about 10 per cent includes mineral rights. Putting these facts in perspective, Nunavut covers more than two million square kilometres and has a majority population of Inuit peoples who have fee simple title ownership to 18 per cent of the total surface rights (317,972 square kilometres) and both surface and subsurface rights to a mere 37,870 (Peters 1999). The remainder is Crown land, where Inuit have unrestricted harvesting rights (Peters 1999) and development companies have unrestricted mineral rights (NRTEE 2001), although both are subject to government regulation. The basic principles of free entry, characterized as guaranteeing a "right to mine," under the *Canada Mining Regulations* exist, except where subsurface minerals are owned by Aboriginal people. With non-Aboriginal-directed development options available for the vast majority of the land, cultural survival and subsistence activities of Aboriginals are vulnerable (NRTEE 2001).

In contrast to the federal government focus on non-renewable resource megaprojects, the Nunavut Land Claims Agreement clearly states as its priorities conservation and subsistence use before commercial use. Inuit Tapiriit Kanatami (2002) describes the role Inuit will play in the next phase of economic and political development throughout the Canadian North:

We cannot, however, assume that this new role [in economic and political development] will be developed at the expense of more traditional activities which characterize our mixed subsistence based economies that are so vital for the long term economic and social health of our communities.

Inuit communities have taken a very different approach to economic development than industry and governments in Canada. With a focus on community and the local level, almost every Inuit community has a marketing co-operative for Inuit carving and print making, and one in seven people consider themselves to be artists (Inuit Tapiriit Kanatami 2004). This is a lucrative activity without a large ecological footprint. As well, to support sustainable development a modest hunter support program has begun in Nunavut. This makes it economically viable for those wanting to pursue a hunting way of life to do so and, by doing so, contribute to the overall well-being of their communities and extended families (NRTEE 2001). NRTEE's consultations suggest widespread agreement among Aboriginal peoples and other northerners that significant powers, along with sufficient money to exercise them effectively, should be transferred from the federal government, based in Ottawa.

There have been a number of positive developments in Nunavut. Noteworthy have been: the establishment and operation of the Nunavut Wildlife Management Board; the harmonization of the bylaws of regional and wildlife organizations and hunters and trappers associations; the expansion of commercial fishing ventures in Nunavut; the revival of the bowhead whale hunt; and the beginning of a modest hunter support program. Recently, the National Contaminants Program and the formation of Canadian Arctic Peoples Against POPS (CAIPAP) have provided vehicles for Arctic people to have some input into environmental research, programs, and regulation that affect them.

CONCLUSION

Changes in the ecosystem, through chemical contamination, resource depletion, and ill-conceived government regulation jeopardize the food security of people in Inuit communities dependent, nutritionally and culturally, on wildlife harvested by hunting, fishing, and trapping. A look at food security of Inuit peoples, focusing on Nunavut, shows the toxic role of outside forces undermining sustainability through cultural imperialism and long-range pollution. The impact of environmental change on food security in the North on Aboriginal people is of heightened importance due to (1) widespread poverty, (2) contamination causing toxic impacts, and (3) government policies restricting access to land and resources. Northern communities, with their isolation, poverty, and limited infrastructure, have a vital need for country foods for food security. Discussing vulnerability only in the positive terms of resilience and capacity places the onus on Aboriginal people to absorb and counteract negative environmental impacts caused by the industrial economy, rather than targeting the source of the problems to demand change.

This review of food security and vulnerability generated the following recommendations:

1 To ensure sustainable Aboriginal communities, a coordinated policy framework should address cumulative effects from local and long-range pollution, prioritizing culture, food security, community life, and subsistence activities. The need to follow the recommendation of the International Joint Commission to ban toxic, persistent chemicals is evident from human and animal contamination in Nunavut.

2 To support sustainable development the federal government should support Inuit community economic development (*e.g.,* carving and print-making co-operatives, hunter support program) with a focus on sustainable livelihoods, rather than megaprojects that deplete non-renewable resources.

3 To ensure meaningful Aboriginal involvement in the decisions that affect them, a devolution of authority over non-renewable resources from the federal government to Nunavut, including sufficient money to exercise them effectively, should occur.

4 To reduce vulnerability of Arctic Aboriginal communities due to limited infrastructure not meeting basic needs, the federal government should set aside a block of funding to be used as a minimum threshold allocation. Basic human rights include the right to food, shelter, and safe drinking water. Infrastructure development, upgrading of water treatment facilities, and increased training in many communities, as well as efforts in local communities to improve sanitation and decrease contamination, are required to protect Inuit communities from water-borne health hazards.

5 To strengthen culture, a pro-indigenous education system is required that teaches Inuktitut and traditional ecological knowledge. Traditional knowledge shared by elders and others should be included in school curricula, in field trips to harvest and trap, in textbooks, and in government decision-making. Presently, the formal school system reinforces the devaluation of local knowledge systems, resulting in the loss of the intergenerational transmission and the erosion of such knowledge.

Assembly of First Nations. 2001. *First Nation Vision for Housing.*
http://www.afn.ca/Programs/Housing/HS/housing.htm (25 January 2004).

Archibald, Linda, and Roda Grey. 2000. *Evaluation of Models of Health Care Delivery in Inuit Regions.* Ottawa: Inuit Tapirisat of Canada.

Battiste, Marie, and J.Y. Henderson. 2000. *Protecting Indigenous Knowledge and Heritage: A Global Challenge.* Saskatoon: Purich Publishing.

Berkes, F., C. Folke, and M. Gadgil. 1995. "Traditional Ecological Knowledge, Biodiversity, Resilience and Sustainability." In *Biodiversity Conservation.* Eds. C. Perrings *et al.* Dordrecht: Kluwer.

Bone, Robert M. 1992. *The Geography of the Canadian North: Issues and Challenges.* Toronto: Oxford University Press.

Borré, K.S. 1994. "The Healing Power of the Seal: The Meaning of Inuit Health Practice and Belief." *Arctic Anthropology* 31(1): 1–15.

Brockman, Aggie. 1997. "When All People have the Same Story, Humans will Cease to Exist." *Protecting and Conserving Traditional Knowledge: A Report for the Biodiversity Conventions Office.* Churchill: Dene Cultural Institute.

Brody, Hugh. 1975. *The People's Land: Whites and the Eastern Arctic.* New York: Penguin.

CACAR. 1997. *Canadian Arctic Contaminants Assessment Report.* Jensen, J., K. Adare, and R. Shearer (eds). Ottawa: Department of Indian Affairs and Northern Development.

CACAR. 2003. *Contaminant Levels, Trends and Effects in the Biological Environment,* edited by A. Fisk, K. Hobbs, and D. Muir. Ottawa: Department of Indian Affairs and Northern Development.

Canada-Nunavut Business Service Centre. 2003. *Nunavut Economic Development Strategy: Building a Foundation for the Future.* Iqaluit: Sivmmutt Economic Development Strategy Group.

Canadian Dietetic Association. 1991. "Hunger and food security in Canada: Official position of the Canadian Dietetic Association." *Journal of the Canadian Dietetic Association* 52: 139–45.

Chambers, L., and C. Conway. 1992. "Sustainable Rural Livelihoods: Practical Concepts for 21st Century." *Institute of Development Studies Discussion Paper 296.* Brighton, U.K.: IDS Publications.

Cohen, Stewart Jay. 1997. *Mackenzie Basin Impact Study Final Report: Summary of Results.* Ottawa: Environment Canada.

Churchill, Ward. 1999. *Struggle for the Land: Native North American Resistance to Genocide, Ecocide, and Colonization.* Winnipeg: Arbeiter Ring Publishing.

Dewailly, É, *et al.* 2000. *Diet Profile of Circumpolar Inuit.* GETIC, Quebec: Université Laval.

Douglas, M. 1992. *Risk and Blame: Essays in Cultural Theory.* London: Routledge.

Egede, I. 1995. "Inuit Food and Inuit Health: Contaminants in Perspective." *Avativut/ Ilusivut Newsletter* 2(1) (December): 1–3.

Environment Canada. 2003. "The Inuit Economy: Sustaining a Way of Life." SOE Fact Sheet No. 94–1. *http://www.ec.gc.ca/soer-ree/English/products/factsheets/94-1.cfm*

Fisk, Aaron T., Karen Hobbs, and Derek C. G. Muir, eds. 2003. *National Contaminants Program. Canadian Arctic Contaminants*

Assessment Report 11: Biological Environment. Ottawa: INAC.
http://www.ainc-National Contaminants Program.gc.ca/ncp/pub/pube.html

Freeman, Milton, Lyudmila Bogolovskaya, Richard A. Caulfied, Ingmar Egede, Igor
I. Krupnik, and Marc G. Stevenson. 1998. *Inuit, Whaling and Sustainability.* Walnut
Creek, CA: Altamira Press.

Furgal, Chris, Sarah Kalhok, Eric Loring and Simon Smith, eds. 2003.
*National Contaminants Program. Canadian Arctic Contaminants
Assessment Report 11: Knowledge in Action.* Ottawa: INAC.
http://www.ainc-National Contaminants Program.gc.ca/ncp/pub/pub_e.html

Gottlieb, Robert. 2002. *Environmental Unbound: Exploring New Pathways for Change.*
Cambridge: MIT Press.

Grandjean, P., P. Weihe, R.F. White, F. Debes, S. Araki, K. Yokoyama, *et al.* 1997.
"Cognitive deficit in 7-year old children with prenatal exposure to methylmercury."
Neurotoxicology Teratology 417–28.

Hanrahan, M. 2002. "Identifying the Needs of Innu and Inuit Patients in Urban
Health Settings in Newfoundland and Labrador." *Canadian Journal of Public
Health* 93(2): 149–52.

Hofrichter, R, ed. 2000. *Reclaiming the Environmental Debate: The Politics of Health
in a Toxic Culture.* London: MIT Press.

Huisman, M., C. Koopman-Esseboom, C. Lanting, C. Van der Paauw, L. Tuinstra,
V. Fidler, N. Weisglas-Kuperus, P. Sauer, E. Boersma, and B. Touwen. 1995.
"Neurological conditions in 18 months-old Children Perinatally Exposed to
Polychlorinated biphenyls and Dioxins." *Early Human Development* 41: 165–76.

Inuit Tapiriit Kanatami. 2002. *The Nine Qulliqs of Inuit Health: An Inuit Health Policy.*
Ottawa: Inuit Tapiriit Kanatami.

Indian and Northern Affairs Canada. 2003. *What Is Sustainable Development?*
Updated 16 January 2003. *http://www.ainc-inac.gc.ca/sd/whatis_e.html*

———. 2004. *Aboriginal Human Resource Development Council of Canada.* Ottawa:
Inuit Tapiriit Kanatami.

Jacobson, J., S. Jacobson and H. Humphrey. 1990. "Effects of in utero exposure to
polychlorinated biphenyls and related contaminants on cognitive functioning in
young children." *Journal of Pediatrics* 116: 38–45.

Kinloch, D., H. Kuhnlein, and D.C.G. Muir. 1992. "Inuit Foods and Diet: A
Preliminary Assessment of Benefits and Risks." *Science of the Total Environment*
122: 247–78.

Kinnon, D. 2002. *Improving Population Health, Health Promotion, Disease Prevention
and Health Protection Services and Programs for Aboriginal People: Recommendations
for NAHO Activities.* Ottawa: National Aboriginal Health Organization.

Krupnik, I., and D. Jolly, eds. 2002. *The Earth Is Faster Now: Indigenous Observations
of Arctic Environmental Change.* Fairbanks, AK: Arctic Research Consortium of the
United States.

Kuhnlein, H.V., O. Receveur, and H. M. Chan. 2001. "Nutrient Benefits of Arctic
Traditional/Country Food." In *Synopsis of Research Conducted under the 2000–2001
Northern Contaminants Program*, edited by S. Kalhok, 56–64. Ottawa: Indian
Affairs and Northern Development.

Kuhnlein, H.V., O. Receveur, H. M. Chan, and E. Loring. 2000. *Assessment of Dietary
Benefit/Risk in Inuit Communities.* Ste Anne-de-Bellevue, Quebec: Centres for
Indigenous Peoples' Nutrition and Environment (CINE).

Lawn, J. 1994. *Air Stage Subsidy Monitoring Program: Final Report.* Vol. 2: *Food Consumption Report.* Ottawa: Department of Indian Affairs and Northern Development.

——. 1995. *Report on the Development of Regional Inuit Food Baskets.* Ottawa: Department of Indian and Northern Affairs.

Lockhart, W.L., G. Stern, P. Roach , M. Evans, M. Ikonomou, G. Low, K. Koczanski, S. Atkins, D. Muir, G. Stephens, M. Hendzel, G. Boila, J. DeLaronde, and S. Friesen. 2003 "Temporal and Spatial Trends of Organohalogen and Heavy Metal Contaminants in Fish from Canadian Arctic Lakes." Canadian Arctic Contaminants Assessment Symposium. Ottawa: National Contaminants Program, INAC.

Marcus, Alan Rudolph. 1995. *Relocating Eden: The Image and Politics of Inuit Exile in the Canadian Arctic.* Hanover: Dartmouth College.

Muckle, G. E. Dewailly, P. Ayotte, J. Jacobson and S. Jacobson. 2001. "Prenatal exposure of the Northern Quebec Inuit infants to environmental contaminants." *Environmental Health Perspectives* 109 (12): 1291–99.

Murphy, Robert, and J.H. Steward. 1956. "Tappers and trappers: parallel process in acculturation". *Economic Development and Cultural Change* 4: 335–53.

NRTEE. 2001. National Round Table on the Environment and the Economy. *State of the Debate: Aboriginal Communities and Non-Renewable Resource Development.* Ottawa: Renouf.

NTE. 1991. Northwest Territories Education. *Inuvialuit Pitquusiit: The Culture of the Inuvialuit.* n.p. NTE.

NTI. 1999. Nunavut Tunngavik Incorporated. *Taking Stock: A Review of the First Five Years of Implementing the Nunavut Land Claims Agreement.* Nunavik: NTI.

Nunavut Planning Commission. 2004. *General Information about Nunavut.* *http://npc.nunavut.ca/eng/nunavut/general.html* (accessed 23 January 2004).

Parajuli, Pramod. 1997. "Discourse on Knowledge, Dialogue and Diversity: Peasant Worldviews and the Science of Nature Conservation." *Worldviews: Environment, Culture, Religion* 1: 189–210.

Peters, Evelyn J. 1999. "Geographies of aboriginal self-government." In *Aboriginal Self-Government in Canada: Current Trends and Issues,* 2nd ed., edited by John H. Hylton, 411–30. Saskatoon: Purich Publishing.

Receveur, O., M. Boulay, C. Miles, W. Carpenter, and H. Kuhnlein. 1996. *Variance in Food Use in Dene/Métis Communities.* Montreal: Centre for Indigenous Peoples' Nutrition and Environment (CINE), McGill University.

Redefining Progress. 2004. *Ecological Footprint.* *http://www.redefiningprogress.org/programs/sustainabilityindicators/ef/methods/components.html* (24 January 2004).

Ridington, Robin. 1998. *Trail to Heaven: Knowledge and Narrative In A Northern Native Community.* Iowa City: University of Iowa Press.

Robertson, G., and H. Gilchrist. 1998. "Evidence for Population Declines Among Common Eiders Breeding in the Belcher Islands, Northwest Territories." *Arctic* 51: 378–85.

Sen, Amartya. 1992. *Inequality Reexamined.* Boston: Harvard University Press.

Shearer, Russell, and Sui-Ling Han. 2003. "Canadian Research and POPS: The Northern Contaminants Program." In *Northern Lights Against POPS: Combating Toxic Threats in the Arctic,* edited by David Leonard Downie and Terry Fenge, 41–59. Montreal: McGill-Queen's University Press.

Sivummut Economic Development Strategy Group. 2003. *Nunavut economic development strategy: building a foundation for the future.* Iqualuit: Canada-Nunavut Business Service Centre. *http://www.cbsc.org/nunavut/english/pdf/ NunavutEconomicDevelopmentStrategy.pdf*

Statistics Canada. 2001 *Aboriginal Population Profile.* *http://www12.statcan.ca/English/Profilo1ab/Details/detailspop1.cfm* (21 August 2003).

———. 1993. *Language, Tradition, Health, Lifestyle and Social Issues, 1991 Aboriginal Peoples Survey.* Ottawa: Minister of Supply and Service.

Szathmary, E.J.E., C. Ritenbaugh, and C.S.M. Goodby. 1987. "Dietary Change and Plasma Glucose Levels in an Amerindian Population Undergoing Cultural Transition." *Social Science and Medicine* 24(10): 791–804.

Tester, Frank , and Peter Kulchyski. 1994. *Tammarniit (Mistakes): Inuit Relocation in the Eastern Arctic, 1939–1963.* Vancouver: UBC Press.

Thouez, J.P., A. Rannou, and P. Foggin. 1989. "The Other Face of Development: Native Population, Health Status, and Indicators of Malnutrition – the case of the Cree and Inuit of Northern Quebec." *Social Science and Medicine* 29(8): 965–74.

Tracy, B.L., and G.H. Kramer. 2000. "A method for estimating caribou consumption by northern Canadians." *Arctic* 53(1): 42–52.

UNEP. 2002. United Nations Environment Programme "Human Vulnerability to Environmental Change." In *Global EnvironmentOutlook 3: Past, Present and Future Perspectives.* UNEP. *http://web.rolac.uneo.mx/geo/geo3/english/pdfs/chapter3_vulnerability.pdf*

UNESCO. 1999. Modern Science and Other Systems of Knowledge. *http://www.unesco.org/science/wcs/eng/#3* (11 August 2002).

Usher, P., G. Duhaime, and E. Searles. 2003. "The Household as an Economic Unit in Arctic Aboriginal Communities, and its Measurement by Means of a Comprehensive Survey." *Social Indicators Research* 61: 175–202.

Usher, P.J., M. Baikie, M. Demmer, D. Nashima, M.G. Stevenson, and M. Stiles. 1995. *Communicating About Contaminants in Country Food: The Experience in Aboriginal Communities.* Ottawa: Inuit Tapirisat of Canada.

Usher, P., and G. Wenzel. 1989. "Socio-economic aspects of harvesting." In *Keeping the Land: A Study of the Feasibility of a Comprehensive Wildlife Support Programme in the Northwest Territories,* edited by R. Ames, D. Axford, P. Usher, E. Weick, and G. Wenzel. Ottawa: Canadian Arctic Resources Committee.

Van Oostdam, J. , S. Donaldson, M. Feeley, N. Tremblay, eds. 2003. *National Contaminants Program. Canadian Arctic Contaminants Assessment Report II: Human Health.* Ottawa: INAC, 2003. *http://www.ainc-National Contaminants Program.gc.ca/ncp/pub/pub_e.html*

Wayland, M. H. Gilchrist, D. Dickson, T. Bollinger, C. James, R. Carreno and J. Keating. 2001. "Trace Elements in King Eiders and Common Eiders in the Canadian Arctic." *Archivia Environmental Contaminants Toxicology* 41: 491–500.

World Health Organization. 1984. "Concepts and Principles of Health Promotion: Report on a WHO Meeting." World Health Organization Regional Office for Europe, Copenhagen, 9–13 July 1984.

Wuttunee, Wanda. 1995. "Paddle Prairie Mall Corporation." In *Northern Aboriginal Communities: Economies and Development,* edited by Peter Douglas Elias, 193–209. North York, ON: Captus Press.

CHAPTER 4

ECOSYSTEM-BASED MANAGEMENT &
MARINE ENVIRONMENTAL QUALITY
IN NORTHERN CANADA

Donald Cobb (Department of Fisheries and Oceans)
Mina Kislalioglu Berkes (University of Manitoba)
Fikret Berkes (University of Manitoba)

INTRODUCTION

Consistent with the provisions of the *Oceans Act* is a commitment to the integrated management (IM) of ocean resources and activities. Integrated management brings together the environmental, economic, and social considerations by planning for sustainable use. As such, there is ongoing interest in selecting ecosystem-based management objectives that ensure the health of the marine environment. Indicators are needed to assess progress toward meeting these ecosystem-based management objectives. Although the Department of Fisheries and Oceans (DFO) has made some progress in the development of a framework for the environmental objectives and indicators, the social and cultural dimensions of these objectives and indicators have not, to any extent, been incorporated into the discussion, and very little of the existing science takes into account local and traditional knowledge. This is not surprising, considering that in fact MEQ (marine environmental quality), and even ecosystem-based management in Canada's coastal waters, are relatively new. Hence, the timing is good to initiate discussion on how to bring together MEQ-related science and local/traditional knowledge. This chapter attempts to provide a starting point for that.

The integration of social and cultural considerations into long-term research in marine ecosystems is a priority. Since humans are part of the ecosystem and human resource use impacts ecosystems, social and cultural dimensions of marine resource use have to be addressed. However, there is relatively little capacity and experience in this area in the DFO, and there has been little research to address questions such as: Do northern peoples have views that are similar to or different from those of scientists regarding MEQ and ecosystem-based management? Is there relevant local and traditional knowledge? Are there ways to develop community-based monitoring approaches for MEQ? What would these monitoring systems look like, and how would they work?

We explore some possible approaches to take into account indigenous knowledge, both for MEQ and, in the broader context, for ecosystem-based management undertaken to manage human activities that may impact ecosystems (and not to manage the ecosystem itself). The objective of the chapter is to examine some of the challenges and opportunities in establishing MEQ indicators in Canada's northern coastal areas and the appropriateness of considering traditional ecological knowledge along with science.

OCEANS ACT, ECOSYSTEM-BASED MANAGEMENT AND MARINE ENVIRONMENTAL QUALITY

Canada's *Oceans Act* (Canada 1997) provided the Minister of Fisheries and Oceans (DFO) the mandate to lead and facilitate in the development of a national oceans strategy. Canada's Oceans Strategy (DFO 2002a) provides the overall strategic framework for Canada's oceans-related programs and policies. The strategy is based on the principles of sustainable development, integrated management (IM), and the precautionary approach. The overarching program to deliver on the *Oceans Act* is the Integrated Management Program. The Policy and Operational Framework for Integrated Management (DFO 2002b) specifies an objectives-based approach to management through the establishment of social, cultural, economic, and environmental objectives in IM plans. Within the context of the IM framework, tools such as Marine Protected Areas (MPA) and Marine Environmental Quality (MEQ) programs attempt to provide a means to achieve and assess the effectiveness of IM plans. In order to achieve ecosystem-based management, those components or functions of the ecosystem which should not be compromised with various ocean uses will be established at a broad, or Large Ocean Management Area (LOMA), scale (*e.g.*, Beaufort Sea). Embedded within, and transferred from ecosystem objectives from the LOMA scale, are MEQ objectives for coastal management (*e.g.*, western Hudson Bay), which are then applied to IM or MPA planning. Integrated management aims at participation by coastal communities and stakeholders, while respecting land claims agreements. Integrated management planning in this context would offer a means to gather scientific and traditional knowledge, set MEQ objectives in a collaborative way, select indicators, monitor, assess, and report. The challenge is how to achieve these objectives in Canada's diverse coastal areas, and particularly in northern Canada with its unique environmental and social setting.

To assist IM planners tasked with implementing ecosystem-based management, DFO held a workshop to develop a framework for ecosystem objectives to be applied at the LOMA scale, and MEQ objectives for use in IM or MPA planning at the local coastal management scale (Jamieson *et al.* 2001). Conservation of species and habitats (the environmental dimension) and the sustainability of human uses (economic and socio-cultural dimensions) were the two overarching objectives

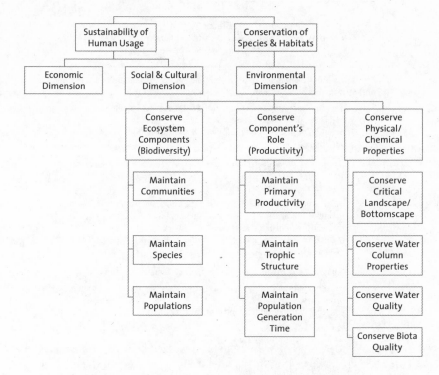

Figure 4.1 Ecosystem objectives from DFO Workshop. (Jamieson *et al.* 2001)

(Figure 4.1). Since this was a science-driven process, workshop participants discussed only the environmental component, and focused on a scientific approach to "unpacking" broad ecosystem objectives into MEQ objectives, targets, and indicators. They did, however, recognize that within IM plans, the other components can be as important. As a result of input from this workshop, the subsequent Policy and Operational Framework for Integrated Management (DFO 2002b) contains characteristics of ecosystems that could be used as a guide to setting ecosystem objectives for a LOMA. Vandermeulen and Cobb (2004) expanded on this list to develop a more complete set of overarching objectives (Table 4.1). They concluded that more work remains to define ecosystem-based management objectives at national and regional scales.

The MEQ program is recognized in s. 32 of the *Oceans Act*, which states that the Minister "may establish marine environmental quality guidelines, objectives and criteria respecting estuaries, coastal waters and marine waters" (Canada 1997). Although the concept of MEQ and associated objectives is contained within the *Oceans Act* as a tool for IM and MPA planning, the Act does not elaborate upon the nature of MEQ, and the intent of the *Oceans Act* regarding MEQ must be interpreted to provide policy and operational direction. The evolution of MEQ over several decades is fully described in Vandermeulen and Cobb (2004). Of particular relevance to our discussion is the broadening of the definition

a) *Maintain enough components (e.g., communities, species, populations) to ensure natural resilience of ecosystems;*

- maintain communities within bounds of natural variability
- maintain species within bounds of natural variability
- maintain populations (genetic diversity) within bounds of natural variability.

b) *Maintain function of each component of ecosystem to allow it to play natural role in food web (i.e., not cause any component of ecosystem to be altered such that it ceases to play its natural role);*

- maintain primary production within the bounds of natural variability
- maintain trophic structure so that individual species/stages can play their natural role in the food web
- maintain mean generation times of populations such that population resiliency is assured.

c) *Maintain physical and chemical properties of ecosystem;*

- conserve critical landscape/bottomscape features and water column properties
- conserve water, sediment and biota quality.

of MEQ. Early federal activity related to MEQ was led by Environment Canada during the 1980s, where it mainly had a pollution focus. In the early 1990s, interdepartmental (DFO and DOE) efforts attempted to broaden the view of MEQ to focus on ecosystem-based management. There was the recognition of a need for consultation and collaboration with stakeholders, and many elements of these efforts are reflected in the DFO IM framework. The concept of MEQ now encompasses ecosystem structure and function and includes such factors as population viability, contaminant and nutrient loading, biodiversity, disease, and physical disturbance. This broadening of the meaning of MEQ is significant, and would make it a valuable tool in setting ecosystem objectives and more specific MEQ objectives within an IM or MPA plan. MEQ objectives are designed to direct management action on environmental issues specific to a particular marine area. MEQ objectives lead to the development of MEQ indicators, with reference points (limits and targets) for management action.

DFO has also an interest in developing a suite of national MEQ indicators in order to address sustainable development issues. Vandermeulen and Cobb (2004) discuss problems and approaches to the development of indicators for use in integrated management. The need for socio-economic and environmental indicators has now become a priority for DFO (Rice 2003). DFO recognizes the need for a more integrative approach to assessing marine environmental quality. For northern indigenous peoples, the use of holistic approaches to environmental change is not a novel one. The use of multi-level, multi-year observing is central to traditional knowledge. Thus the development of indicators and approaches may in fact gain strength from both the integrative approaches being attempted in Western science and the holistic traditional knowledge of northern indigenous peoples.

So although DFO's approach to MEQ has an environmental focus, the selection of MEQ objectives, indicators and reference points should be accomplished with contributions from both scientists and stakeholders involved in the IM planning process. Local and traditional knowledge has an invaluable contribution to make to the setting of MEQ objectives in the north, as northern indigenous people's concept of the ecosystem is a broad one. Traditional knowledge has been successfully applied to understanding environmental change (McDonald *et al.* 1997) and assessing environmental impacts of development (Sadler and Boothroyd 1994), and these will be further explored in the following sections. Moreover, local coastal communities should be involved in the development and delivery of monitoring programs. There are opportunities to work collaboratively with scientists through participatory environmental monitoring and research, and these programs are discussed in this chapter and others in the book.

NORTHERN VIEWS ON CULTURE AND ENVIRONMENT

There are a number of different views of environment and culture among northern indigenous peoples, and these views share some common elements. Indigenous languages have words that usually get translated into English as "land" but carry other meanings as well. To them, land is more than a physical landscape; it encompasses the living environment, including humans (Berkes *et al.* 1998). For example, the term used by the Dene groups, such as the Dogrib, ndé (ndeh), is usually translated as "land." But its meaning is closer to "ecosystem" because it conveys a sense of relations of living and non-living things on the land. However, ndé differs from the scientific concept of ecosystem in that it is based on the idea that everything in the environment has life and spirit (Legat *et al.* 1995). Similarly, the Cree of the James Bay area use the word "aschkii" (askii), which is often translated as "land." However, it is a comprehensive concept of the environment because it refers to plants, animals, and humans as well as the physical environment. The Western James Bay Cree consider that "the Indians go with the land" as part of "land's dressing" in the sense that it is the presence of humans that makes the land complete (Preston *et al.* 1995).

The environment has always been the source of livelihood and basis of culture for northern indigenous peoples. In the contemporary world, it continues to be crucially important in the mixed economy of northern communities in sustaining social relationships and distinctive cultural characteristics of a group. The environment helps maintain social identity and provides a source of social values, such as sharing. Traditional knowledge, ethics and values, and cultural identity are transferred to succeeding generations through the annual, cyclical repetition of activities on the land, from berry gathering to whaling (Freeman 1993). Any loss of resources, or the health of the environment, has the potential to damage indigenous cultures through the loss of social relations of production, socialization of children, land stewardship ethics, and traditional knowledge.

Traditional community feast in Rankin Inlet, Nunavut. Photo by Steve Newton, 2002.

Traditional knowledge or traditional ecological knowledge is a body of knowledge built up by a group of people through generations of living in close contact with nature. The working definition we have used for traditional ecological knowledge is "a cumulative body of knowledge, practice and belief, evolving by adaptive processes and handed down through generations by cultural transmission" (Berkes 1999: 8). We use the term indigenous knowledge and traditional knowledge in the broader sense to mean knowledge specific to an area or culture, and traditional ecological knowledge when the knowledge is of ecological nature (not all traditional knowledge is). These are dry definitions compared to what Aboriginal people themselves have to say about traditional knowledge. For example, when native participants in a conference in Inuvik were asked to describe traditional knowledge, there was consensus on the following meanings: practical common sense; teachings and experience passed through generations; knowing the country; rooted in spiritual health; a way of life; an authority system of rules for resource use; respect; obligation to share; wisdom in using knowledge; using heart and head together.

Traditional knowledge, as a "way of life," has been part of indigenous culture for millennia. But it has become part of a shared northern culture and politics only relatively recently. The Government of Northwest Territories (GNWT) policy recognizes that "aboriginal traditional knowledge is a valid and essential source of information about the natural environment and its resources, the use of natural resources, and the relationship of people to the land and to each other" (GNWT 1993). The policy may be seen as a way of implementing self-government: "… the GNWT has adopted what is probably the first formal traditional knowledge policy in Canada, in an attempt to improve democratic representation in the North by moving the policies and practices of territorial government closer to reflecting the values and needs of all northern residents" (Abele 1997).

Traditional knowledge has a place in the *Oceans Act* through Canada's Oceans Strategy (COS). The strategy addresses the integrated management of activities in coastal, marine, and estuarine waters of Canada, and contains language that provides for the inclusion of "bodies established under land claims agreements" and other stakeholders, specifically mentioning "affected aboriginal organizations, coastal communities and other persons." In the Arctic region, this means that the operational frameworks for *Oceans Act* implementation must be consistent with land claims agreements, providing for co-management with land claims organizations and for the incorporation of traditional ecological knowledge into decision making.

The land claims agreements bring traditional ecological knowledge into the forefront for environmental management in the North. Traditional knowledge may be seen as a key mechanism for participatory approaches in environmental research and policy. This is true for fisheries and marine mammal co-management under such bodies as the Fisheries Joint Management Committee (*e.g.*, Beaufort Sea Beluga Management Plan) and the Beaufort Sea Integrated Management Planning Initiative (Fast *et al.* 2001, and Berkes and Fast chapter, this volume). Environmental change, detected through local observations and traditional knowledge, has been shown to enrich the understanding, for example, of Arctic climate change (Riedlinger and Berkes 2001). Hence, traditional knowledge has become a key mechanism for implementing participatory management in a number of areas of resource and environmental management, including monitoring (Berkes *et al.* 2001).

There are two areas in which northern views of culture and environment are particularly important for our thinking about MEQ and ecosystem-based management. The first is that local observations and place-based research are important for understanding ecosystems for ecosystem-based management purposes because ecosystems are complex and information is needed at multiple scales. Scientific models dominate the discussion of MEQ and ecosystem-based management in general. But can these models provide the whole answer? One may argue that scientific models, without local observations of change, are limited in their explanatory power. As well, models do not directly address the major human and ecological impact of environmental change, which is not so much about mean change but about the local impact of environmental quality changes on such factors as food, nutrition, and culture. The shortcomings of global and regional models can be seen, for example, in the case of climate change research in which local observations and understanding are needed to supplement these models (Krupnik and Jolly 2002).

Second, dealing with indicators based on traditional knowledge requires a major shift in thinking. Land claims agreements in the Arctic have produced a new model of governance in which communities, regional organizations, and governments share power. Hence, community-based research is significant in bringing community objectives to the forefront. However, working with indigenous peoples and dealing with local and traditional environmental knowledge is not always easy to do. Developing new models of community-based research that do justice to local observations and facilitate sharing of knowledge is a challenge. For example, community-based participatory research and monitoring to better understand near-shore sea ice was a key recommendation from a workshop on sea ice variability and change (Barber *et al.* in press), where local hunters and trappers from the Inuvialuit Settlement Region and sea ice scientists gathered to share the two ways of knowing about sea ice. Such an approach marks a shift away from expert-knows-best science and toward acceptance of traditional knowledge (and civil science in general) as a source of legitimate knowledge (Berkes 1999). We explore such challenges further, after a consideration of how science and traditional knowledge can be used together in the area of contaminants research.

CONTAMINANTS: SCIENCE AND TRADITIONAL KNOWLEDGE
Indicators as Used in Science

A great deal of work has been done in the past twenty years regarding northern contaminants, their chemistry, biology, and human health implications, as documented in the two massive Canadian Arctic Contaminants Assessment Reports (Jensen *et al.* 1997; NCP 2003; see also Myers *et al.* chapter, this volume). The Northern Contaminants Program is the main source of this large database. The Program has had direct input from northern indigenous groups since 1989, when its technical committee was expanded to include five Aboriginal parties (see chapter by Myers *et al.* for more detail). The inclusion of northern voices helped establish a participatory process for setting objectives, and oriented the program to address northern concerns. However, integrating Aboriginal views and approaches into contaminants research has been slow. In this section, we explore how science and local knowledge can be used together in searching for indicators of environmental quality.

We use the area of contaminants as an illustration of possible approaches for bringing together the two kinds of knowledge. Our discussion is exploratory, acknowledging that the two kinds of knowledge are in fact different. Indicators provided by science and traditional knowledge are different in the way they have been arrived at, but are similar in showing the state of well-being of the environment. Hence, we look at the use of scientific and local ecological indicators: What are they? How similar or different are they? How or when can they be brought together? To do this, first we consider the use of indicators in toxicology. Second, we investigate how northern indigenous peoples consider environmental signs and signals, and the kinds of indicators that may be identified by them.

In the broadest sense, indicator is a general term that is used to denote a relatively simple signal of complex trends and conditions or states, often

Inuvialuk woman harpooning a beluga whale near East Whitefish Station in the Mackenzie Delta. Photo by Fisheries Joint Management Committee, 1998.

obtained from various sources (Meadows 1998). The most common use of indicators is to document signs of change and to monitor trends of change. Various environmental conditions, such as pollution, express their effect on biological systems in a variety of ways. Biomarkers are indicators that signal some kind of change in an organism or system in response to a stressor. A biomarker can be a change in the amount or production of a particular chemical, such as an amino acid, a protein, or an enzyme. If the biomarker is a gene, then the term genetic marker is used. Biomarkers are detected by specific tests and techniques.

Biomarkers are generally measured at the biochemical level in tissue extracts or at the cellular level by examining change in cells. For example, the heavy metal exposure biomarker for fish, metallothionein, normally occurs in tissues in trace amounts; metallothionein is widely distributed in the animal kingdom. Exposure to heavy metals induces metallothionein production in fish. Heavy metals are not the only chemicals that have the ability to induce metallothionein, but this effect is much greater in the case of heavy metals (20–50 fold increases) than for other chemicals that induce metallothionein production (for example, for glucocorticoids that cause a 2–4 fold increase). Hence, metallothionein is a good biomarker. Its levels in fish tissues are measured with spectrophotometers of high sensitivity, and evaluations are made about the level of heavy metal exposure (Klaverkamp *et al.* 2002).

For broader responses at the level of organism to the same stressor, the term biomarker is not used. Instead, one refers to clinical signs and pathology. These may include parameters such as changes in body weight, external changes in skin, body, fur, eyes, mucus membranes; changes in respiratory, circulatory, or nervous systems; and changes in activity and behaviour patterns. Such changes may be expressions of multiple causes rather than changes specific to a stressor. Hence, they are not as valuable as biomarkers for diagnostic purposes. Biomarkers are called biological indicators (bioindicators) when they are causally linked to ecologically relevant end points. Bioindicators typically reflect cumulative effects of stressors at individual, population, and community levels. As such, they do not indicate clear-cut cause-and-effect relationships to specific contaminant exposure (ESD 2003).

The term "bioindicator" is commonly used in pollution research and biological effect monitoring. In recent studies, the trend is toward the use of a number of broader and less specific biological or ecological indicators (Wrona and Cash 1996). In fact, the larger the system under consideration, the less specific (but the more inclusive) are the indicators. There is a difference between the reductionistic approaches of analytical sciences using chemical and cytological methods commonly employed in toxicology, on the one hand, and the holistic approaches of ecosystem monitoring studies, on the other (Suter 2001).

The toxicological approach focuses on identifying and quantifying a specific effect: What amount of a certain toxicant produces what level of undesirable effect? By contrast, ecosystem monitoring studies focus on documenting effects without necessarily identifying or quantifying the precise nature of causality. One could argue that the holistic approach to ecosystem monitoring is on a continuum, partway between the reductionistic approach of toxicology and the holistic approach of traditional ecological knowledge. Further, one can make the argument that current multi-level, multi-year ecosystem health or ecosystem integrity studies are in some ways more similar to the Aboriginal way of observing and understanding of the world than are toxicological studies.

Indicators in Traditional Ecological Knowledge

What is the nature of the "indicators" used in traditional ecological knowledge? The term and concept of environmental quality indicators have no direct translation in most northern indigenous languages. However, many indigenous people who are knowledgeable about the land do recognize and monitor various environmental signs and signals. These may be related to changing seasons, the timing of harvesting activities, abundance of animals, health of animals, and unusual patterns and deviations from the norm (Table 4.2). Such "indicators" may be chosen on the basis of shared culture, values, and issues important for that community, and reflect the knowledge and experience of current and previous generations. This accumulated experience with the environment may be used to detect trends, for example, in fish catches. Indigenous fishers are experts in keeping mental track of the catch per unit of effort and judging trends, after allowing for year-to-year variations (Berkes 1999, Chapter 7). If, year after year, fewer fish are caught per unit of effort, fishing at the same locations with similar nets, then this is an indication of declining productivity or some kind

Table 4.2

ENVIRONMENTAL SIGNS AND SIGNALS
USED BY SOME NORTHERN INDIGENOUS HUNTERS

Signs and signals	Description	Reference
Changing seasons	Noting changes in sea ice, winds, snow cover, temperature, etc. for reading the weather and predicting hunts	Krupnik and Jolly (2002)
Signs and signals for harvesting	Environmental cues for harvesting, *e.g.*, whitefish are spawning when tundra changes colour (Chisasibi, James Bay)	Berkes, unpublished field notes
Catch per unit of effort to track abundance	Monitoring harvest success, usually per unit of time, *e.g.*, catch per net per day	Berkes (1999)
Monitoring health of animals by noting body condition	Observation of fat in certain parts of the body to judge health of big game, small game, birds, fish	Kofinas *et al.* (2002); Berkes (1982)
Noting unusual patterns in distribution and abundance	Unusual occurrences of species in an area, *e.g.*, unfamiliar species, strange distributions, breeding failure	Jolly *et al.* (2002)
Monitoring biophysical change by noting extremes	Detecting change by noting, not averages, but extremes and major deviations from the norm, *e.g.*, in sea-ice cover and thickness	Nichols *et al.* (2004); Jolly *et al.* (2002)
Noting changes in environmental quality	Detecting change through taste of fish and game, observations of pathological conditions	(see Table 4.3)

of environmental change. Evaluation of indicators over time allows users to receive feedback from the ecosystem, enabling them to assess various aspects of the system. For example, a catch of burbot with shrivelled, discoloured livers may mean that something in the environment is causing this or something in the water may have changed (Lockhart *et al.* 1987).

Table 4.3 provides a sample of local indigenous observations on environmental quality changes, including the burbot example noted above. The table is compiled from examples recorded in the Canadian Arctic Contaminants Assessment Report (Jensen *et al.* 1997). Note that local observations may be useful not only in detecting abnormal body conditions (liver, body deformity, small eggs) but also abnormal taste and consistency, parasitism, poor condition (lower body fat content), and abnormal behaviour, as in the example of altered spawning behaviour. Locally used indicators can be quite specific. For example, the mesentery fat content of fish caught in nets would mean the fish are in poor condition, hence not suitable for consumption in James Bay. When the reservoir of the LG 2 dam was being built, Cree fishers were checking the

Table 4.3
EXAMPLES OF COMMUNITY OBSERVATIONS REGARDING CONTAMINANTS AND OTHER SUSPECTED NEGATIVE ENVIRONMENTAL QUALITY CHANGES

Observation	Community and/or area
Decrease in the quantity and size of whitefish and trout eggs	Yukon First Nations
Changes in texture and consistency of fish flesh	Yukon First Nations
Altered migratory behaviour in spawning salmon, upstream travel distance reduced	Yukon First Nations
Changes in fish flesh quality and fish numbers	Dene Nation
Fish (burbot, *Lota lota*) with spotted, shrivelled or discoloured livers	Dene, Mackenzie River area
Increases in deformities of fish and other animals	Dene Nation
Thinner marine fish, reduced firmness of flesh	Tuktoyaktuk area
Pacific herring (*Clupea harengus*) with white spots in flesh and altered taste	Tuktoyaktuk area
Trichinosis in walrus associated with suspected negative environmental influences	Sanikiluaq, Hudson Bay
Sores on the insides (*i.e.*, body cavity) of seals	Avativut report, Nunavik and Labrador

condition of the *Coregonus* species in the estuary of the La Grande River to monitor the health of the fish (Berkes 1982; Olsson *et al.* 2004). Some indicators show an understanding of ecological interactions and effects of key variables acting together (*e.g.*, sea ice and wind). For example, the observation of skinny ringed seal pups was connected to lack of sea ice south of Sachs Harbour (Nichols *et al.* 2004).

In the case of contaminants as well, the Inuit do not appear to make linear, cause-and-effect connections as usually made in Western science. Rather, they see environmental change and observations such as those in Table 4.3 as empirically connected (O'Neil *et al.* 1997). Among the Inuit and in many other northern cultures, systematic generalizations regarding cause-and-effect relationships are in general regarded negatively. According to the Inuit worldview, making simplifications and generalizations of complex phenomena is "childish" and without sense (without *ihuma*) (Omura 2002). As pointed out in the chapter by Myers *et al.*, these considerations suggest that the problem in communicating contaminants information to the Inuit is not exclusively a translation problem. The poor communication is not only due to lack of suitable Inuktitut (Inuit language) terminology. It is in part due to differences between the Inuit worldview and the Western one which emphasizes cause-and-effect relationships.

Similarly, the Inuit concept of wellness and sickness is holistic, as the following Inuit quotations from Nunavik (northern Quebec) taken from O'Neil *et al.* (1997, 32, 33) illustrate: "We just keep finding again and again that everything is interlocked. Everything is intertwined. Everything is not neat [like] with [scientific] classification. The world does not work like that to Inuit people. Do your labelling but we see this whole. So let us cherish this [Inuit] knowledge."

The diagnosis of a sick animal is also holistic: "The Inuit know what animals are sick or when they are not sick because they know it even without samples because they have been hunting it for years and years" (O'Neil *et al.* 1997, 32). Nevertheless, Inuit hunters do make reference to specific signs that tells them that an animal is not well and should not be eaten. The following refer mainly to seals: animals with *manimiq* (lumps), skinny animals, discoloured bones, abnormal liver, bumps, and blue-coloured spots in the intestines. The problem may be with the meat, the behaviour of the animal, or its outward appearance. The health of the animal would be in doubt if "it did not look normal" (O'Neil *et al.* 1997). Assessing the "wellness" of an animal by appearance and behaviour is rather similar to a medical doctor assessing the wellness of a person by looking at clinical signs, such as weight, skin colour, eyes, breath, and so on. In the case of the Inuit, the signs of wellness are read continuously and cumulatively, and that is perhaps why an experienced hunter has a good sense about the state of health of an animal. The Inuit hunter's logic is similar to some integrated scientific approaches to stressors in which anomalies are noted and quantified as a percentage and used as a component of an index of biological integrity (Tong 2001).

We have argued that the Inuit (and other northern indigenous) views of contaminants and health are holistic. However, there are specific signs and signals of environmental quality that native hunters monitor within a context of holistic understanding. Returning to the question of the kinds of indicators that may be identified by the two ways of looking at the world, Table 4.4 provides a listing of contaminant-related indicators as used in toxicology and as may be used by indigenous observers. The table is exploratory and is not meant to imply that the two kinds of knowledge systems make similar observations using similar methodologies. They do not. The cultural context of the observations are obviously also different. Rather, our point is that each set of indicators, as used by the two knowledge systems, assesses environmental conditions and wellness in its own way.

At the chemical, biochemical, and cellular levels, toxicology uses many indicators. Local observations and traditional knowledge are generally not very useful at this scale – except that some northern indigenous people are apparently able to taste and smell some contaminants, or effects of contaminants, in animal tissues. At the organismic, population, and community levels, however, local observations can provide a great deal of information. There are some effects, such as physiological changes, that would not be observable to hunters. However,

Table 4.4
EXAMPLES OF INDICATORS OF CONTAMINANT-RELATED EFFECTS, AS MAY BE IDENTIFIED BY WESTERN SCIENCE AND BY TRADITIONAL KNOWLEDGE

	Scientific knowledge	Traditional knowledge
Chemical/biochemical level		
Metallothionein levels (*e.g.*, Cd, Pb, Hg)	+	–
Inhibition of enzyme and protein synthesis in liver, kidney, brain (*e.g.*, Hg)	+	–
Contaminants in tissues, sediments, water	+	–?
Cellular level		
Adenomas, nuclear, mitochondrial, cytological changes in	+	–
Structural changes in cells	+	–
Organismic level		
Structural alteration in fish epidermal mucous	+	+
Tumours	+	+
Lesions related to parasites	+	+
Parasitic infestation	+	+
Reduction in sperm viability	+	–
Changes in survival of larvae and fry	+	–
Growth rate by size (from catch data)	+	+
Growth rate by age (*e.g.*, otolith data)	+	–
Body condition	+	+
Muscle firmness, mesentery fat	–	+
Physiological changes (*e.g.*, osmoregulation)	+	–
Visible neurophysiological changes (*e.g.*, swimming)	+	+
Other behavioural changes	–	+
Population and community level		
Abundance (numbers, biomass)	+	+
Fecundity; sex ratio in catches	+	+
Reproductive life span	+	–?
Age at maturity	+	–?
Genetic diversity	+	–
Community change	+	+

Sources: Compiled from various sources, Attrill and Depledge (1997), Lockhart *et al.* (1992) and Muir *et al.* (1992) for scientific knowledge; Jensen *et al.* (1997), O'Neil *et al.* (1997) and McDonald *et al.* (1997) for traditional knowledge. See also Tables 4.2, 4.3, 4.5, and 4.6.

such physiological effects may express themselves as behavioural effects, and the Inuit are experts on reading those. There are other indicators that may be noted by indigenous observers and not normally studied by science. These include the use of mesentery fat of fish as an indicator of health (Berkes 1982) and observations of a range of animal behaviours.

The implication of Aboriginal worldviews for the question of environmental quality and health indicators is that a high degree of indicator specificity is not a sought-after characteristic on its own. A holistic worldview favours a large number of less specific (and probably multicausal) indicators used simultaneously as a suite, giving the community feedback on many aspects of the environment. At the same time, it gives them a more complete and holistic picture of the environment. Unlike common scientific indicators, local indicators do not produce formalized generalizations. This gives community-based indicators built-in adaptability, that is, they would be readily modified with changing conditions and thus be flexible.

The approach of using a broad suite of simpler indicators, instead of a few detailed and costly ones, is finding favour in Western science as well (Wrona and Cash 1996; Kislalioglu *et al.* 1996). It is increasingly recognized that the use of a few indicators, no matter how well chosen and researched, may be inadequate in reflecting complexity.

LESSONS FROM SOME INDIGENOUS KNOWLEDGE SYSTEMS

The use of traditional knowledge for MEQ and ecosystem-based management is relatively new. However, the potential contribution of traditional knowledge is considerable, given local environmental expertise guided by generations of experience. The knowledge held by indigenous experts enables local scale understandings of impacts and changes in environmental quality, and can be used as a guide for research and application. Documenting this knowledge is only the first step. Recognizing and including local expertise requires building relationships between scientists and communities – those who are studying change and those who are experiencing it. Table 4.5 provides some examples of community observations regarding environmental change and ecosystem-based management in the Arctic. Many of these examples are climate-related but illustrate the nature of ecosystem changes that are being observed by communities and are therefore relevant for community-based monitoring.

A number of community-based monitoring projects have been or are currently being conducted in the North (see chapters by Manseau *et al.* and by Parlee *et al.*). Some are directly related to marine environmental quality, while others are related to other components of the environment. Valuable lessons can be learned from these projects regarding the application of traditional and scientific knowledge to environmental quality. One of these projects is the Tariuq (Ocean) community-based monitoring program in Aklavik and Tuktoyaktuk, NT, carried out since 2000 (Cobb *et al.* 2003). Its objective is to understand the health

Table 4.5

EXAMPLES OF COMMUNITY OBSERVATIONS REGARDING ENVIRONMENTAL CHANGE AND
ECOLOGICAL LINKAGES AS MAY BE RELEVANT TO THE BROADER DEFINITION OF MEQ

Observation	Community/area	Reference
Warming trends affecting fish populations, Arctic char looking unhealthy, smaller size	Rankin Inlet, NWT	1
Earlier spring arrival, shallower rivers, poorer Arctic char flesh quality	Rankin Inlet, NWT	1
More polar bears on land and near communities perhaps because of thinning of ice on floe edge	Whale Cove, Arviat, others	1
Occurrence of fish species not normally known in the area (*e.g.*, salmon in Beaufort Sea), related to climate change	Tuktoyaktuk; Sachs Harbour	2, 4
Lack of ringed seals because of absence of sea ice habitat	Sachs Harbour; Whale Cove	4,1
Coastal erosion increasing because of more wave action due to longer ice-free seasons	Tuktoyaktuk; Sachs Harbour	3, 4
Changes in wind direction affecting Pacific herring harvesting	Tuktoyaktuk Harbour	5

Sources: (1) DFO (2001); (2) Chiperzak and Cockney (2000a); (3) Chiperzak and Cockney (2000b); (4) Jolly *et al.* (2002) ; (5) A. Kristofferson, DFO, pers. comm.

of the marine ecosystem, using community-selected indicators and monitors. The strength of the program results from the working group, which consists of elders, youth, and experienced hunter and trapper members. The DFO sits on the working group and provides a conduit to input from other agencies and scientists on a required basis. This arrangement allows community concerns to be expressed through the selection of indicators and having teams of monitors consisting of youth and an experienced (and often an elder) hunter and trapper. Monitors use scientific methods to carry out sampling, and tissues samples are analyzed for a variety of contaminants. Marine environmental quality objectives are established through workshops. The scientific objectives are to conserve populations and species and quality of water and biota. The social and cultural objectives were established as the use of traditional knowledge; sharing of knowledge and awareness; training and capacity-building; and providing information for decision-making.

A second example is the traditional knowledge project of the Lutsel K'e Dene First Nation (Parlee *et al.* 2001). The Lutsel K'e Dene notion of healthy fish involves observations regarding five points: (1) size and shape: visual assessment of the length/weight ratio, with attention to deformities; (2) population and diversity:

whether the species that are supposed to be present in a place are present there; (3) fatness of fish: fat around the internal organs is a sign of good fish health and good water quality; (4) cleanliness and healthy organs: infections, parasites, and deformities are signs of poor health; and (5) colour and texture of the flesh: firm texture and appropriate colour, for example, darker red meat in the case of lake trout (Parlee *et al.* 2001, p. 33). The Lutsel K'e Dene traditional knowledge project covers indigenous concepts of the health of the environment and the health of the community, which are considered to be interrelated. Hence the overall social objective of a healthy Dene way of life encompasses indicators of healthy fish and other animals, such as caribou. The results regarding fish are interesting because the five-point list, completely produced by Dene hunters and elders, is concise but comprehensive. It suggests a range of indicators that can actually be used for monitoring change.

A third example is the Inuit Bowhead Knowledge Study, carried out under the provisions of the *Nunavut Land Claims Agreement*. The Inuit concept of a healthy ecosystem is clearly articulated in the report: animals "remain healthy and abundant only if they were harvested and treated with respect" (Hay *et al.* 2000). The social and cultural objectives were to follow traditional rules of respect; to share food and never to waste it; and to renew traditions of bowhead whale hunting. The traditional knowledge study was based on Inuit monitoring, and the main sources of information were (1) the frequency of sightings, (2) trends in bowhead group size, and (3) observations of cows with calves (Hay *et al.* 2000). These indicators of a healthy and recovering bowhead population are likely to be considered suitable by scientists as well. Inuit hunters are assessing changes by mentally tracking changes in these indicators over the years. The indicators can be made consistent with science by quantifying them. However, there is a notable discrepancy in objectives. While the scientific objectives were about conserving populations and species, the Inuit objectives were about Inuit-bowhead relationships and access to the resource. The two sets of objectives may be reconciled by aiming for the long-term sustainability of the resource.

A fourth example is *Voices from the Bay*, a report on the traditional environmental knowledge of Inuit and Cree in the Hudson Bay region (McDonald *et al.* 1997). The primary objective of the study was to assess region-wide environmental change related to cumulative impacts of hydroelectric development. It was a remarkable project, initiated and carried out by indigenous people themselves, documenting what communities said about changes occurring in their environment, combining these local observations into a regional whole, and using this information as a baseline in the face of additional hydroelectric development. The report makes a holistic assessment of observed changes, including those related to contaminants and climate change, as well as to hydro development impacts. Table 4.6 is an attempt to capture the Cree and Inuit notions of respect as a starting point, followed by concepts of healthy human-environment relations and signs and signals of problems.

Table 4.6

CREE AND INUIT VIEWS FROM THE HUDSON BAY AREA RELATED TO A HEALTHY ENVIRON-
MENT. SELECTION OF ITEMS COMPILED FROM MCDONALD *ET AL.* (1997)

Concept of respect (pp. 1, 4, 5)

- Knowledge of the land from ancestors
- Co-existence with the environment
- Respect for the land tied to a healthy environment

Concept of healthy human-environment relations

- Knowledge of seasonal cycles (p. 25)
- Ability to anticipate change by watching animals (p. 25)
- Knowledge of long-term population cycles, *e.g.*, walrus (p. 42), beaver (p. 43), beluga (p. 87)

Wellness indicators (p. 43)

- Seasonal fat thickness
- Condition of the liver
- Meat colour
- Fur condition
- Behaviour of the animal

Signs and signals of problems

- Changes in sea-ice (pp. 30, 31 and throughout)
- Polynyas (open-water areas) freezing over (pp. 30–31)
- Changes in currents (pp. 30, 31 and throughout)
- Changes in weather (p. 28)
- People cannot predict weather and seasons anymore (p. 29)
- Taste of snow and rainwater has changed (p. 30)
- Changes in colour, composition and taste of snow (p. 27)
- Taste of land animals and plants changing (p. 27)
- Changes in animal migrations, *e.g.*, flyways of geese shifted east in James Bay (pp. 46, 84, 86)
- Behaviour of animals changing, *e.g.*, polar bears lost fear of humans and dogs (p. 91)
- Species disappearing, *e.g.*, Rankin Inlet area, fish and ringed seals (p. 27)
- Change in fish condition, *e.g.*, Deception Bay, northern Quebec
- Change in fish size, *e.g.*, Great Whale area, Arctic char and trout (p. 26)
- Change in fish meat quality, *e.g.*, James Bay, Harrikanaw River sturgeon (p. 83)

The indigenous expertise leading to Voices was built stepwise, starting with the La Grande hydro development of the 1970s. The hydro project produced unexpected impacts in the La Grande estuary, and the Cree learned to use scientific styles of monitoring (*e.g.*, coring for ice thickness), in addition to their own traditional monitoring (*e.g.*, judging safety of ice by colour and the sound of tapping stick) (Berkes 1988). The Cree and Inuit used their own knowledge of sea ice, currents, and animal and plant distributions to assess regional-scale change. They used signs and signals (*e.g.*, changes in sea ice and currents) as well as their knowledge of ecological relations to produce a comprehensive evaluation. For example, the Cree observed that changes in the freshwater-saltwater balance not only affected fish distributions but also marine grasses that, in turn, affected the distributions of geese feeding on them. The Inuit of Sanikiluaq reported winterkills of eider ducks and reduction in polynyas (open-water habitat for wintering eider) associated with post-1970 changes in currents and sea ice. Subsequently, Robertson and Gilchrist (1998) provided scientific follow-up and cross-verification of these observations.

CONCLUSIONS

The framework for ecosystem and MEQ objectives and indicators developed at the DFO workshop (Jamieson *et al.* 2001) (Figure 4.1) was an important step for moving forward with the implementation of IM and MPA activities under Canada's Oceans Strategy. The environmental objectives remain to be further defined as objectives, targets, and indicators. As part of the IM node within the Oceans Management Research Network, we have begun to examine the framework and its implementation in northern Canada. Our intent was not to redesign the current framework, rather to examine how the social dimension inherent in northern societies can contribute to and enhance the present framework, and subsequent work on MEQ objectives and indicators.

A number of factors led us to conclude that there is likely to be overlap and interaction between indicators of the environmental and social components of ecosystem and MEQ objectives in their application to IM planning in northern Canada. We conclude this because northern indigenous peoples have a unique perspective on their environment, based on societal, cultural, and spiritual ties to the land. Moreover, the use of traditional knowledge, which has been transferred from generation to generation through time-honoured use of the land, is integrated into all aspects of society, community conservation plans, and land claims resource co-management arrangements. Northern indigenous peoples have their own perceptions about what constitutes a healthy marine ecosystem, and they have a holistic approach to understanding environmental change.

Based on the four examples of recent and ongoing monitoring studies provided in this chapter, we conclude that northern indigenous peoples have criteria as to what constitute suitable indicators (signs and signals) of environmental change. This comes from many generations of observing seasonal patterns of flora and fauna used for subsistence foods, and travelling across the land and sea ice. These indigenous criteria should be examined and attempts made to validate the observations. The Arctic Borderlands Ecological Co-op has attempted to consolidate several years of community observations gathered by interviewing community members (Kofinas *et al.* 2002).

There are several examples of the use of the two kinds of combined knowledge to improve resource and environmental management (Berkes 1999). For example, in the Inuit Observations of Climate Change project, Riedlinger and Berkes (2001) developed a conceptual framework for linking science-based research with local knowledge. The framework was articulated through five interrelated convergence areas; that is, research areas that could facilitate collaboration and communication between scientists and local experts. These convergence areas were the use of traditional knowledge (i) as local scale expertise; (ii) as a source of climate history and baseline data; (iii) in formulating research questions and hypotheses; (iv) as insight into impacts and adaptation in Arctic communities; and (v) for long-term, community-based monitoring.

The five areas highlight the ability of local experts to address the complexities of Arctic environmental research at spatial and temporal scales often under-represented in Western science. All five areas are applicable to environmental quality indicators. For example, indigenous hunters can and do note changes in the environment and deviations from the normal, as in the abnormal burbot liver example (Lockhart *et al.* 1987) and in the northern Quebec seal examples (O'Neil *et al.* 1997). Animals that "do not look normal" is an expression typical of the results of continuous indigenous monitoring of the environment, or continuous visual "sampling" as scientists might put it, and the use of mental reference points to make an assessment of whether an animal is sick or not. The logic is similar to the scientific one, but the observation is holistic and the discourse does not fit well with scientific discourse.

What are the implications for these explorations for MEQ indicators? We conclude that probably no single approach to setting ecosystem and MEQ objectives, or selecting and monitoring indicators will be suitable in the North. The vast territory, harsh environment, and high costs of research make long-term monitoring programs a challenge. We suggest that the judicious use of traditional ecological knowledge and community monitoring provides the most effective way of moving forward. The concept of mutual learning rather than attempting to "integrate" local knowledge into science might provide a "weight of evidence" approach to environmental change. Collaborative or participatory approaches to setting ecosystem objectives, selecting indicators, and monitoring are likely to provide the best path forward. More research is needed in order to develop the most effective models to bring scientists and those with ecological knowledge together to move along this path. Each land claim settlement region has its own structure, and facilitating the adequate engagement within each region is important and challenging.

The strengths contained within community-based monitoring seem to indicate that this is the best way of overcoming this challenge. The various approaches of ongoing community-based monitoring programs are examined elsewhere in this book (Parlee *et al.*, this volume). Each has its strength and weakness, but common to all are the challenges of sustaining long-term monitoring. Sustaining long-term funding, maintaining interest and relevance in the programs, maintaining capacity within both the communities and scientific agencies to dedicate the time needed to accomplish community-based monitoring, shifting policies, etc., are all challenges that must be overcome in order to make a successful monitoring program. More research is required to further develop and refine this approach for use both locally and throughout the North in the implementation of Integrated Management, and more specifically in helping to "unpack" ecosystem and MEQ objectives. Is it possible to develop a common suite of indicators for use in different regions of the Arctic, or will they have to be tailored to suit the individual settlement regions?

REFERENCES

Abele, F. 1997. "Traditional ecological knowledge in practice." *Arctic* 50: i–iii.

Attrill, M.J., and M.H. Depledge. 1997. "Community and population indicators of ecosystem health: targeting links between levels of biological organization." *Aquatic Toxicology* 38: 183–97.

Barber, D.G., L. Barber, L. Carpenter, D. Cobb, and R. DeAbreu. In press. Sea Ice Variability and Climate Change –"Two ways of Knowing"– identifying research priorities for the future. Can. Manuscr. Rep. Fish. Aquat. Sci. *http://www.umanitoba.ca/geography/ceos/arcticworkshop/index.html*

Berkes, F. 1982. "Preliminary impacts of the James Bay hydroelectric project, Quebec, on estuarine fish and fisheries." *Arctic* 35: 524–30.

——— . 1988. "The intrinsic difficulty of predicting impacts: lessons from the James Bay Hydro Project." *Environmental Impact Assessment Review* 8: 201–20.

——— . 1999. "Sacred Ecology." *Traditional Ecological Knowledge and Resource Management.* Philadelphia and London: Taylor & Francis.

——— , M. Kislalioglu, C. Folke, and M. Gadgil. 1998. "Exploring the basic ecological unit: Ecosystem-like concepts in traditional societies." *Ecosystems* 1: 409–15.

——— , J. Mathias, M. Kislalioglu, and H. Fast. 2001. "The Canadian Arctic and the Oceans Act: The development of participatory environmental research and management." *Ocean & Coastal Management* 44: 451–69.

Canada. 1997. *Oceans Act.* Statutes of Canada, 1996, c. 31. In force 31 January 1997.

Chiperzak, D., and M. Cockney. 2000a. Results from the Tuktoyaktuk Community Marine Ecosystem Health (MEH) Workshop, January 25–26, 2000. Fisheries and Oceans Canada, Oceans Programs Division, Winnipeg.

——— . 2000b. Results from the Aklavik Community Marine Ecosystem Health (MEH) Workshop, March 6–7, 2000. Fisheries and Oceans Canada, Oceans Programs Division, Winnipeg.

Cobb, D, S. Newton, and K. Cott. 2003. Tuktoyaktuk and Aklavik Tariuq (Ocean) Community-Based Monitoring Program, 2001–2002 Monitoring Summary Report. Fisheries and Oceans Canada, Oceans Programs Division, Winnipeg.

DFO. 2001. Fisheries and Oceans Canada Trip Report: Churchill, Kivalliq and Iqaluit Community Tour, March 7–19, 2001. Fisheries and Oceans Canada, Oceans Programs Division, Winnipeg.

——— . 2002a. *Canada's Oceans Strategy: Our Oceans, Our Future.* Ottawa.

——— . 2002b. Policy and operational framework for integrated management of estuarine, coastal and marine environments in Canada, Ottawa.

ESD. 2003. Bioindicators of aquatic ecosystem stress. Environmental Sciences Division, Oakridge National Laboratory. *http://www.esd.ornl.gov/programs/bioindicators* (September 2003)

Fast, H., J. Mathias, and O. Banias. 2001. "Directions toward marine conservation in Canada's Western Arctic." *Ocean & Coastal Management* 44: 183–205.

Freeman, M.M.F. 1993. "The International Whaling Commission, small-type whaling, and coming to terms with subsistence." *Human Organization* 52: 243–51.

GNWT. 1993. *Traditional knowledge policy.* Government of the Northwest Territories, Yellowknife.

Hay, K., D. Aglukark, D. Igutsaq, J. Ikkidluaq, M. Mike. 2000. Final Report of the Inuit Bowhead Knowledge Study. Nunavut Wildlife Management Board, Iqaluit.

Jamieson, G., R. O'Boyle, J. Arbour, D. Cobb, S. Courtenay, R. Gregory, C. Levings, J. Monro, I. Perry, and H. Vandermeulen. 2001. Proceedings of the national workshop on objectives and indicators for ecosystem-based management. Sydney, BC. 27 February–2 March 2001. Fisheries and Oceans, Canadian Science Advisory Secretariat, Proceedings Series, No. 2001/09.

Jensen, J., K. Adare, and R. Shearer, eds. 1997. Canadian Arctic Contaminants Assessment Report. Northern Contaminants Program. Indian and Northern Affairs Canada, Ottawa.

Jolly, D., F. Berkes, J. Castleden, T. Nichols, and the community of Sachs Harbour. 2002. "We can't predict the weather like we used to: Inuvialuit observations of climate change, Sachs Harbour, western Canadian Arctic." In *The Earth is Faster Now: Indigenous Observations of Arctic Environmental Change*, edited by I. Krupnik and D. Jolly, 92–125. Fairbanks, AK: Arctic Research Consortium of the United States.

Kislalioglu, M., E. Sherer, and R.E. McNicol. 1996. "Effects of cadmium on foraging behaviour of lake charr, *Salvelinus namaycush*." *Environmental Biology of Fishes* 46: 75–82.

Klaverkamp, J.F., C.L. Baron, B.W. Fallis, K.G. Ranson, K.G. Wautier, and P. Vanriel. 2002. "Metals and metallothionein in fishes and metals in sediments from lakes impacted by uranium mining and milling in northern Saskatchewan." *Can. Tech. Rep. Fish. Aquat. Sci.* 2420: v + 72 p.

Kofinas, G. with the communities of Aklavik, Arctic Village, Old Crow, and Fort McPherson. 2002. "Community contributions to ecological monitoring." In *The Earth Is Faster Now: Indigenous Observations of Arctic Environmental Change*, edited by I. Krupnik and D. Jolly. Fairbanks, AK: Arctic Research Consortium of the United States.

Krupnik, I., and D. Jolly, eds. 2002. *The Earth Is Faster Now: Indigenous Observations of Arctic Environmental Change*. Fairbanks, AK: Arctic Research Consortium of the United States.

Legat, A., S.A. Zoe, and M. Chocolate. 1995. "The importance of knowing." In *NWT diamonds project environmental impact statement*. Vol. 1, Appendices. Vancouver: BHP Diamonds.

Lockhart, W.L., D.A. Metner, D.A.J. Murray, and D.C.G. Muir. 1987. "Hydrocarbons and complaints about fish quality in the Mackenzie River, Northwest Territories, Canada." *Water Pollution Research Journal of Canada* 22: 616–25.

———, R. Wagemann, B. Tracey, D. Sutherland, and D.J. Thomas. 1992. "Presence and implications of chemical contaminants in the freshwaters of the Canadian Arctic." *The Science of the Total Environment* 122: 165–243.

McDonald, M., L. Arragutainaq, and Z. Novalinga. 1997. *Voices From The Bay: Traditional Ecological Knowledge of Inuit and Cree in the Hudson Bay Bioregion*. Ottawa: Canadian Arctic Resources Committee.

Meadows, D.H. 1998. Indicators and Information Systems for Sustainable Development. Hartland Four Corners, VT: The Sustainability Institute. *http://www.sustainer.org* (January 2004)

Muir, D.C.G., R. Wagemann, B.T. Hargrave, D.J. Thomas, D.B. Peakall, and R.J. Norstorm. 1992. "Arctic marine ecosystem contamination." *The Science of the Total Environment* 122: 75–134.

NCP. 2003. Canadian Arctic Contaminants Assessment Report II. Northern Contaminants Program. Indian Affairs and Northern Development, Ottawa.

Nichols, T., F. Berkes, D. Jolly, N.B. Snow, and the Community of Sachs Harbour. 2004. "Climate change and sea ice: Local observations from the Canadian western Arctic." *Arctic* 57: 68–79.

Olsson, P., C. Folke, and F. Berkes. 2004. "Adaptive co-management for building resilience in social-ecological systems." *Environmental Management* 34: 75–90.

Omura, K. 2002. Problems of use of traditional ecological knowledge in resource management: a case from Canadian Arctic. Papers of the Monbukagakusho International Symposium, National Museum of Ethnology, Osaka, Japan.

O'Neil, J.D., B. Elias, and A. Yassi. 1997. "Poisoned food: cultural resistance to the contaminants discourse in Nunavik." *Arctic Anthropology* 34: 29–40.

Parlee, B., M. Basil, and N. Casaway. 2001. Traditional knowledge in the Kache Kue Study Region. Submitted by the Lutsel K'e Dene First Nation. West Kitikmeot Slave Study Society, Yellowknife.

Preston, R.J., F. Berkes, and P.J. George. 1995. Perspectives on sustainable development the Moose River Basin. Papers of the 26th Algonquian Conference: 378–93.

Rice, J. 2003. "The role of indicators in integrated coastal management." *Ocean and Coastal Management Journal* 46(3–4).

Riedlinger, D., and F. Berkes. 2001. "Contributions of traditional knowledge to understanding climate change in the Canadian Arctic." *Polar Record* 37: 315–28.

Robertson, G.J., and H.G. Gilchrist. 1998. "Evidence of population declines among common eiders breeding in the Belcher Islands, Northwest Territories." *Arctic* 51: 378–85.

Sadler, B., and P. Boothroyd. 1994. A background paper on traditional ecological knowledge and modern environmental assessment. Vancouver, UBC Centre for settlement studies. Canadian Environmental Assessment Agency and International Association for Impact Assessment.

Suter, G.W. II. 2001. "Applicability of indicator monitoring to ecological risk assessment." *Ecological Indicators* 1: 101–12.

Tong, S.T.Y. 2001. "An integrated exploratory approach to examining the relationships of environmental stressors and fish responses." *Journal of Aquatic Ecosystem Stress and Recovery* 9: 1–19.

Vandermeulen, H., and D. Cobb. 2004. "Marine environmental quality: a Canadian history and options for the future." *Ocean & Coastal Management* 47: 243–56

Wrona, F.J., and K.J. Cash. 1996. "The ecosystem approach to environmental assessment: moving from theory to practice." *Journal of Aquatic Ecosystem Health* 5: 89–97.

CHAPTER 5

INTEGRATED MANAGEMENT PLANNING IN CANADA'S WESTERN ARCTIC:

AN ADAPTIVE CONSULTATION PROCESS

Helen Fast, Doug B. Chiperzak, Kelly J. Cott,
and G.M. Elliott (Department of Fisheries and Oceans)

In collaboration, stakeholders learn and therefore change. While heterogeneity and diversity may be impediments to dialogue, they are also an immense source of creative potential. Collaboration leads to the reconciliation of diverse frames of reference, and therefore to the transformation of agents' mindsets, and thus, indirectly to the modification of the original setting.

— Paquet and Wilkins 2002, 9

THE REGIONAL CONTEXT

The Inuvialuit Settlement Region (ISR) lies in the Canadian western Arctic region (see Figure 5.1). Created with the signing of the Inuvialuit Final Agreement (IFA) in 1984, the ISR covers 906,430 square kilometres. It includes four distinct geographic regions: the Beaufort Sea, the Mackenzie River Delta, the Yukon North Slope, and the Arctic islands. The Mackenzie Delta includes lake, swamp, and river channels covering 35,000 square kilometres. The population of the region in 2003 was about 5,600 people, including 3,300 Inuvialuit.

The marine environment of the ISR includes a permanently ice-covered region, a seasonally ice-covered region, and a coastal area influenced by the mixing of salt water and fresh water from the Mackenzie River. The continental shelf of the Beaufort Sea is quite narrow, nowhere exceeding 150 kilometres offshore. The average depth of the shelf is less than 65 meters, and ranges from around 10 meters in the Mackenzie Delta to 600 meters around Amundsen Gulf. The shelf seas and ice edges provide food for the Inuvialuit and other top predators. The Beaufort Sea marine region has a large population of polar bear, ringed and bearded seals, the largest summer feeding population of bowhead whales, and perhaps the world's largest summering stock of beluga whales.

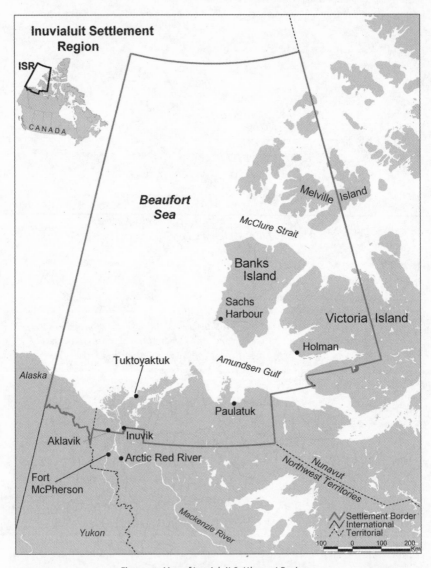

Figure 5.1 Map of Inuvialuit Settlement Region.

The region is rich in non-renewable hydrocarbon resources. Hydrocarbon exploration in the Canadian Beaufort began in the late 1960s, fuelled in part by the discovery of oil at Prudhoe Bay, Alaska, in 1968. The period of exploration which followed lasted approximately twenty years and resulted in a number of significant oil and gas discoveries both in the nearshore and offshore areas of the Beaufort Sea. A proposal to construct a pipeline from the Mackenzie Delta was tabled in the 1970s. The Federal Government responded by hold-ing an inquiry to assess what impacts would occur with the construction of a pipeline. The Report of the Mackenzie Valley Pipeline Commission (Berger 1977) recommended that comprehensive land use planning be undertaken to address resource use conflicts identified during the commission's hearings, and further, that part of the area of West Mackenzie Bay should become a beluga sanctuary. The commission also recommended a ten-year moratorium on the construction of the pipeline in order to allow time to settle land claims in the Mackenzie Valley. The recommendation for a moratorium coincided with a fall in oil prices. Hydrocarbon exploration activities in the Beaufort Sea were subsequently scaled back and ultimately shut down.

Six years later the Task Force on Northern Conservation was established to provide advice to DIAND (Department of Indian and Northern Affairs) concerning the development and implementation of a comprehensive conservation policy for northern Canada (DIAND 1984a). The recommendations tabled by the Task Force emphasized the need for marine conservation management and planning initiatives, including a comprehensive network of land and/or water areas subject to special protection, taking into account local knowledge and uses of the area. The IFA which was signed in the following year provided legislative support to those recommendations.

The three goals of the IFA are:

a) to preserve Inuvialuit cultural identity and values within a changing northern society;
b) to enable Inuvialuit to be equal and meaningful participants in the northern and national economy and society; and
c) to protect and preserve the Arctic wildlife, environment and biological productivity. (DIAND 1984a, x).

Under the IFA the Inuvialuit Regional Corporation (IRC) was given responsibility for managing the compensation and benefits received by the Inuvialuit. The Inuvialuit Game Council (IGC) was given responsibility for representing the collective Inuvialuit interest in wildlife. The Fisheries Joint Management Committee (FJMC) was given the responsibility to assist Canada and the Inuvialuit in administering the rights and obligations relating to fisheries under this Agreement, and to provide advice to the Minister of Fisheries and Oceans Canada in carrying out his responsibilities for the management of fisheries. The Wildlife Management Advisory Council (NWT) (WMAC) with representation from Canada, the Government of the Northwest Territories, and the Inuvialuit, was created to give advice to the appropriate minister on request, on all matters relating to wildlife policy and the management, regulation and administration of wildlife habitat and harvesting in the western Arctic region.

The Wildlife Management Advisory Council (NWT) and the FJMC drafted the Inuvialuit Renewable Resource Conservation and Management Plan (IRRCMP) (FJMC and WMAC 1988). This plan lays out a long-term strategy for the conservation and management of fish and wildlife in the Inuvialuit Settlement Region. Soon after, efforts initiated earlier by the Department of Fisheries and

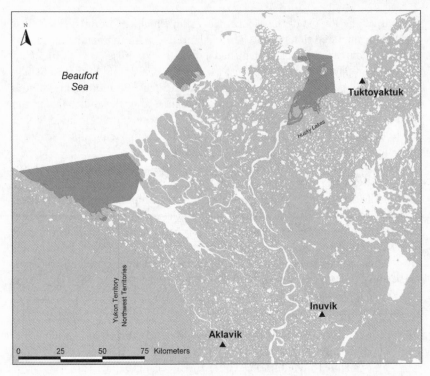

Figure 5.2 Map of the Zone 1(a) areas.

Oceans (DFO) toward the development of a beluga management plan were concluded with drafting of the Beaufort Sea Beluga Management Plan (BSBM Plan) (FJMC 2001). FJMC assumed responsibility for implementing the plan. Parties to the plan included the FJMC, the six community Hunters and Trappers Committees, and DFO. Consistent with the themes and goals of the IRRCMP, the purpose of this plan was to ensure the responsible and effective long-term management of the beluga resource by the Inuvialuit and the Department of Fisheries and Oceans.

Management issues addressed under the BSBM Plan include the following: oil, gas, and mining exploration, production and related development including dredging drilling, seismic and sounding surveys, island/camp maintenance, vessel movements, helicopter and fixed-wing flights, and ice-breaking, shipping routes, port development, possible future commercial fisheries development, contaminant levels in marine waters and mammals, a developing tourism industry, a myriad of regulators, transboundary issues, subsistence hunting practices and traditional values closely related to the beluga harvest, and climate change.

Compliance with the BSBM Plan is voluntary. The document was intended to provide clarity for industrial and other users wishing to conduct activities in the Beaufort Sea. Authors of the BSBM Plan classified the estuarine and marine waters into four management zones. These zones were consistent with the values and wishes of Inuvialuit communities. Areas zoned as levels two and three allowed for development that would not adversely affect the beluga or their habitat. Zone four was used to classify international waters. Beluga management issues here are an international responsibility.

Areas classified as Zone 1a and 1b were regarded as areas needing special protection, and strict limits were placed on the types of activities allowed. Three areas known to be important beluga habitat in the Mackenzie River estuary were identified as Zone 1(a) areas. These areas were also important to maintaining the local subsistence economy. In order to protect these interests, the BSBM Plan states that "in the review of any development proposal, Zone 1 is to be considered a Protected Area [and] the oil and gas industry should not be permitted to explore for resources within or on the shores of any Zone 1 waters nor to produce hydrocarbons or construct/operate any type of facility" (FJMC 2001, 13). The Zone 1(a) areas include a large section of Kugmallit Bay, an area bounded by Kendall, Pelly and Garry Islands (hereafter referred to as Kendall Island) and a large area in the northern portion of Shallow Bay (also known as Mackenzie Bay) (Figure 5.2). The total area covered by these three Zone 1(a) areas is 1,716 square kilometres.

A GROWING NEED FOR INTEGRATED MANAGEMENT

In the late 1990s interest in oil and natural gas exploration in the Mackenzie Delta and Beaufort Sea resurfaced. The legacy left in the Beaufort Sea offshore area by earlier exploration activities included eleven Exploration Licences, one Production Licence, and thirty-two Significant Discovery Licences. (A Significant Discovery License [SDL] is granted by Indian and Northern Affairs Canada and gives the licence holder the rights to the petroleum resources within the SDL, as well as the right to apply for a production licence whenever they choose to do so.) The area covered by these licences is 10,096 square kilometres. While the economic potential offered by the resurgence of activity was generally welcomed, the potential for negative environmental effects was of concern to community members who depended on the natural resources in the region for food, and whose culture and traditional way of life depended on their continued use of the land and sea.

The continued protection of the three Zone 1(a) areas is an important issue. In addition to their ecological importance, these areas are culturally significant to the Inuvialuit and an important source of food. Traditional fishing and whaling camps have long been established in each of the three Zone 1(a) areas. Inuvialuit from the communities of Aklavik, Inuvik, and Tuktoyaktuk (Figure 5.2) are primary users of the areas. Tuktoyaktuk is situated on the coast while Inuvik and Aklavik are located in the Mackenzie Delta. The Inuvialuit of Aklavik are the primary users of the Shallow Bay Zone 1(a) area. The Inuvialuit of Inuvik use both the Kendall Island and Kugmallit Bay Zone 1(a) areas for traditional harvesting. Residents of Tuktoyaktuk harvest beluga primarily in the Kugmallit Bay Zone 1(a) area.

Driftwood is used to make smokehouses for fish along the Beaufort Coast, Inuvialuit Settlement Region. Photo by B. Spek, Department of Fisheries and Oceans.

The types of environmental impacts likely to occur with renewed exploration and development of hydrocarbon resources in the Mackenzie Delta-Beaufort Sea will be significant. These activities will also potentially bring major employment opportunities for northerners. They include a large increase in ship movement and barge traffic through the region. A shipping corridor exists through Kugmallit Bay. This shipping corridor is essential for the supply of goods and materials for coastal communities both in the ISR and in parts of western Nunavut. The corridor also plays an important role logistically for oil and gas activities in the ISR. Several Significant Discovery Licences exist within the Kendall Island Zone 1(a) area. During periods of intense activity in previous decades it was not uncommon to see an average of a hundred vessels of all types in Kugmallit Bay at any given time – including barges, platforms, and supply vessels. Dredging activities will also increase.

The shorebases that will be built to support offshore activities are known to produce localized impacts on the marine environment. For example, Tuktoyaktuk Harbour and McKinley Bay served as staging areas for offshore drilling that was carried out in the Beaufort Sea during the 1970s and 1980s. Studies have shown that some of the highest hydrocarbon concentrations in the Arctic occur in Tuktoyaktuk Harbour and McKinley Bay. These hydrocarbons appear to originate primarily from chronic fuel spills and runoff from work-yards (AMAP 1998). Tuktoyaktuk Harbour plays an important role in the shipment of goods throughout the region and beyond. The Northern Transportation Company Limited (NTCL), the largest shipping company in the region, has a large docking and staging facility in Tuktoyaktuk Harbour. During the earlier hydrocarbon exploration days this harbour was the main base for companies operating in the Beaufort Sea.

Beluga summering in the Beaufort Sea migrate through areas where oil and gas production and transportation activities are proposed for the future. They concentrate in areas where mining (gravel removal), deep water port development and shipping could affect water regimes, water quality, and food availability. Such activities could affect beluga either directly (underwater noise, oil spills) or indirectly (changes in stability or integrity of ice, timing of breakup, chronic hydrocarbon contamination of food species). The aggregations of beluga whales during the summer months is a large draw for tourism activity in and around the Zone 1(a) areas. Since whale harvesting is also conducted during the summer, the Beaufort Sea Beluga Management Plan prohibits tourism activities such as whale watching in the area from spring breakup (normally in July) until August 15. As well, the Zone 1(a) areas are important for migrating anadromous fish such as inconnu, Arctic cisco, and, in the Mackenzie Bay section only, Dolly Varden char (North/South Consultants Inc. and Inuvialuit Cultural Resource Centre 2003, 2004).

Considering the magnitude of possible development scenarios, members of the FJMC and Inuvialuit beneficiaries expressed concern regarding the absence of legally enforceable mechanisms available under the BSBM Plan. The lack of scientific knowledge that could be used to assess the relative sensitivity of marine mammals and their habitat to disturbance by various activities in the Zone 1(a) areas was identified as another management concern. Finally, industry and others had repeatedly requested simplification of the maze of legislation and regulation which currently governs management decision-making processes in the region. Recognizing that a time of major change was imminent, the Inuvialuit Regional Corporation (IRC), the Inuvialuit Game Council (IGC), the Fisheries Joint Management Committee (FJMC), the Department of Fisheries and Oceans (DFO), and industry represented by the Canadian Association of Petroleum Producers (CAPP) met to consider whether the recently passed *Oceans Act* (Canada 1997) could be used to facilitate implementation of a planning process that would balance development and community interests in the months and years to come.

With passage of the *Oceans Act* (Canada 1997), Canada had become one of the first countries in the world to make a legislative commitment to a comprehensive approach for the protection and development of oceans and coastal waters. The *Oceans Act* calls for wide application of the precautionary approach to the conservation, management and exploitation of marine resources. It also recognizes the significant opportunities offered by the oceans and their resources for economic diversification and the generation of wealth for the benefit of all Canadians, particularly those in coastal communities. To achieve these commitments, the Act calls on the Minister of Fisheries and Oceans to lead and facilitate the development of plans for integrated management. The concept of integrated management as described in the *Oceans Act* includes collaborative planning and management of human activities to minimize conflict among users. This planning process respects existing divisions of constitutional and departmental authority, and does not abrogate or derogate from any existing Aboriginal or treaty rights.

In 1999 the Inuvialuit management and co-management bodies, DFO, and industry agreed to follow the model outlined in the *Oceans Act* and collaborate on the development of integrated management planning for marine and coastal areas in the Inuvialuit Settlement Region. This agreement is called the Beaufort Sea Integrated Management Planning Initiative (BSIMPI). The BSIMPI Senior Management Committee (SMC) represents the interest groups that formed the initiative. The committee is responsible for overseeing the development of a management planning process for ocean-related activities in the Beaufort Sea. One of its first actions was to form a Working Group to implement effective collaboration on ocean management efforts in the region. Representation on the Working Group mirrors that of the SMC, with the addition of a member from DIAND.

The Senior Management Committee and Working Group are not formal co-management bodies; however, the balanced representation on these committees is consistent with the principles of co-management outlined in the IFA. This ensures that the Inuvialuit have a strong leadership voice in the decision-making process. Administrative, technical, and communication support for the Senior Management Committee and Working Group is provided through the BSIMPI Secretariat, which consists of regional DFO Oceans Program staff and the independent Chair of the BSIMPI Working Group. The secretariat works to ensure that other organizations, governments, and communities with an interest in ocean use and management are brought into the BSIMPI process by inviting them to participate in selected WG meetings. As well, the secretariat keeps these groups informed of BSIMPI activities and progress, and ensures that any issues, comments, and recommendations are brought back to the Working Group and, if appropriate, the Senior Management Committee (Figures 5.3 and 5.4). The BSIMPI Secretariat is responsible for leading consultations, often with the participation of Inuvialuit Working Group members.

SMC members agreed during their first meetings that balancing both the conservation and development interests in the BSBMP Zone 1(a) areas was a high priority. It was understood that developing a working relationship between BSIMPI partners would take time, and that building a shared sense of trust was paramount to long-term success. It was agreed that given the extremely complex, dynamic, and unproven environment in which the BSIMPI had been created, that the BSIMPI Working Group would be asked to focus on one major task. Doing so would provide the opportunity for participants to identify shared interests and develop mutual understanding and respect for one another's values. Consistent with this intention, and consistent with National DFO Oceans' priorities at the time, the BSIMPI WG was directed to conduct an evaluation of the merits of establishing a MPA in the Zone 1(a) areas. The work commenced in early in 2001.

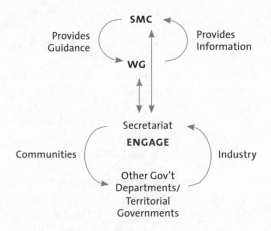

Figure 5.3 Oceans Governance – BSIMPI.

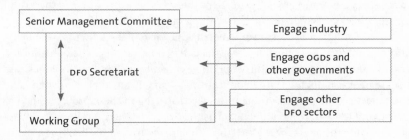

Figure 5.4 Organizational structure of the BSIMPI.

THE CONSULTATION PROCESS

As specified in the National Framework for Establishing and Managing Marine Protected Areas (DFO 1999), and consistent with the philosophy espoused in the *Oceans Act*, the BSIMPI Working Group initiated assessments of the ecological, economic, social, and cultural environment of the proposed MPA, as well as of the technical merits of the proposal (North/South Consultants Inc. 2002; North/South Consultants Inc. 2001; Kavik-AXYS Inc. 2002; Elliott 2002). The purpose of these assessments was to provide baseline information needed by the communities, industry, government, and others to evaluate the proposed MPA against the stated criteria for an MPA (Canada 1996). Section 35(1) of the *Oceans Act* defines an MPA as an area of the sea that ... has been designated ... for special protection for one or more of the following reasons:

(a) the conservation and protection of commercial and non-commercial fishery resources, including marine mammals, and their habitats;

(b) the conservation and protection of endangered or threatened marine species, and their habitats;

(c) the conservation and protection of unique habitats;

(d) the conservation and protection of marine areas of high biodiversity or biological productivity; and

(e) the conservation and protection of any other marine resource or habitat as is necessary to fulfil the mandate of the Minister [of Fisheries and Oceans].

The assessment reports indicated that the Zone 1(a) areas met the requirements for a Marine Protected Area as set out in the *Oceans Act*.

The next task was to conduct a consultation process with ISR communities, industry stakeholders, and government organizations. The objective of this process was to determine the level of support and interest among the Inuvialuit, government, and industry for the idea of creating a Marine Protected Area (MPA) in the three Zone 1(a) areas. If sufficient interest was expressed in proceeding, further direction would be sought on implementing the MPA.

The consultation process ramped up as the various assessments were being drafted, vetted, and finalized. The level, extent, and depth of the consultation process undertaken exceeded that of other consultations processes which had to date been conducted in the communities. This meant that secretariat staff and BSIMPI Working Group members participating in these processes had to design and test working models even as they were conducting the actual work of evaluating the areas for MPA status. This proved to be a challenging task, requiring a capacity for flexibility and adaptation and a high tolerance for change. Though the process has proven dynamic and often unpredictable, principles established by BSIMPI to guide the process lent a sense of stability and focus to these efforts. These principles included the following:

- recognition of Inuvialuit rights established under the IFA;
- respect for the views of all parties;
- commitment to building consensus;
- the ongoing use of local, traditional, and scientific knowledge to inform the evaluation; and
- the adoption of transparent, timely, and coordinated procedures.

Three separate consultation strategies were developed to meet the unique needs of the communities, industry, and government. There were common elements across all three strategies which ensured that the consultations would ultimately bring the interest groups to the same level of understanding. Early stages of the consultation process sought to ensure that as many participants as possible would achieve a common basic level of understanding of various aspects of process. This included knowledge about BSIMPI, its membership, its purpose, and what options were available for meeting with that group. Explaining the role of BSIMPI provided an opportunity to discuss Canada's *Oceans Act*, and the expectation that ocean resources should henceforth be managed collaboratively with coastal communities, industry, government, and other interested parties. The concept of integrated management as elaborated in Canada's *Oceans Act* and how it could be applied in the Inuvialuit Settlement Region was discussed. Finally, the opportunity to protect marine areas of special interest with a regulation leading to a Marine Protected Area was presented.

Once these concepts had been communicated, discussed, and understood, it became possible to move to elaborate on the specific process that could lead to the establishment of an MPA in the three Zone 1(a) areas of the BSBM Plan, and to discuss the implications of establishing an MPA. The purpose of these discussions was to ensure that communities, government agencies, and industry stakeholders would have an opportunity for a thorough and open dialogue considering the proposed MPA and how it could be established.

The concerns, views, and desires expressed throughout the process were documented. This record of information fed into future meetings and guided the thinking as plans for the establishment of a MPA advanced and were formalized into draft documents. This living document was shared with the Inuvialuit management and co-management bodies, interested government agencies, and industry stakeholders. The BSIMPI Working Group was kept informed of progress achieved, and special meetings were called as necessary to work through potential conflicts as soon as they became apparent.

Consultations were supported by other BSIMPI activities. For example, Working Group members were provided opportunities to visit the Zone 1(a) areas. These field trips proved invaluable in increasing the awareness of these areas for members based outside the Inuvialuit Settlement Region. The trips also served to increase the visibility of BSIMPI and helped build a foundation of trust with the people camping at these sites. Other supporting activities included annual Oceans Day celebrations and youth retreats in the communities of Aklavik and Tuktoyaktuk. A community-based monitoring program designed to monitor fish health and abundance in these two communities also served to profile the practical relevance of ocean management to the daily lives of coastal residents.

Community Consultations

Formal community consultations are documented in Table 5.1. Of the six Inuvialuit communities, Aklavik, Inuvik, and Tuktoyaktuk are the ones directly affected by the proposed MPA. For this reason consultations were conducted more frequently in these communities. The communities of Paulatuk, Holman, and Sachs Harbour were consulted initially; however, they opted to let the communities active in the Zone 1(a) areas represent the overall community interests

Table 5.1

RECORD OF BSIMPI CONSULTATIONS FALL 2001 TO MARCH 2003[*]
(does not include informal presentations, telephone communications or
other non-face-to-face correspondence)

Date	Group Consulted	Summary of Information
Fall 2001	• Northern Canadian Marine Advisory Council includes Major shipping companies in North as well as Canadian Coast Guard and Dept. of Transportation	Attended annual meeting and provided information on BSIMPI and process underway.
Fall 2001	• Dept. of Environment • Other Federal Dept. if NWT Federal Council • Renewable Resources, Wildlife and Economic Development-GNWT	Overview of Canada's *Oceans Act* and BSIMPI
Fall 2002	• Industry	Overview of Canada's *Oceans Act* and Strategy, Who and what is BSIMPI? Value added of BSIMPI to Industry
Fall 2002	• Inuvik Hunters and Trappers Committee • Tuktoyaktuk Hunters and Trappers Committee • Tuktoyaktuk Elders Committee • Inuvik Community Corporation • Inuvik Elders Committee • Aklavik Community Corporation • Aklavik Hunters and Trappers Committee	Phase 1 Presentation: Introduction to BSIMPI, What is a Marine Protected Area? What is Canada's *Oceans Act*? How is BSIMPI evaluating the BSBMP Zone 1(a)? Overview of BSIMPI's Consultation Strategy
Fall 2002	• Northern Canadian Marine Advisory Council: includes Major shipping companies in North as well as Canadian Coast Guard and Dept. of Transportation	Information sharing (see Fall 2002 industry and community meeting descriptions above).
Fall 2002	• Dept. of Renewable Resources, Yukon Gov't • Dept of Environment • Other Federal Departments in the NWT Federal Council • Geophysical Forum – various government departments both Federal and Territorial	Canada Oceans Strategy and BSIMPI, information sharing
Winter 2002	• DFO Regional Oceans Coordinating Committee	BSIMPI and proposed Marine Protected Area
Fall-Winter 2002–03	• Individual meetings with Petroleum Companies: Devon Canada, Aanadarko Canada and Suncor Energy • Northern Environmental Managers Group • Meetings, Telephone updates with Tourism Companies: Arctic Nature Tours and Ookpik Tours	Information sharing (see Fall 2002 industry and community meeting descriptions above).

Date	Group Consulted	Summary of Information
Winter 2003	• Aklavik Hunters and Trappers Committee • Aklavik Community Corporation • Aklavik Elders Committee • Tuktoyaktuk Elders Committee • Tuktoyaktuk Community Corporation • Tuktoyaktuk Hunters and Trappers Committee	Review of Phase 1 Presentations, Phase II: Review of Assessment Report Results, Overview of next steps, Active Dialogue on Issues facing MPA.
Winter 2003	• Inuvik Hunters and Trappers Committee • Inuvik Elders Committee • Inuvik Community Corporation • Public Meeting in Tuktoyaktuk • Public Meeting in Inuvik • Public Meeting in Aklavik	Update on process and results of Assessment Reports
Winter 2003	• National Energy Board • Environment Canada and Canadian Wildlife Service	BSIMPI, proposed MPA and regulatory responsibilities
March 4, 2003	• Joint Meeting: All Hunters and Trappers Committees, Community Corporations and Elders Committees	Brought all organizations together to discuss views, and determine support for continuing the process of evaluating the MPA.
March 6, 2003	• BSIMPI Working Group meeting with Tuktoyaktuk Community Corporation	Addressed specific concerns of the Community Corporation as brought up during March 4 meeting.
March 20, 2003	• BSIMPI Senior Management Committee Meeting	Letters of conditional support for continuing the evaluation process from the community organizations were tabled.

in the consultation process. (See Table 5.1 for a summary of BSIMPI consultations beginning in the fall of 2001 to March 2003.) Organizations represented on the BSIMPI were encouraged to have their members attend and participate in community consultations. Their active participation helped to reduce, or even on occasion eliminate, concerns that the process was a DFO-driven initiative. As well, industry participation allowed the communities to hear industry voice its support for and/or concerns regarding various issues directly. In turn, their participation allowed industry to hear directly from the community regarding local concerns and expectations. An extra benefit of having industry participate in community meetings was the capacity to answer and apply expertise to industry-specific questions immediately.

To ensure effective consultation with the communities, the strategy had to allow for varying interests and changing priorities between and within the communities. This required a multi-phased process. During each phase a series of meetings was held with various subgroups within each community. These typically included the following: the Hunters and Trappers Committees; the Community Corporations; the Elders' Committees; and a general public meeting. The Hunters and Trappers Committees manage fish and wildlife issues; the Community Corporations represent economic interests; and the Elders' Committees ensure that the traditional knowledge and history of the area is taken into account when decisions are made. All meetings were documented, and the notes taken were sent back to the respective groups for their approval.

The strategy proposed for involving a given community in the evaluation process was presented to each of the various groups during the first information-sharing meeting. At that time community members were asked to give their opinions on the strategy and to comment on whether or not they felt it was a viable strategy for achieving the consultation goals. In response to community requests, an additional round of consultation meetings was organized to bring together the Hunters and Trappers Committees from all three communities to share their views. Similarly, the Community Corporations and the Elders' Committees from the three communities were given the opportunity to meet and to share their views with their peers.

Industry Consultations

Three industry sectors were identified as working in and/or around the Zone 1 (a) areas. These sectors included oil and gas, transportation (marine and air), and tourism. The industry consultation plan had to be flexible and adaptive to allow for varying levels of interest and capacity among the companies in these different sectors. Industry interest and capacity were highest in the oil and gas sector and lowest in the tourism sector. Initial contact with these groups included phone calls and the distribution of printed materials documenting BSIMPI activities and the provision of background information. The level of response to these contacts helped assess the level of interest of individual companies in participating further. Meetings were subsequently scheduled as required. The

Beluga muktuk (blubber) about to be prepared, Inuvialuit Settlement Region. Photo by P. Cott, Department of Fisheries and Oceans.

industry representative on the BSIMPI Working Group proved integral to engaging individual oil and gas companies with interests in the area. Subsequent individual consultations were conducted as required based on the level of interest expressed by the company. Those with significant interest were invited to make presentations to the BSIMPI Working Group. Information gaps identified by industry were addressed through reports, workshops, or the identification of research needs whenever possible. Once again, the results from industry consultations were shared with other interested groups.

Government Consultations

Government consultations were held at both the federal and territorial level. Federally, consultations occurred within DFO and with other federal departments. Within DFO, an existing Regional Oceans Coordinating Committee provided a forum for the consultations. Frequent updates were also given to staff of local DFO offices. Since the level of responsibility for the marine environment varies widely among the different federal departments, in general the level of consultation was directly related to that level of responsibility. If a department wished to have a higher level of engagement this was arranged. Meetings with the membership of the Northwest Territory Federal Council were used to provide information and determine levels of interest of the broad suite of federal departments. Departments which had larger regulatory or mandated responsibilities within the marine environment were engaged more frequently and actively.

Governments of the Northwest Territories and Yukon have limited regulatory responsibility in the Beaufort Sea. Formal consultations with territorial government agencies were completed as needed. Informal updates on the MPA assessment process were often worked into discussions on other issues of mutual interest such as climate change, contaminants, and/or land/ocean interactions.

EVALUATION OF THE CONSULTATION PROCESS

The overt purpose of the extensive three-year consultation process described was to assess the level of support among the Inuvialuit, government, and industry for the idea of creating a Marine Protected Area in the Zone 1(a) areas of the Beaufort Sea Beluga Management Plan. This was to be achieved by ensuring that communities, government agencies, and industry stakeholders were fully informed of BSIMPI, the *Oceans Act*, integrated management, and the concept of Marine Protected Areas and its application to the Zone 1(a) areas. Indicators of knowledge and understanding are difficult to measure. There was, however, an observable shift from uncertainty and lack of trust to a willingness to participate constructively. The authors attribute this to a better understanding of the issues. Though the decision that resulted from the consultation process was to proceed with the steps necessary to establish a Marine Protected Area, the success of the consultation process should be determined by assessing whether or not the original objectives of the consultation strategies were met, not whether the decision was to proceed with a MPA.

To ensure transparency of the consultation process, detailed records were kept and verified with the consulted parties. The verification process ensured that the secretariat had an accurate understanding of the views expressed and that the information they passed on to the other organizations and to the BSIMPI Working Group was reflective of the views and concerns brought forward through the consultations. This ongoing process of sharing information was an essential role of the secretariat. For example, the secretariat presented documentation of the consultations to the Working Group, and in turn, when the Working Group presented their recommendation to the Senior Management Committee, they were able to present this together with the documented results of the consultations. By presenting their recommendation in this manner, the Working Group felt confident that accurate information was being incorporated into the decision-making process. This also served as a demonstration of transparency in the consultation process, as organizations were able to see that their comments and concerns were tabled with the decision-making bodies during the decision-making process.

A larger purpose of the process was to develop effective working relationships between members of the BSIMPI. There was a desire to use this exercise to build a shared sense of trust that would allow the BSIMPI to begin broader integrated management planning for the marine waters of the ISR. Measures used to evaluate progress achieved toward building effective working relationships included a willingness of community groups and others to participate constructively in the consultation process; a willingness to contribute resources to the effort, whether cash or in-kind; and finally a willingness to accept that the information exchanged was fair and accurate.

An increased willingness to participate in the process was demonstrated when organizations increased their effort and level of contribution to the process, and when organizations which had not previously been involved sought to participate. As the consultation process continued over time, the benefits of participation became more apparent and initially reluctant partners recognized that the process facilitated their ability to represent their interests. In many cases the result was a gradual shift toward becoming a more willing partner. Willingness to participate also became more evident as the number of rumours and highly negative or inflammatory statements such as "an MPA would mean a pull-out of hydrocarbon activities in the Beaufort Sea" were reduced.

Over time, constructive participation in the process grew. Early consultation meetings were occasionally antagonistic and used as a platform to raise other issues not related to BSIMPI. As understanding of the process and why it had been initiated increased, meetings became more focused and more constructive in nature. BSIMPI in turn developed a greater understanding of the issues of concern to the various stakeholders and determined how to communicate more effectively with various sectors.

An increased willingness to contribute resources, especially in terms of in-kind support, became apparent as the process continued. At the community level, in-kind support translated into assisting in the organization of community meetings. For industry and other organizations support included dedicating staff for two-to-three-day periods for meetings and workshops, and covering costs associated with attending those meetings. Government agencies and industry contributed financially to several workshops conducted by the BSIMPI.

Developing trust between BSIMPI and the various stakeholders was not without its challenges. As the *Oceans Act* is relatively new legislation, its implementation and the ramifications of its implementation were not well known or understood. In addition, organizations and communities were concerned that BSIMPI was really a DFO- or government-led process rather than a partnership with the Inuvialuit. Three factors that likely contributed to this perception are: the BSIMPI Secretariat is made-up of DFO staff; funding for BSIMPI initiatives is primarily from DFO; and the legislation under which the MPA evaluation process was being undertaken is a DFO responsibility. The Inuvialuit are often skeptical of new government-led initiatives because of past negative experiences. The enactment of the gun registry and the inability of the Inuvialuit to change a national park boundary are two recent examples of government-led initiatives that caused concern and contributed to the perception that "government will do what it wants despite what the communities say."

BSIMPI has tried to counter this sense of mistrust in a number of ways. Representatives from each organization in BSIMPI participated in the consultation meetings in communities. Having adequate representation assisted in fostering the sense that BSIMPI was a partnership and not just DFO with a different name. This participation was particularly beneficial when the Inuvialuit partners emphasized their role in BSIMPI, defended BSIMPI as a partnership and indicated their support for the BSIMPI process. The presence of industry at meetings further helped address this issue. DFO also provided assurances that it would continue to respect and work with the co-management arrangements established by the IFA, and further that if an MPA were created in the ISR it would be managed through a co-management body. Being able to relay this message to the communities in the form of a written assurance from the Minister of Fisheries and Oceans was imperative to the process of building trust.

Canada's *Oceans Act* and its associated policies provided the regulatory and policy framework in which to initiate BSIMPI and its associated consultation process. In an effort to achieve a consensus-based decision regarding the designation of an MPA, BSIMPI designed and implemented consultation strategies that were specific to the needs of the Inuvialuit, government, and industry. BSIMPI created a forum that included key interested parties in the decision-making process, thereby providing the opportunity to contribute directly to the management of ocean resources.

BSIMPI created a tailored consultation process that has proven effective in the ISR. This process addressed the needs of organizations with a diverse range of interests and overcame challenges resulting from the diversity of the participants. BSIMPI made a commitment to lead a participatory process in which all parties would be involved in the decision-making process at all levels. As a result of the extensive consultation process adopted, all parties will be able to come to a mutually agreed-upon decision regarding creation of an MPA.

Reflecting on other lessons learned during the process is important for informing further work in the western Arctic and elsewhere. For example, it proved important to ensure that key individuals and organizations were able to participate regularly and at key points in the consultation process. Extra attention was given to accommodating the constraints faced by these individuals in order to ensure the process would not subsequently be delayed because of perceived or real gaps in support for the process.

Staff regularly reviewed basic understandings and agreements reached at all meetings. Since there was routinely a high turnover among participants, this helped to bring newcomers up to date. It also served to quell false rumours or expectations about how the results of the process would be used or the MPA implemented. Staff maintained liaison functions with members of key organizations and agencies so that they would be able to anticipate whether organizations, groups, or individuals had intentions of raising unrelated or personal agenda items and so derailing the intended discussions, whether deliberately or otherwise. If so, they would come prepared to keep discussions on course. Another major concern in the relatively small communities of the ISR is consultation fatigue. Efforts were made to coordinate the consultation efforts with other scheduled meetings and to ensure the meetings were well organized and well managed.

Developing collaborative partnerships has been a key component of the consultation process adopted by BSIMPI. Berkes *et al.* (2001) describe the term participatory as referring to the inclusion of local groups, land claimants and other stakeholders in the decision-making process. Collaborative (participatory) processes, especially those focusing on communities, have become widely used in a variety of sectors around the world (Ananda and Herath 2003; Berkes *et al.* 2001; Wells and White 1995; and Wiseman *et al.* 2003). Much of the literature on participatory processes focuses on public or community participation (Fenton *et al.* 2002; Kaza 1988; Wells and White 1995). Although the importance of

Seismic operations in the Mackenzie Delta, winter 2002. Near Tuktoyaktuk, NT.
Vibrator Energy Source and Recording Equipment. Photo by Pete Millman, Devon Canada Corporation.

community consultations cannot be overstated, broader consultation that includes industry, government bodies and other interested parties is required.

Mitchell (1997) refers to collaborative partnerships as those in which real decision-making power is shared with the intent to achieve mutually compatible objectives. These benefits are apparent in the BSIMPI process, and the basic benefits of a participatory process described by Mitchell (1997) have been achieved. For example, there has been a better definition of issues or problems. Communities, Inuvialuit management and co-management bodies, industry, and government agencies, including DFO, have a better understanding of the complexities of balancing conservation and development in the complex offshore environment of the Beaufort Sea. Access to information and understanding beyond a single realm such as science has been achieved. Over the past three years the BSIMPI has provided an effective conduit for exchanging information between communities, industry, and government. The lack of scientific knowledge has been acknowledged and traditional knowledge has been accessed to achieve a better environmental understanding. Ultimately, the objective of integrated management is to influence human behaviour. This is the realm that has been advanced through the BSIMPI consultation process.

The identification of alternative solutions which are acceptable to all parties is another benefit of the participatory process discussed by Mitchell. The BSIMPI process has brought pro-development interests, conservation interests, and political interests to one table. The uncertainty in terms of understanding the natural systems and/or the risk assessment of action vs. inaction has created an atmosphere of mutual understanding and a willingness to respect alternative views. The BSIMPI process described has led to respect for the wishes of Inuvialuit coastal communities, and it is that group which will largely influence the final recommendation to the Minister of Fisheries and Oceans on whether to proceed with the MPA or not. In this regard, then, Mitchell's observation that a sense of ownership over the plan or solution will facilitate implementation will also characterize the process described in this chapter. Completion of a participatory consultation process can also reduce the potential for future conflicts, lead to greater acceptance of the end result, provide acceptable solutions to problems, build social capital, reduce regulatory offences through voluntary compliance, and be viewed by all involved as an acceptable process through which possible new initiatives could be launched. Ideally, when the process is completed, participants will feel comfortable that their views and ideas have been utilized in the decision-making process and have a sense of ownership and control over the decisions taken.

BSIMPI has taken many positive steps toward attaining the goal of establishing good working relationships and developing trust. These relationships will be tested as BSIMPI moves into broader integrated management initiatives. As well, it is to be expected that management organizations will experience changes in membership and that individuals new to the process will require time and effort to be integrated into the mindset that has been achieved to date. It is anticipated that many of the difficulties that have been encountered and overcome to date will re-emerge as the process continues. They will require ongoing management.

The consultation process followed was not without difficulties. The BSIMPI Secretariat identified issues such as misunderstandings, delays, and contradictory expectations as challenges throughout the process. It was particularly important that BSIMPI's level of decision-making power was clearly communicated repeatedly to ensure that expectations were not exaggerated or underestimated. This also helped to emphasize that processes and bodies established under the IFA would not be diminished. The varying interests, cultural backgrounds, levels of education and technical expertise, methods of communicating and interpreting information, values, and expectations of the groups involved in the process often contributed to the challenges mentioned above. Expectations regarding the length and speed of the process also had to be addressed. Some voiced concerns that the process was moving too quickly, while others voiced concerns that the process was moving too slowly. The secretariat addressed these concerns as they were raised by reviewing as often as needed the steps in the assessment process, and by modifying the speed of the process as appropriate. Significant progress was made in alleviating these difficulties; however, continual management of these issues was required.

Despite the challenges, the BSIMPI consultation strategies have contributed to strengthening partnerships between stakeholders, engaging Canadians in oceans related decisions in which they have a stake and developing inclusive and collaborative oceans governance in the ISR. The process followed and the strategies adopted support modern ocean management as outlined in the *Oceans Act* and Canada's Oceans Strategy (DFO 2002). The principles used to develop BSIMPI's consultation strategies for this process can be adapted and used to facilitate participatory decision making in other multi-stakeholder environments.

Paquet and Wilkins (2002) comment that "socio-political factors are as important as the dynamics of the natural system in the governance of oceans" (p. 21), and further that "in a turbulent environment, organizations can only govern themselves by becoming capable of learning both what their goals are, and the means to reach them, *as they proceed*" (28). The consultation process described in this chapter illustrates the truth of these observations. The process demanded adaptability, in order to meet the needs of those being consulted and to address changes in priorities and issues as they arose. Those being consulted came from different cultural backgrounds, different levels of technical expertise, different interests, motivations, and mandates. Our limited scientific knowledge of the area dictates that likely impacts on the environment of further oil and gas exploration in the Beaufort Sea are at best a guess. Likely cost/benefit outcomes for proceeding with a Marine Protected Area are equally difficult to ascertain. Given these levels of uncertainty, socio-political factors are driving the initiative.

The consultation processes that have been developed over the past three years have promoted learning and resilience among communities, industry, and government departments. Canada's Oceans Strategy has been adapted to the regional context and implemented with considerable success. A model for effective local coordination of information and decision making has been developed and tested. Core principles that can serve to guide further ocean management efforts in the Inuvialuit Settlement Region have proven viable in a dynamic, complex, and multi-faceted context.

REFERENCES

AMAP (Arctic Monitoring and Assessment Programme). 1998. AMAP Assessment
Report: Arctic Pollution Issues. Arctic Monitoring and Assessment Programme,
Oslo, Norway. xii+859 pp.

Ananda, J., and G. Herath. 2003. "Incorporating Stakeholder Values into Regional
Forest Planning: A Value Function Approach." *Ecological Economics* 45(1):75–91.

Berger, T. R. 1977. Northern Frontier Northern Homeland: the report of the
Mackenzie Valley Pipeline Inquiry, 2 vols. Minister of Supply and Services Canada,
Ottawa.

Berkes, F., J. Mathias, M. Kislalioglu, and H. Fast. 2001. "The Canadian Arctic and
the *Oceans Act*: the development of participatory environmental research and
management." *Ocean & Coastal Management* 44: 1–19.

Canada. 1997. *Oceans Act*, S.C. 1996, c. 31 (also c. O-2.4).
http://canada.justice.gc.ca/FTP/EN/Laws/Chap/O/O-2.4.txt

DFO (Fisheries and Oceans Canada). 1999. National Framework for Establishing and
Managing Marine Protected Areas.

DFO (Fisheries and Oceans Canada). 2002. Canada's Oceans Strategy. Fisheries and
Oceans Canada.

DIAND (Department of Indian and Northern Affairs). 1984a. Report of the Task Force
on Northern Conservation.

DIAND (Department of Indian and Northern Affairs). 1984b. The Western Arctic
claim: the Inuvialuit Final Agreement, Ottawa.

Elliott, G. 2002. Technical assessment: proposed Beaufort Sea Marine Protected Area.
Fisheries and Oceans Canada.

Fenton, D.G., P. Macnab, J. Simms, and D. Duggan. 2002. Developing Marine
Protected Area Programs in Atlantic Canada – a Summary of Community
Involvement and Discussions to Date. Proceedings of the fourth international
conference on Science and Management of Protected Areas, Wolfville, Nova Scotia,
14–19 May 2000. Science and Management of Protected Areas Association. 1413–26.

FJMC (Fisheries Joint Management Committee). 2001. Beaufort Sea Beluga
Management Plan, Amended Third Printing. Inuvik, Northwest Territories.

FJMC (Fisheries Joint Management Committee) and WMAC (Wildlife Management
Advisory Committee). 1988. Inuvuialuit Renewable Resource Conservation and
Management Plan, Northwest Territories.

Gray, R.H. 1995. "Protected Areas: Public Outreach and Involvement." In Proceedings
of the Second International Conference on Science and the Management of
Protected Areas. Dalhousie University, Halifax, Nova Scotia, Canada 16–20 May
1994.

Kavik-AXYS Inc. 2002. In association with Gustavson Ecological Research
Consulting. Socio-economic assessment of the proposed Beaufort Sea Marine
Protected Area Final Report. Submitted to Fisheries and Oceans Canada.

Kaza, F. 1988. "Community involvement in marine protected areas." *Oceanus* 31(1):
75–81.

Mitchell, B. 1997. *Resource and Environmental Management*. London: Longman.

North/South Consultants Inc. and Inuvialuit Cultural Resource Centre for Fisheries
and Oceans Canada. 2003. Ecological Assessment of the Beaufort Sea Beluga
Management plan Zone 1(a) as a Marine Protected Area of Interest.

———. 2004. Overview of the Coastal Marine Ecosystem of the Southeastern Beaufort Sea in the vicinity of the Mackenzie River estuary.

Paquet, G., and K. Wilkins. 2002. Ocean governance: An inquiry into stakeholding. Centre on Governance, University of Ottawa.

Stephenson, W.R. 2002. The top ten human dimension initiatives in a protected area ecosystem management program. Proceedings of the Fourth International Conference on Science and Management of Protected Areas, Wolfville, Nova Scotia, 14–19 May 2000. Science and Management of Protected Areas Association. 681–91.

Wells, S., and A.T. White. 1995. "Involving the community." In *Marine Protected Areas: Principles and Techniques for Management.* Edited by Susan Gubbay. London: Chapman & Hall.

Wiseman, V., G. Mooney, G. Berry, and K.C. Tang. 2003. "Involving the General Public in Priority Setting: Experiences from Australia." *Social Science & Medicine* 56(5).

CHAPTER 6

MARINE STEWARDSHIP & CANADA'S OCEANS AGENDA IN THE WESTERN CANADIAN ARCTIC:

A ROLE FOR YOUTH

Michelle Schlag (University of Manitoba)
Helen Fast (Department of Fisheries and Oceans)

INTRODUCTION

Stewardship is a term being used increasingly by resource industries, government agencies, and community activists to describe their philosophy of resource use. Most stewardship literature focuses on the care, protection, and monitoring of terrestrial resources. Over the past decade, however, Canada has moved to assume management responsibilities for an exclusive economic zone encompassing 2.9 million square kilometres of estuarine, coastal, and marine waters. The *Oceans Act* (1997) outlines objectives for the management of Canada's ocean resources. Canada's Oceans Strategy (Government of Canada 2002a) is the policy being used to guide implementation of the Act. It identifies an important role for stewardship in fulfilling Canada's marine management obligations.

Implementation of Canada's Oceans Strategy along the western Arctic coastline in the Inuvialuit Settlement Region is being led by a co-management group called the Beaufort Sea Integrated Management Planning Initiative (BSIMPI) Working Group. During its first year of operation the Working Group identified the need to involve Inuvialuit youth in their work as a priority. There is a relatively small number of people involved in co-management processes in the North. Those who do participate tend to be of middle age or older. There remains very little time to involve and train the younger generation to assume these responsibilities.

The purpose of this research was to develop a strategy that will foster increased Inuvialuit youth interest and participation in ocean stewardship activities. A qualitative approach was used to conduct the research. The lead author of this chapter, a young person herself, conducted all the fieldwork for this research. She found that most youth were interested in talking to her about many aspects of stewardship. These discussions were supplemented by further interviews with resource managers, elders, teachers, parents, and others. Together they provided a coherent and in-depth understanding of the training, educational, and other stewardship opportunities available in the various communities. Youth also expressed their views of these opportunities, and what opportunities they would like to have made available to them. Key findings include evidence that: (a) few opportunities are available to the 1,700 youth in the region; (b) youth share the community's view that they are ill-prepared to assume resource management responsibilities from their elders; (c) youth feel they are being shortchanged in terms of the quality of formal education they are receiving and in terms of being taken on the land in order to gain the skills and knowledge possessed by their elders; and (d) youth want to play meaningful roles in their society.

This chapter outlines a strategy for addressing the research findings and identifies BSIMPI as having a key role to play in promoting the knowledge gathered from this study in ways that will encourage the larger community to assume responsibility for its implementation.

STEWARDSHIP AND CANADA'S MARINE RESOURCES

Stewardship is a term being used increasingly by resource industries, government agencies, and community activists to describe their philosophy of resource use (Laynard and Delbrouck 1994; Environment Canada 1996; Biodiversity Convention Office 2001; Government of Canada 2002c; CAPP 2003). Consistent across the range of definitions used by these groups is an ethic of caring for the earth and assuming responsibility for preserving, protecting, and even restoring the environment. Included in some perspectives is responsibility for protecting the country's economic and social fabric as well as the environment (CAPP 2003). The notion of stewardship is commonly understood to include an obligation to ensure a healthy environment for present and future generations (CWS et al. 1995; Laynard and Delbrouck 1994; Knight and Landers 1998; Lerner 1993). Implicit is the acknowledgement of individual and personal responsibility for positive action. Government and industry both claim to seek to engage citizens and endorse business practices that protect the natural environment while enhancing the quality of life (CAPP 2003; Government of Canada 2002b). The Government of Canada, the provinces, and the territories, through Canada's Stewardship Agenda, recognize stewardship as a key conservation tool and are committed to further supporting and encouraging stewardship activities (Government of Canada 2002c). The research undertaken for this chapter was funded by Fisheries and Oceans Canada, and provides evidence of a genuine attempt to involve northern youth in the national stewardship agenda.

Stewardship activities typically include advocacy, conservation, monitoring, research, fundraising, and working co-operatively with other stakeholders and interest groups. Lerner (1993) noted that individuals who become involved in these activities tend to be humanistic intellectuals such as teachers, social workers, other service professionals, and students. She describes these people as resourceful, high-minded, self-sacrificing, and socially conscious individuals. They become involved in stewardship activities because they have concerns about an issue and perceive that it will not be addressed to their satisfaction by government or others. In many cases individuals and community groups are more effective in addressing issues than government might have been, since they can respond to circumstances more quickly, are independent of constraints characteristically imposed on government agencies, and tend to have a more intimate knowledge of their local landscape characteristics (IWCO 1998; Lerner 1993).

People who become involved in stewardship activities enjoy a range of benefits. These benefits include learning more about their natural environment, meeting other people with similar values and interests, having fun, and taking pride in having contributed to the well-being of their community and the natural environment. Once people become involved in stewardship activities they tend to stay involved. Reasons for this include the shared feelings of solidarity, strength, camaraderie, and empowerment associated with their efforts. Communities benefit directly from their citizens' efforts and indirectly by a growing capacity for self-reliance (CWS et al. 1995; Lerner 1993).

Most stewardship literature focuses on the care, protection, and monitoring of terrestrial resources (Dallmeyer 2003; Knight and Landers 1998; Lerner 1993). Over the past decade, however, Canada has moved to assume management responsibilities for an exclusive economic zone encompassing 2.9 million square kilometres of estuarine, coastal, and marine waters. Almost one-quarter of Canada's total population lives in these coastal regions. The management complexities associated with these additional responsibilities are illustrated by the numerous ocean-related conventions to which Canada is a party. These include shipping, fisheries, biodiversity, pollution, climate change, anadromous stocks, and safety of life at sea (Government of Canada 1997).

The *Oceans Act* (1997) outlines economic, social, and environmental objectives for the management of Canada's ocean resources (Government of Canada 2003). Canada's Oceans Strategy (Government of Canada 2002a) is the policy that is being used to guide implementation of the Act. It identifies an important role for stewardship in fulfilling Canada's marine management obligations. Specifically, it recognizes that involving Canadians in the stewardship of their marine resources is key to its implementation. In this document, stewardship is defined as "acting responsibly to conserve the oceans and their resource for present and future generations" (Government of Canada 2002a, 20).

MARINE STEWARDSHIP IN CANADA'S WESTERN ARCTIC
The Inuvialuit Settlement Region (ISR)

The Inuvialuit Settlement Region (ISR) is located in Canada's western Arctic (see Figure 5.1). It includes the northern portion of the Mackenzie Delta, the Beaufort Sea, Banks Island, and the western portion of Victoria Island. The Inuvialuit, a group of Inuit, largely inhabit the area. (See Table 6.1 for a summary of population numbers in the ISR communities.)

Table 6.1
INUVIALUIT SETTLEMENT REGION COMMUNITY POPULATION NUMBERS

Community	Total population	Inuvialuit population	Inuvialuit under age of 29	
			Population	% of Inuvialuit population
Aklavik	700	350	179	51
Inuvik	3,200	1,200	624	52
Paulatuk	270	270	154	57
Sachs Harbour	120	120	65	54
Holman	400	400	212	53
Tuktoyaktuk	940	940	526	56
Total	5,630	3,280	1,760	54

There are also Gwich'in First Nation people in Inuvik and Aklavik, and non-native people in Inuvik. The small communities in this region offer few employment and/or entrepreneurial business opportunities. The traditional economy of the ISR includes marine subsistence hunting and fishing. The Inuvialuit have for centuries harvested a variety of whales, seals, and marine fish for food for household consumption. Harvesting occurs on the sea, sea ice, and the seashore. Substantial benefits are associated with sharing food, retaining long-standing cultural practices, and integrating young people into work roles and the community including reducing the need for a cash income. The private sector cash economy includes marine shipping, marine-related arts and crafts, marine-related tourism, research, and significant offshore oil and gas potential (G.S. Gislason & Associates and Outcrop Ltd. 2003).

Subsistence harvesting is evident in all the ISR communities, but is strongest in the smaller communities that have few wage employment opportunities (Ayles and Snow 2002). The area is governed by a comprehensive land claims agreement titled the Inuvialuit Final Agreement (IFA) entered into by the Government of Canada and the Inuvialuit and signed in 1984. It recognizes Inuvialuit rights to specific surface and subsurface lands. Monetary compensation and participation in resource development, harvesting, and management arrangements are also covered in the agreement (Notzke 1995; Government of Canada 1984).

Marine and freshwater resources in the ISR are managed through two administrative structures: those created by federal legislation and the co-management arrangements created under the IFA (Muir 1994). In the ISR, co-management is an institutional arrangement between the Government of Canada and the Inuvialuit. It outlines the specific roles, rights, powers, and obligations of parties to the co-management arrangements as they pertain to the management of renewable resources (Berkes *et al.* 2001; Government of Canada 1984). The Inuvialuit have an advisory role only concerning development of non-Inuvialuit lands in the ISR (Notzke 1995).

Communities of the ISR

Of the six communities in the ISR, two are inland (Aklavik and Inuvik) and four are coastal (Paulatuk, Sachs Harbour, Holman and Tuktoyaktuk) (see Figure 5.1). The community of Aklavik is located in the Mackenzie Delta. Wage employment in the community is largely associated with the local government and the oil and gas industry. Inuvik is located on the western branch of the Mackenzie River. It was built in the 1940s as a regional centre to replace Aklavik. This community has most of the wage-employment opportunities in the ISR. It is also home to Aurora College and the Aurora Research Institute (ARI). Subsistence hunting, fishing, and harvesting continue to be important in community life (Ayles and Snow 2002).

The community of Paulatuk is located at the mouth of the Hornaday River. Subsistence harvesting is the main activity as there is almost no cash economy. Paulatuk is the gateway to Tuktuk Nogiat National Park and has become a focus for mineral exploration. The community of Sachs Harbour is located on Banks Island. It is the smallest community in the ISR. Aulavik National Park headquarters are located there. The only non-governmental employment opportunities available in the community are through tourism, sport hunts, and the commercial harvest of muskoxen. The community of Holman is located on the west coast of Victoria Island. It is a very traditional community known for its printmaking artwork. Seal hunting played a large role in the economy of Holman until the anti-sealing movement caused a downturn in the market. This community has very close ties with the eastern Arctic. The community of Tuktoyaktuk is situated on the coast of the Arctic Ocean. The only deepwater port in the region is located here, and the area is the focus of onshore and offshore oil and gas exploration. Though Tuktoyaktuk is a coastal community, its subsistence harvesting activities extend inland into the Mackenzie Delta (Ayles and Snow 2002).

The total population of the six communities is approximately 5,600. Of this number, about 3,300 are Inuvialuit. Over half of the Inuvialuit population is under twenty-nine years of age (see Table 6.1).

Implementation of Canada's Oceans Strategy in the ISR

Implementation of Canada's Ocean's Strategy along the western Arctic coastline is being led by a co-management group called the Beaufort Sea Integrated Management Planning Initiative (BSIMPI) Working Group. During its first year of operation the Working Group identified the need to involve Inuvialuit youth in their work as a priority. Members realized that their long-term obligation for the careful and responsible management of marine resources in the region would soon depend on the active involvement of the younger generation.

The need to involve youth in regional management responsibilities was a concern shared by others. Community leaders had observed almost ten years earlier that a small pool of middle-aged qualified and willing northerners who sat on resource management committees and councils were getting older and would need replacements in the near future. Though they recognized that a logical choice for replacements was the next generation, they perceived that youth lacked interest in assuming this role (Notzke 1995). Inuvialuit elders have more recently expressed concern that youth lack the necessary skills and knowledge to assume resource management responsibilities assumed by the Inuvialuit under the IFA. They note that there has been a loss of connection between young people and the natural environment. Elders fear that this loss has resulted in a lack of knowledge of the land, a loss of culture, and an unwillingness to assume responsibility for the natural environment (NRTEE 2001). Steps taken to try to address these concerns have been few and they have had very limited success.

Research Purpose

The purpose of this research was to develop a strategy that will foster increased Inuvialuit youth interest and participation in oceans stewardship activities in the ISR. The specific research objectives were:

- To assess and evaluate trends related to the level of Inuvialuit youth participation in oceans stewardship activities;
- To identify components of a successful Inuvialuit youth oceans stewardship strategy;
- To draft a strategy for increasing Inuvialuit youth participation oceans stewardship activities in the ISR.

This chapter summarizes the research findings, documents a proposed strategy to engage youth, and concludes with lessons learned and implications for the future.

Research Methods

The researcher is a young person. As a result, northern youth considered her a peer, saw her as a role model, and wanted to spend social time together. The level of communication and mutual understanding possible was thus enhanced by her age. A qualitative approach was used to conduct this research, as the subject matter does not lend itself to quantitative analysis. All six ISR communities were included in the study in order to capture any differences between coastal and delta communities. The research methods included a preliminary site visit, participation, youth focus groups, and expert interviews. A young Inuvialuit beneficiary was hired as a research assistant. The assistant introduced the researcher to the Inuvialuit communities, culture, and most importantly to other youth. Ninety-one individuals participated in the study. This included Inuvialuit youth (between the ages of fourteen and twenty-nine), elders, parents, past and present high school teachers, and local resource managers.

A preliminary site visit to the communities of Inuvik and Tuktoyaktuk was conducted in July 2002. The purpose of this visit was to prepare for the fieldwork by obtaining a scientific research licence from the ARI and to begin meeting with people from with the various communities. Primary information was collected while visiting the six communities between November 2002 and January 2003.

Participation was used to build relationships with people in the communities. While conducting the research, I lived with different families and participated in the family's daily activities. These included things like washing dishes, preparing food, watching television, snowmobiling etc. The families I lived with also gave me advice on the Inuvialuit culture. I used this information to guide how I conducted subsequent interviews and focus groups. I spent many of my evenings and weekends socializing with local youth. This allowed me to develop numerous friendships and acquire a depth of knowledge and understanding that would not have been possible if I had relied only on focus groups and formal interviews.

The purpose of the focus groups was to gain insight into Inuvialuit youths' perspectives, perceptions, motivations, and understanding of the concept of oceans stewardship and management. I facilitated focus groups in the five high schools in the ISR. Though the community of Sachs Harbour does not have a high school, the youth from this community were included in the Inuvik focus group. When large groups of students were involved a teacher and/or my research assistant helped with the facilitation.

I conducted interviews with Inuvialuit elders, parents, past and present high school teachers, local resource managers, including government people and Hunter and Trapper committees (HTCs), as well as with youth environmental stewardship program administrators from outside the ISR. These interviews were semi-structured with open-ended questions.

I transcribed, coded, and categorized the information I had gathered after I had completed the fieldwork portion of my research. Next I conducted a content analysis to extrapolate themes and patterns. I organized the information for further analysis using Atlas.ti software. Finally, I synthesized and evaluated my findings. After preparing a summary report, I returned to the ISR communities to present my findings to the research participants and the communities at large. This verification of findings occurred between 2 May and 28 May 2003. This step helped to ensure the trustworthiness and validity of information, findings, and conclusions of the study (Leedy and Ormrod 2001; Grenier 1998).

INUVIALUIT YOUTH INVOLVEMENT IN OCEANS STEWARDSHIP

When I approached Inuvialuit youth about this research, most were interested in talking with me about many aspects of stewardship. Taken together, these discussions provided me with a coherent and in-depth understanding of the training, educational, and other opportunities available. I also learned what the general views of youth are in regard to these opportunities and what opportunities they would like to have made available to them. Perhaps most significant was the clear expression of a desire to learn and a wish to assume meaningful roles in their communities.

Inuvialuit youth share the community's lack of confidence about their capacity to take over from their elders. The roots of their lack of confidence are evident

Oceans Day parade, July 2002, youth banner, Tuktoyaktuk, NWT.
Photo by Michelle Schlag, Department of Fisheries and Oceans, 2002.

and stem from chronic feelings of abandonment by their community and the larger society. They note a lack of opportunity to learn the traditional values and customs of their grandparents' generation, a lack of expectations that they will be productive citizens, a lack of role models to inspire them, and being given an inferior education as factors that have limited their active involvement. Combined with the serious social ills that are common in many homes, the result has been low self-confidence, little initiative, and a low quality of life.

Opportunities and Limitations

There are few opportunities in the ISR for youth to get involved in oceans stewardship either through 'formal' (*e.g.,* scientific research, laboratory work, monitoring), or 'traditional' oceans activities (*e.g.,* resource use – hunting, fishing, harvesting). Participating in traditional activities is problematic for a number of reasons. Cost is one factor. Large families is another. Often there simply is not room for all to be taken on the land. Time on the land also conflicts with classroom requirements. As well, community leaders are often not aware of the interest youth have in participating. Further, the number and type of opportunities available vary by community. Most 'formal' opportunities are available in Inuvik, Aklavik, and Tuktoyaktuk, while most 'traditional' opportunities appear to occur in the smaller communities of Sachs Harbour, Paulatuk, and Holman. Current stewardship opportunities available to youth include the

Fisheries Joint Management Committee (FJMC) Student Mentoring Program, The Tariuq Monitoring Program, scientific research, "Oceans 11" – a new Arctic marine science curriculum, "Oceans Day" festivities and workshops, cultural camps, and informal traditional activities (*i.e.*, going out on the land and ocean with family). The Department of Fisheries and Oceans (DFO) and the Inuvialuit Government are involved to different degrees in most of the oceans steward-ship programs in the ISR. Parks Canada (PC) offers stewardship programs for younger children in the ISR and supports the FJMC Student Mentoring Program by providing work placements for participants. At present there are no other oceans stewardship programs in the ISR administered by other federal, territo-rial, and municipal government agencies, nor non-governmental organizations. Current opportunities are described more fully below.

The FJMC Student Mentoring Program is run annually in partnership with the DFO. It is designed to give students an introduction to fisheries science and resource management. This program has been running in the community of Inuvik since 1996. Given the technical nature of most jobs in this field, the pro-gram is intended to encourage youth to continue their schooling so that they can become the future scientists and resource managers of the ISR (FJMC 2001). The program provides participants with summer jobs in which they are given placements with resource managers and scientists in different agencies in the ISR and at the DFO Freshwater Institute in Winnipeg. The program typically attracts three to four students from Inuvik. Students from the other ISR com-munities have previously participated in the program. In recent years, however, the program has focused on youth in Inuvik.

The Tariuq Monitoring Program is a community-based pilot project funded by the DFO through its Oceans Program. The program operates in the communities of Tuktoyaktuk and Aklavik. Participants include representatives from DFO, the Hunters and Trappers (HTC), elders' and youth committees. Program participants monitor fish abundance and health, as well as water temperature. The objective of this program is to provide baseline information that will allow researchers to monitor changes in coastal and anadromous fish over time (BSIMPI 2002). Students who participate in this program learn to apply monitoring techniques, which are helpful in assessing fish populations, an important component of the freshwater and marine ecosystem.

Youth occasionally have an opportunity to assist with scientific research. When working for research scientists, young people can learn a variety of skills such as sampling and monitoring techniques, conducting interviews, and helping to organize and facilitate meetings. These infrequent opportunities are open to youth from any of the ISR communities, depending on where the research is being conducted.

The Oceans 11 Arctic marine science curriculum was developed by the DFO Oceans Program in recognition of the need for Arctic science and traditional ecological knowledge (TEK) curriculum material. The Departments of Educa-tion of the Northwest Territories, Nunavut, and the Yukon participated in the

Sachs Harbour, Banks Island, NWT, in summer. Photo by Fikret Berkes.

development of these materials. The course was piloted to grade eleven students in Inuvik, Holman, and Paulatuk in the winter of 2003. If deemed successful by the Beaufort Delta Education Council, it will be expanded and offered in Aklavik and Tuktoyaktuk in the near future.

Oceans Day is a national event intended to celebrate and remember the importance of oceans to Canadians (DFO 2003). Oceans Day celebrations and activities have been hosted by the DFO and community organizations in the ISR for the past two years. These events have included specific activities for youth. In 2002, a youth retreat was held in Tuktoyaktuk. Three youth from each of the ISR communities were brought together for this event. They learned about the ocean ecosystem and oceans-related careers. During Oceans Day 2003 in Aklavik, youth were taken camping. The community responses over the past two years have been so positive that the event will likely be held annually, and with greater support from local management agencies.

The Community Corporations and/or Brighter Futures run cultural camps in all six communities. The Community Corporations try to improve the so-cial, cultural, and economic well-being of beneficiaries in each ISR community. Brighter Futures, developed by Health Canada in co-operation with First Nation and Inuit communities, is a program designed to improve the mental, physical, and social health of children, families, and communities (IRC 2003). The goal of these camps is to teach youth respect for the land and ocean, to help them develop a sense of connection to the land, and to pass on TEK to younger genera-tions. Between ten and twenty-five youth from each community are taken out on the land each summer. During this time, they are taught traditional hunting, fishing, harvesting, and food preparation skills. Some of the communities also offer cultural camps in the winter. In this way youth learn about seasonal difference on the land. Many youth said they would like to spend more time out on the land with their families. However, families are typically large and this often means only the very young can go because of limited equipment and high costs of supplies. In addition, going out on the land can mean missing tests and exams and no allowances are made for this at school (Youth and Parents, pers. comm. 2002/2003).

In addition to the very limited number of opportunities to become involved in oceans stewardship, youth lack awareness of those that are available in their home community or in other ISR communities. Even on those occasions when they learn of a program and want to get involved, they don't know whom to contact. Youth are also not aware of oceans-related careers. Consequently, they do not consider pursuing careers as, for example, biologists, fisheries officers, or resource managers. Though scientists and ocean managers are present in the ISR communities, they are not visible. Many are from southern Canada and are not integrated into the Inuvialuit community. Youth, in turn, are not given information about their work, or about its importance to their community (Youth, pers. comm. 2002/2003).

Further complicating access to information about stewardship opportunities is poor communication between agencies, groups, and the schools. This lack of collaboration has led in some instances to turf wars and competition for a small pool of youth participants. It has also caused concern among leaders of educational institutions. The perception is that agencies prefer to offer their own programs rather than working collaboratively to provide the best opportunities to youth (Resource Manager 2002). Schools typically are not consulted on program developments to ensure that they meet the needs of the school and youth. Instead, educators are often approached with ready-made programs, which may not fit with the school curriculum or the school year. For example, while the schools are asked to promote and suggest students for the FJMC Student Mentoring Program, the program is inconsistent with the school year in the coastal communities, and therefore, youth from these communities are excluded from participating (Teacher, pers. comm. 2002).

Opportunities to become involved in ocean-related activities are obviously too limited, recognizing that there are over seventeen hundred Inuvialuit youth who might wish to participate. Interest and enthusiasm appear to be strongest among those who live in the coastal communities. Youth in these communities spoke of a personal relationship to the ocean that is the result of living next to it every day. This relationship was not as evident among youth in Inuvik and Tuktoyaktuk, as they did not mention having a connection to the ocean. Some youth from Inuvik said that they had not even seen the ocean. Nonetheless, almost all opportunities to participate in formal oceans activities are based in the communities of Inuvik, Aklavik, and Tuktoyaktuk.

A Role for Education

Educational standards in the ISR are low and getting lower (Teachers and Resource Manager, pers. comm. 2002; Vodden 2001). Even at current levels, local resource management professionals are commenting that high school graduates lack the capacity and proficiency to demonstrate basic skills such as reading and writing. There are also concerns that educational institutions in the ISR are under pressure to increase their completion rates and are modifying programs and advancing students who lack basic literacy skills (Resource Managers and Teacher, pers. comm. 2002). Inuvialuit youth recognize that they are receiving an inferior education. They want to be given a higher quality of formal education.

Youth who do attend school get little support or encouragement from their families or communities to continue (Youth, pers comm. 2002). Parents told me that they sometimes feel alienated from the educational system. They commented that they might not encourage their children to attend school regularly and graduate because they themselves have not graduated from high school. Alternatively, they may have had bad experiences in school when they were children (Farrow and Wilman 1989; Parent, pers. comm. 2002). In Inuvik and Tuktoyaktuk, this indifference is reinforced by the fact that people can often make a decent living without a formal education.

Another underlying concern shared by youth, elders, and parents is most strongly felt in the smaller coastal communities. This concern is that even if youth do become educated, there will be no local job opportunities available to them (Youth, Elders, and Parents, pers. comm. 2002/2003; Purich 1992). The assumption that youth will have to relocate to find employment, or end up underemployed or unemployed if they remain in their home community, is consistent with Condon's 1987 findings. Alternatively, elders and parents do not see a formal education as important for youth who want to be active in traditional ocean activities and local co-management processes. Rather, it is felt that their time would be better spent on the land than in school (Elders, pers. comm. 2002/2003).

The need for positive role models in their own age group was stressed repeatedly by youth (Youth, pers. comm. 2002/2003). Role models are important because youth look to their communities for examples of success. Youth would be encouraged to make better choices by having the chance to see that other young people like themselves have made good lifestyle and career choices (FedNor 1994; City of Calgary 2002). Inuvialuit youth also want mentors who could help them learn about oceans-related career possibilities in the ISR. At present they feel they are not getting the information they need to make good decisions (Youth, pers. comm. 2003).

A Role for Traditional Knowledge – A Long Tradition of Stewardship

Aboriginal people have typically adapted their needs to the capacity of the surrounding environment and have had a relationship of reciprocity and balance with other living things (Booth and Jacobs 2001). Lerner (1993) observes that many

Aboriginal people have an in-depth knowledge of the environment. Aboriginal people are acting as stewards, she says, when they have a personal relationship with the land and animals, and when they use hunting, fishing, and harvesting methods that demonstrate respect for the environment. Aboriginal people were able to harvest animals and fish without depleting resources by using specialized equipment (*e.g.*, Chisasibi Cree used certain gill net sizes to catch specific age and species of fish), and by having control over which species are harvested, when they are harvested, and what size is harvested (Lerner 1993).

The Inuvialuit have historically been stewards or caretakers of their coastal and marine resources. Much of their present knowledge has been passed down over time from one generation to the next generation (Inuit Tapirisat Katami 2003). They have tried to maintain a balance between sustainable resource use and conservation based on their experience, traditional knowledge, and respect for their environment (Fast *et al.* 2001, 184; McDonald *et al.* 1997). The deeply held connection that many Inuvialuit continue to have to coastal and ocean resources leads them to be concerned with protecting their natural environment for future generations (NRTEE 2001).

It is largely elders who hold the traditional knowledge of the ISR at the turn of the twenty-first century. There is general agreement among elders, community leaders, parents, teachers, and youth that this knowledge is not being passed on to the present generation. This loss of knowledge has many implications for the future of this region. By spending time on the land, youth develop an ethic of respect and a sense of connection to the land, the ocean, the animals, and each other. They also develop a strong sense of place through experiencing the land on a daily basis. The ethos of caring for the community, working together for the common good, becoming self-reliant and confident of one's abilities, as well as developing leadership skills and knowing the land, can only be acquired through personal experience (FJMC 2000). This generation's ability to use and maintain ocean resources for food and to sustain its cultural values and traditions will directly affect its quality of life (Fast *et al.* 2001). Youth are going to inherit the Earth, and they must be equipped to take care of it for future generations.

Elders are highly respected and knowledgeable members of the community. They have the responsibility to teach youth about the importance of the ocean, and to pass on the community's body of traditional knowledge and shared values (Elder, pers. comm. 2003). Parents are also responsible for ensuring that their children learn to respect the ocean and that they learn the traditional knowledge of their elders (Parent, pers. comm. 2002). The Inuvialuit consider respect for wildlife and the environment as being essential to understanding their place in the world. Elders believe that problems arise out of a lack of respect (Elders and Parents, pers. comm. 2002/2003). Industry, government, and community groups such as the hunters and trappers committees are also seen by youth, elders, and the committees themselves as having a responsibility to teach youth about ocean ecology and management.

All study participants recognized the importance of youth acquiring traditional knowledge if they are to be stewards of the ocean. Inuvialuit elders felt that TEK is extremely important because one cannot be a good manager or caretaker of the ocean without that knowledge. Though most participants thought that it would be ideal for youth to possess both traditional and formal skill sets (competency in reading, writing, and Western science), many elders felt that youth could gain competency in formal skills without graduating from high school (Elders, pers comm. 2003).

A Role for Local Resource Management Professionals

Resource management professionals (*i.e.,* government staff) in the ISR tended to have a pessimistic attitude toward youth. Many see youth as a lost cause and have concluded that there is no point in trying to involve them. Past efforts to involve youth have proven disappointing and there is not an inclination to continue trying. Some felt that it would be more beneficial to target a younger age group and work with them to instill an ethic of social responsibility to protect the environment (Resource Managers, pers. comm. 2003). Youth told me that they are aware of the negative attitudes of resource management professionals and said that this attitude can be a barrier to their participation. In some agencies attitudes varied among staff, with some more willing than others to take the initiative to involve youth. Many resource management professionals fail to recognize the value and importance of involving youth and simply see it as 'slowing them down.' Some resource management professionals expressed that they felt obliged to involve youth but personally were not interested working with them (Resource Managers, pers. comm. 2003). Youth are aware of the negative attitudes toward them held by some professionals in the region. This further lowers their self-esteem and adds to their sense of hopelessness (Youth, pers. comm. 2003).

In sum, many youth said that they often feel abandoned and forgotten. Youth in communities other than Inuvik expressed that they feel forgotten by other Canadians and abandoned by their own people in Inuvik, where their government is based. Their communities are isolated, and they too feel personally isolated from other Inuvialuit and the rest of Canada (Youth, pers. comm. 2002). Youth in the communities other than Inuvik felt especially abandoned because of the lack of opportunities available to them. Some even felt as if they were being punished for staying in their home communities, where opportunities are very few (Youth, pers. comm. 2003).

THE WAY AHEAD: A STRATEGY FOR ENGAGING INUVIALUIT YOUTH IN MARINE STEWARDSHIP

The proposed strategy for engaging Inuvialuit youth in marine stewardship outlined in this section is based on the research findings. The proposed strategy identifies key activities that need to be undertaken in the communities. Implementation of this strategy will require the collaborative efforts of the BSIMPI, community leaders and members, educators, elders, and local resource managers. The strategy addresses all six Inuvialuit communities and focuses on the issues of participation, the need for expanded and development of new programs, better communication, the need for higher educational standards, and the urgent need to take youth on the land for extended periods. The strategy also outlines roles for BSIMPI, educators, HTCs, elders, local resource managers, the Inuvialuit Regional Corporation (IRC), and industry. What is being proposed requires a new way of thinking and doing things related to youth. If it is fully implemented youth will rise to the challenge of being tomorrow's resource managers. The benefits achieved will exceed their active participation in marine stewardship.

Strategy Components

1 *Continue to participate in existing opportunities*

Youth should continue to participate in existing marine stewardship opportunities as described earlier. Existing opportunities should be regularly evaluated by both program administrators and participants to ensure that they meet the needs of youth and the administering organization. Where possible, existing programs such as the FJMC Student Mentoring Program should be expanded to provide opportunities to engage youth in all ISR communities.

2 *Develop new stewardship opportunities and encourage participation*

Youth, in collaboration with local resource managers, should develop new opportunities for participation. This process should be facilitated by a Marine Stewardship Youth Coordinator. The goal here is to strive to create and fill this position with a youth. The idea has been discussed by the FJMC in the past and they would agree. Ideally, the position would be filled by someone who has previously participated in some of the programs. New opportunities should provide youth with hands-on, participatory experiences that are both interesting and challenging (Youth, pers. comm. 2003; FedNor 1994). Youth should be involved in program design to ensure that opportunities are interesting and attractive to youth. Youth should also be provided incentives and recognition for their participation in stewardship (Environmental Program Administrator, pers. comm. 2003). These measures will increase the likelihood of youth staying involved in stewardship programs.

3 *Increase youth awareness of stewardship opportunities*

Increasing awareness of stewardship opportunities and how to get involved is critical to the success of a strategy to engage youth in oceans stewardship. Currently, only individual program providers tend to have information about their program. One-stop shopping to access information would be helpful because youth are less likely to seek out information if they must go to a number of different locations. Information should be located in convenient places easily accessible to youth, such as schools and youth centres (FJMC 2000). It is very important that a concerted effort be made in the smaller coastal communities to increase awareness of opportunities. A variety of media, such as television, newspapers, newsletters and Web sites, could be used to inform youth about opportunities.. Involving youth in the design and development of promotional materials will help ensure their effectiveness (FedNor 1994).

4 Improve communication between groups

Currently, there is a fragmented approach to providing stewardship opportunities to youth due to a lack of communication between agencies, groups, and schools. A fragmented approach increases the likelihood of duplication and gaps of programs (NRTEE 2001). Better cross-communication between these groups when developing programs will help ensure that there is no duplication of opportunities and that gaps will be filled. This approach will also help ensure that programs are relevant to course curriculum and that they are consistent with the school year. A coordinated approach requires that groups work collaboratively to make certain that youth are provided with the best opportunities possible, and that competition for the same pool of youth participants is reduced. The appointment of a Marine Stewardship Youth Coordinator would be a very positive step. This coordinator would be responsible for collaborating with interested parties, developing new initiatives, and promoting existing and future programs.

5 Offer youth a high quality formal education

Schools in the ISR need to improve their standards, and Inuvialuit youth should be encouraged to complete their formal education and pursue post-secondary education. Educational institutions must have rigorous academic standards to ensure that youth have the necessary skills and knowledge to participate in stewardship and take on leadership roles in their communities (NRTEE 2001). Watering down and modifying programs results in a system that does not adequately prepare youth for life in either the North or the South (Castellano et al. 2000).

The schools should aim to teach youth about ocean ecology, governance, and the importance of the ocean to the Inuvialuit culture through a variety of initiatives. Such initiatives include guest speakers, marine-related projects and science fairs, attending marine related conferences, offering the Oceans 11 curriculum in Aklavik and Tuktoyaktuk, and providing marine-related extra curriculum activities where there is sufficient student interest.

6 Increase youth capacity in traditional skills and knowledge

Inuvialuit youth must also increase their capacity in traditional skills and knowledge. Youth gain these skills and knowledge through experiencing the land and ocean first-hand and learning from family and elders. Therefore, more on-the-land and ocean programs should be offered. The skills and knowledge gained will help provide youth with an understanding of the environment and respect for the land and ocean, and help develop a sense of connection to the earth. It is important that opportunities offered to youth be culturally relevant. Youth need to understand how what they are doing and learning is relevant to them and the Inuvialuit culture. On-land programs could begin to compensate for the decline in traditional methods of handing down TEK.

Strategy Implementation – The Roles and Responsibilities

BSIMPI has a key role to play in implementing this strategy for engaging Inuvialuit youth in marine stewardship. Along with other interested parties, this group will be responsible for promoting the knowledge gathered from this study in ways that will encourage community leaders, elders, educators, resource managers, and others in identifying ways to make the changes needed to ensure the effective implementation of the strategy. BSIMPI is the ideal group to carry out these functions because it represents a cross-section of interested parties in the ISR.

Educators are responsible for teaching Inuvialuit youth about ocean ecology and management. The Oceans 11 course is a take-off point for this and could be used to good advantage in assuming this responsibility. The schools should also make an effort to ensure that stewardship opportunities are available to youth from all six ISR communities.

Members of HTCs should play a much greater role in teaching and engaging youth in marine stewardship at the local level. The HTCs acknowledge this responsibility and suggested that they could invite youth to attend HTC meetings, host workshops, and conduct school presentations to teach youth about marine stewardship and the importance of the marine environment to the Inuvialuit (Resource Managers, pers. comm. 2002/2003).

Elders' committees also have a very important role to play in implementing this strategy. Elders have the responsibility to pass on their knowledge of the land and animals on to younger generations. Some suggested initiatives elders' committees could pursue to engage youth in marine stewardship include hosting field trips to take youth out on the land and teach them traditional skills and knowledge, and conducting workshops in the community on marine-related issues and traditional skills. The Oceans 11 curriculum has an experiential component that brings elders into the classroom and takes youth into the community to meet with elders. This curriculum is being implemented in four of the Inuvialuit communities in 2003–2004.

Local resource managers have perhaps the most important role in implementing the strategy because they currently provide the majority of marine stewardship opportunities to youth. This group includes DFO, PC, Department of Resources, Wildlife and Economic Development (RWED), and the FJMC. Each of these agencies has taken steps to engage youth in resource management activities. They are capable of doing much more. The FJMC should expand its Student Mentoring Program to include all ISR communities. Agencies should make regular school presentations and host or sponsor community events such as Oceans Day to raise the profile of oceans and awareness of the work of that organization. These agencies can also provide competitive summer employment opportunities, assist youth in attending marine related conferences, and involve youth in scientific research.

The IRC, including its community corporations and brighter futures programs, has a crucial role to play in the implementation of the strategy. The IRC can show leadership in supporting existing stewardship opportunities and in developing new initiatives to engage youth. The IRC, as the umbrella governing body in the ISR, must take steps to ensure that youth in the ISR receive a high-quality education. The IRC also has a responsibility to encourage youth to complete and pursue post-secondary education.

Industry has a role to play in the successful implementation of this strategy. Youth have stated that the low skill requirements and relatively high wages for some industry positions can discourage them from completing or returning to school. Industry should therefore pursue initiatives that encourage youth to stay in school and complete their education (Youth, pers. comm. 2002/2003). This can be achieved through hiring policies that encourage youth to complete their education and pursue a post-secondary education, providing scholarships to support students' pursuit of post-secondary education, and conducting presentations at schools about marine stewardship and the benefits of staying in school.

LESSONS LEARNED AND IMPLICATIONS FOR THE FUTURE

There are a number of important lessons to be learned from this research. Perhaps the most important is that a great opportunity to involve Inuvialuit youth in marine stewardship is being missed. Youth in the region have expressed a desire to be involved in stewardship activities and to learn more about the marine environment. To date, there have been limited opportunities for youth involvement. There have also been important lessons learned regarding formal and informal education. It is evident that higher education attainment levels leads to a better quality of life. This includes access to increased income and being able to make better life choices (City of Calgary 2002; NRTEE 2001; Castellano *et al.* 2000). Another lesson learned is that most ISR youth are not being given the chance to spend time on the land even though they want this very much. Spending time on the land allows youth to acquire traditional knowledge, develop moral values of respect for the earth, self-confidence, and a sense of pride and connection with their culture. Acquiring this knowledge will allow youth to become not only better stewards of the land but also leaders of their communities.

This research focused on Inuvialuit youth. The findings regarding education and engaging youth are similar to research findings about Aboriginal youth in Nunavut and northern Ontario (Makokis 2000; FedNor 1994). Although a northern Aboriginal example was used, many of the findings may be applicable elsewhere in Canada. Continued research on engaging youth in stewardship activities in other Aboriginal and northern communities is needed.

REFERENCES

Ayles, Burton, G. and N. Snow. 2002. "Canadian Beaufort Sea 2000: the environmental and social setting." *Arctic* 55(suppl. 1): 4–17.

BSIMPI (Beaufort Sea Integrated Management Planning Initiative). 2002. *Tariuq Fact Sheet.*

Berkes, Fikret, J. Mathias, M. Kislalioglu, and H. Fast. 2001. "The Canadian Arctic and the Oceans Act: the development of participatory environmental research and management." *Ocean and Coastal Management.* 44: 451–69.

Biodiversity Convention Office. 2001. *Working Together: Priorities for Collaborative Action to Implement the Canadian Biodiversity Strategy.* Environment Canada.

Booth, Annie L., and H.M. Jacobs. 2001. "Ties That Bind: Native American Beliefs as a Foundation for Environmental Consciousness." In *Environmental Ethics.* New Jersey: Prentice Hall Inc.

CAPP. (Canadian Association of Petroleum Producers). 2003. *Stewardship initiative. http://www.capp.ca/?V_DOC_ID=1*

Castellano, Marlene, L. Davis, and L. Lahache, eds. 2000. *Aboriginal Education – Fulfilling the Promise.* Vancouver: UBC Press, 2000.

City of Calgary. 2002. *Respecting our youth. http://www.gov.calgary.ab.ca/community/publications/respecting_our_youth*

Condon, Richard. 1987. *Inuit Youth Growth and Change in the Canadian Arctic.* New Brunswick, NJ: Rutgers University Press.

CWS (Canadian Wildlife Service), Department of Fisheries and Oceans, Forest Renewal, British Columbia's Watershed Restoration Program. 1995. *Community Stewardship: A Guide to Establish Your Own Group.* Fraser Basin Management Program.

Dallmeyer, Dorinda G., ed. 2003. *Values at Sea: Ethics for the Marine Environment.* Athens, GA: University of Georgia Press.

DFO (Department of Fisheries and Oceans). 2003. *Oceans Day. http://www.pac.DFO-mpo.gc.ca/oceans/Outreach/oceantalk/background_e.htm*

Environment Canada. 1996. *Report on Federal Actions: Wildlife Conservation and Sustainable Use (1995–2000): Implementing the Canadian Biodiversity Strategy.*

Farrow, Malcolm, D. Wilman, eds. 1989. Self-determination in Native Education in the Circumpolar North. Government of NWT, Department of Education.

Fast, Helen, J. Mathias, and O. Banias. 2001. "Directions toward marine conservation in Canada's Western Arctic." *Ocean and Coastal Management.* 44: 183–205.

FedNor (Federal Economic Development Agency in Northern Ontario). 1994. *Accessing Our Future: Aboriginal Youth on Economic Development in Northern Ontario.* Industry Canada.

FJMC (Fisheries Joint Management Committee). 2001. *Student mentoring program. http://www.fjmc.ca*

———. 2000. *Beaufort Sea Renewable Resources for our Children.* Conference Summary Report. Inuvik.

G.S. Gislason & Associates and Outcrop Ltd. 2002. *The Marine-Related Economy of NWT and Nunavut.* Consulting report.

Government of Canada. 1984. *The Western Arctic Claim: The Inuvialuit Final Agreement.* Indian and Northern Affairs Canada.

————. 1997. *Towards Canada's Oceans Strategy: Discussion Paper.* Ottawa: Fisheries and Oceans Canada.

————. 2002a. *Canada's Oceans Strategy.* Ottawa: Supply and Services Canada.

————. 2002b. Policy and Operational Framework for Integrated Management of Estuarine, Coastal and Marine Environments in Canada. Ottawa: Supply and Services Canada.

————. 2002c. Canada's Stewardship Agenda: Naturally Connecting Canadians. Ottawa: Supply and Services Canada.

————. 2003. *Oceans Act. http://laws.justice.gc.ca/en/0-2.4/*

Grenier, Louise. 1998. *Working With Indigenous Knowledge.* Ottawa: International Development Research Centre.

IWCO (Independent World Commission on the Ocean). 1998. *The Ocean Our Future.* United Kingdom: Cambridge University Press.

Inuit Tapirisat Katami. 2003. *Traditional knowledge.* *http//www.tapirisat.ca/english_text/itk/departments/enviro/tek/*

IRC (Inuvialuit Regional Corporation). 2003. *Inuvialuit corporate group.* *http://www.irc.inuvialuit.com/*

Knight, R. L., and P. Landers, eds. 1998. *Stewardship Across Boundaries.* Washington, DC: Island Press.

Laynard, N., and L. Delbrouck, eds. 1994. *Stewardship 94 – Revisiting the Land Ethic – Caring for the Land.* Victoria: Ministry of the Environment.

Leedy, P.D., and J.E. Ormrod. 2001. *Practical Research Planning and Design, 10th Edition.* Upper Saddle River, NJ: Merrill/Prentice Hall.

Lerner, Sally, ed. 1993. *Environmental Stewardship. Studies in Active Earthkeeping.* Waterloo: University of Waterloo.

Makokis, P.A. 2000. *An Insider's Perspective: The Drop Out Challenge for Canada's First Nations.* Ph.D. dissertation. University of San Diego.

McDonald, M., L. Arragutainaq, Z. Novalinga. 1997. TEK *of Inuit Cree in the Hudson Bay Bioregion – Voices from the Bay.* Prepared for the Canadian Arctic Resources Committee and Environment Committee of the Municipality of Sanikiluaq.

Muir, A.K. Magdalena. 1994. *Comprehensive Land Claim Agreements in the* NWT*: Implications for Land and Water Management.* Canadian Institute of Resources Law.

NRTEE (National Roundtable on the Environment and Economy). 2001. Aboriginal Communities and Non-Renewable Resource Development. Ottawa: Renouf Publishing.

Notzke, C. 1995. *The Resource Co-management Regime of the Inuvialuit Settlement Region. In Northern Aboriginal Communities – Economies and Development.* North York: Captus Press.

Purich, D. 1992. *The Inuit and Their Land: The Story of Nunavut.* Toronto: James Lorimer & Co.

Vodden, Keith. 2001. *Inuvialuit Final Agreement Economic Measure: Evaluation. Final Report.* Consulting report.

SECTION II
RESPONDING & ADAPTING TO NEW CHALLENGES

CHAPTER 7

A PLACE FOR TRADITIONAL ECOLOGICAL KNOWLEDGE IN RESOURCE MANAGEMENT

Micheline Manseau (Parks Canada
& University of Manitoba)
Brenda Parlee (University of Manitoba)
G. Burton Ayles (Canada/Inuvialuit Fisheries
Joint Management Committee)

INTRODUCTION
The Rise of Traditional Ecological Knowledge

As land claims are settled in the northern regions of Canada, further legislative, regulatory or policy requirements are established to ensure that the knowledge, practices, and beliefs of Aboriginal people are protected and included in resource management. It is a policy in Canada that traditional ecological knowledge (TEK) be considered and incorporated into resource management (Usher 2000). National, provincial, and territorial institutions have committed to the understanding and use of TEK (Posey 1999). In some cases the commitments aim to respect, preserve, and promote the use of TEK in managing natural resources (Canada 1995), and in other cases, the intent is toward the integration or harmonization of TEK with other sources of knowledge (Inuit Circumpolar Conference 1992; CFFS 1997).

New co-management institutions have also been created across the North and have the potential to be important vehicles for the inclusion of TEK in resource management (Pinkerton 1989, Berkes 1997). It is argued that such arrangements are key to a successful incorporation of TEK in providing opportunities for communities, governments, and other stakeholders to work together on an ongoing basis, facilitating communication and learning between parties that were conventionally in resource management conflict (Kendrick 2000, Berkes *et al.* 2001). Co-management institutions can be arenas for exchanges of ideas on natural systems, where interactive and mutual learning takes place (Kendrick 2000; 2003). They also often present a redistribution in the balance of power; departing from state-central structures for more equal partnerships, institutionalized joint decision-making or legislated decision-making power at the local level. The change in power dynamics is often critical in the linking of state and especially indigenous knowledge systems (*e.g.,* McCay and Acheson 1987, Pinkerton 1989, Berkes *et al.* 1991, Pomeroy and Berkes 1997, Nadasdy 2003).

The use of traditional knowledge in resource management can bring significant changes to conventional, state management structures. It has the potential to affect the fundamental assumptions of science-based wildlife management (Berkes 1998). In contributing different values and perspectives, TEK can also influence management objectives. Traditional management systems or stewardship practices rarely aim at large-scale land changes, control, or fixed levels of animal population sizes. They place humans inside the unit being managed and incorporate random interconnections and the idea of time as cyclical into their management decisions (Tippett 2000). As a result, the rules and practices also often reflect and respond to the structure and distribution of an animal population rather than aiming at a finite number or toward a given population size (Acheson *et al.* 1998, Berkes and Folke 1998).

The cultural basis of traditional knowledge can also modify the structures and procedures prevailing in management institutions. While management responsibilities often reside with biologists or managers, specific regulation of activities making impacts on the environment is usually vested in traditional authority (Ruddle 1994). This varies according to the social organization, and authority can reside with family groups, senior hunters (Berkes 1998), secular or religious leaders, specialists or rights holders (Ruddle 1994). Indigenous 'management' is not a discrete function within indigenous societies but is practised within the context of the larger cultural system (Tippett 2000). These systems often show a high degree of unification of conception and execution (LaDuke 1994).

This chapter aims at documenting how the strong and high-level commitments to TEK made by Aboriginal groups, academics, and governmental and non-governmental organizations involved in natural resource management have led to an increased use of TEK in the decision-making process. We examine different initiatives from northern Canada to acquire some insights into the mechanisms by which traditional ecological knowledge is contributing to resource management. We explore the following questions:

1   What roles does legislation play in ensuring that TEK is used in resource management?
2   What roles do management institutions play in facilitating the communication, interpretation, and inclusion of TEK in decision making?
3   What roles do the communities play in capturing, collating, and converting TEK for the purpose of resource management?

The first initiative presented in the chapter is from the Inuvialuit settlement region, Northwest Territories; the Fisheries Joint Management Committee (FJMC) is a co-management institution created under the *Inuvialuit Final Agreement* (1986). The work of the FJMC provides useful information on how traditional knowledge is used in the management of fisheries and other marine resources by redefining marine health indicators and by influencing data gathering as well as the analysis and interpretation of results (Day 2002, Harwood *et al.* 2002). The second initiative is from Lutsel K'e, Northwest Territories; this case is a community-based institution established primarily to document and use TEK in resource management in response to local concerns about the environmental and socio-economic impacts of mining exploration and development in the region. Environmental assessment and regulatory requirements in place for the development of diamond mines provided the foundation as well as the resources for the community to carry out ongoing environmental and community health monitoring. The third initiative is from Quttinirpaaq National Park, Nunavut. In this case, legislative requirements for the co-operative management of the Park and the inclusion of traditional knowledge were set out in the *Nunavut Land Claims Agreement* (1993) and further defined in an Inuit Impact and Benefit Agreement (1999). The work of the co-management board reveals how trust between government officials and local community members creates a forum for sharing traditional and scientific knowledge to meet park planning and management goals. Other useful information is also found in the work of the Alaska Beluga Whale Committee, Alaska, and the forest management planning activities of the Wet'suwet'en First Nation in British Columbia and the Innu Nation in Labrador.

THE PLACE OF TEK IN RESOURCE MANAGEMENT –
LESSONS FROM THE CANADIAN NORTH
Management of Wildlife Species –
Canada/Inuvialuit Fisheries Joint Management Committee

Since the ratification of the *Western Arctic Claims Settlement Act* (Inuvialuit Final Agreement) in 1984, the fish and marine mammal resources of the western Arctic have been managed co-operatively by the Canada/Inuvialuit Fisheries Joint Management Committee (FJMC) with its partners the federal Department of Fisheries and Oceans (DFO), the Hunters and Trappers Committees of the Inuvialuit Settlement Region (Figure 5.1), and the Inuvialuit Game Council (for more information on the co-management process, see Bailey *et al.* 1995, Green and Binder 1995, Ayles and Snow 2002 and *http://www.fjmc.ca*).

The Inuvialuit Final Agreement and the various acts and regulations that are under the aegis of the Minister of Fisheries and Oceans define the responsibilities of the FJMC. In general they are (1) to assist Canada and the Inuvialuit in administering the rights and obligations related to fisheries under the Inuvialuit Final Agreement, (2) to assist the minister in carrying out his or her responsibilities for the management of fisheries and marine mammals in the Inuvialuit Settlement Region, and (3) to advise the minister on all matters relating to fisheries in the region.

The establishment of the FJMC has led to significant changes in the co-operative management of fish and marine mammals in the Inuvialuit Settlement Region and the use of TEK in all aspects of decision making. The following sections describe how TEK and the knowledge of local resource users is being integrated into the ongoing activities of the committee, projects operated by the committee and in the long term strategic planning of the committee.

The FJMC is composed of two members appointed by the Inuvialuit, two members appointed by the government of Canada, and a chair appointed by the members. It is supported by a resource biologist and a secretariat that also provides support to the other Inuvialuit co-management bodies (Green and Binder 1995). The committee has five formal meetings annually and also interacts through regular conference calls, special projects, workshops, and meetings. Regular meetings are formally structured with agenda, minutes, and motions on decision items, but discussions and interactions are informal. The two Inuvialuit members are always currently active hunters or fishers or elders with significant experience on the land, and they bring those experiences to the meetings. The contribution of the two federal government members has primarily been to provide input from scientific and government resource management perspectives.

A recent study examining co-management in western Canada (Iwasaki-Goodman 2004) identified several key elements in the meetings: discussion until a consensus was reached; respect for differences in opinions among members; recognition of the importance of their own role in resolving conflicts in resource management between hunters and fishers and the government; a strong sense of alliance and friendship between committee members; and recognition that the goals of the FJMC are long-term. The relationships within the committee are extended to relationships with other individuals, communities, and agencies. The input of traditional knowledge by hunters and fishers and the contribution of their knowledge of the resource and plans for its use are ensured by winter meetings in the communities to discuss community problems and priorities for the upcoming year. In addition, traditional knowledge is brought to the decision-making process through workshops and meetings on specific issues and joint field projects that involve community members as well as outside researchers. Relations with DFO are maintained by having a DFO representative attend each meeting as an active participant in all discussions and by having DFO biologists, scientists, and managers regularly attend meetings to discuss science, fisheries management, and oceans management issues and to receive community input on TEK. Representatives from other government departments regularly come to the meetings, as do non-governmental organizations and industry.

Programs and activities operated or financially supported by the FJMC include: monitoring fish and marine mammal harvests; establishing management and fishing plans for key species, *e.g.,* Beluga whales (*Delphinapterus leucas*), inconnu (*Stenodus leucichthys*), Arctic char (*Salvelinus alpinus*), and developing population and stock assessments of Arctic char and other species in order to establish safe

harvest levels. The FJMC also carries out genetic studies to help identify stocks from different streams or areas, conducts studies of contaminants to ensure animals are not being harmed, and sets up traditional knowledge studies of many animals and areas. The most important area of TEK input is on resource harvesting issues. The subsistence harvest of beluga whales is a critical aspect of Inuvialuit culture and traditions. The Beaufort Sea Beluga Management Plan, first completed in 1991 and scheduled for revision in 2003, is a plan to ensure the sustainability of this harvest and is a priority initiative for the FJMC, the Hunters and Trappers Committees, and DFO (FJMC 2001). The plan addresses objectives for sustainable harvests, conservation, and protection, including the development of beluga management zones that provide guidelines for industry (primarily oil and gas) development activities, tourism as it relates to beluga, and the bylaws of the Hunters and Trappers Committees relating to beluga harvesting. The plan was developed with full input from hunters and fishers.

The beluga monitoring program is an important activity under the umbrella of the plan. It is based on the traditional knowledge of generations, but incorporates the evolving local knowledge of current conditions. The purpose of the program is to document the size and trend of harvesting activities and to obtain the data necessary to assess the health of the beluga population and the impact of the harvest on the stock. The present beluga monitoring program is a result of three decades of evolution and development, but it is based on at least five hundred years of Inuvialuit harvests of beluga in the Beaufort Sea (McGhee 1974). Current monitoring takes place at seven locations along the Beaufort Sea coast and includes all traditional hunting areas and involves four communities. DFO and the FJMC provide the results back to the Hunters and Trappers Committees, schools, and community members through meetings, workshops, and posters. The results are also published in scientific papers (e.g., Stern and Ikonomou 2001, Harwood et al. 2002) and exchanged with Alaskan hunters.

Despite the close co-operation between scientists, managers, and hunters and fishers, and despite legislation and mechanisms to facilitate the input of TEK into resource management decision making, conflicts do arise. Initial aerial surveys of beluga whales in the Canadian Beaufort Sea resulted in published estimates of only about seven thousand animals. Based on their personal observations over the years, the hunters were convinced that the population was much larger. The authors of the original report agreed that the number was an underestimate. They had been very careful to point out that their study results represented the number of whales in the onshore waters of the Mackenzie River estuary and had not been corrected for submerged whales that were not visible and that there were whales beyond those waters (Norton and Harwood 1986). Nevertheless, despite the hunters' concerns, the population estimate of seven thousand animals was published and repeatedly used in other fora. The estimate of 7000 animals ultimately became the focus of discussions on the health of the population (Weaver 1991). Based on these estimates, hunting quotas were proposed, opting for a close to harvest. It is probable that only the reality of the rights assigned within the Inuvialuit Final Agreement prevented the establishment of what would have been arbitrary quotas. The latest aerial survey was conducted in 1992. It gave an index of stock abundance of 19,629 animals, but again, this number did not account for whales under the water during the aerial counts or for whales outside the survey area (Harwood et al. 1996). Subsequent analysis has resulted in a population estimate of over 39,000 animals (Harwood and Smith 2002). In retrospect, the hunters' traditional knowledge was correct: there were many more whales than the original survey revealed.

Traditional ecological knowledge is also critical for the long-term planning of the FJMC. By the end of the 1990s, co-management was well established in the Inuvialuit Settlement Region. With its partners, the FJMC had addressed many of the initial challenges associated with the implementation of co-management arrangements, and the FJMC members began to consider future directions. The result was a vision and strategic plan to guide the committee's objectives over the coming decade (FJMC 2003). Inuvialuit members of the FJMC have used the process to ensure that their beliefs in both the way the committee should operate, and in the way that the fish and marine mammal resources should be managed, would be expressed and communicated to future generations. This is exemplified by one of the fundamental principles of the FJMC: "Committee actions will endeavour to ensure that fish and marine mammals are treated with respect during any harvesting, scientific study, or other use of the resource." Specific

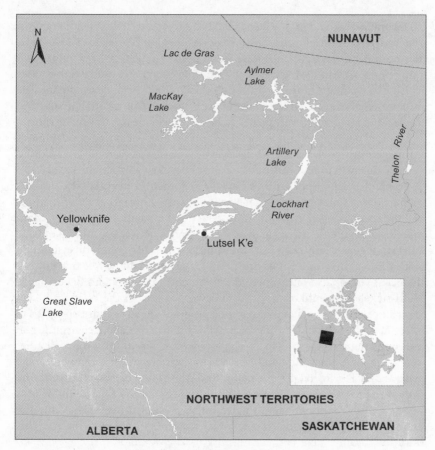

Figure 7.1 Lutsel K'e area.

reference to TEK is found in one of the vision statements: "In the FJMC vision for the future all marine, anadromous and freshwater fish, and marine mammals stocks of the Inuvialuit Settlement Region will be managed and conserved for the wise use and benefit of present and future generations through the use of sound scientific and traditional knowledge of the renewable resources of the Region and their ecosystems."

Management of Industrial Activities – Diamond Mining and the Lutsel K'e Wildlife, Lands and Environment Office

Since the turn of the century, the Lutsel K'e Dene First Nation (Figure 7.1) and other northern communities have witnessed significant non-renewable resource development in their territory, the result of Canadian government efforts to "modernize and develop" the region.

A legacy of environmental and social problems has resulted from many of these projects, including the Talston River hydroelectric project (Bielawski in

Collaboration with Lutsel K'e Dene First Nation 1992) as well as gold, uranium, and lead-zinc mining in the Great Slave Lake region (Parlee 1998). When indicator minerals for diamonds were found in the Lac de Gras area in 1989 and a mine proposed in 1995, government, mining companies, and communities hailed it as the beginning of a different period of industrial development. The recognition of Aboriginal rights under the Canadian Constitution in 1982 and the negotiation of self-government and devolution in the Northwest Territories had created new opportunities for Aboriginal people, including opportunities for integrating traditional knowledge into the assessment, planning, monitoring, and management of non-renewable resource development activities.

Some of these opportunities have been forged in negotiated contracts such as impact and benefit agreements between Aboriginal organizations and the corporations proposing industrial developments. Such contracts are "manifestations of the struggle to achieve balance between local and outside interests, they generally contribute to lasting local benefits, greater diversification of local economies, and better prospects for sustainability while minimizing the negative impacts of resource development" (O'Reilly and Eacott 1998). Multi-stakeholder initiatives were also developed such as the West Kitikmeot Slave Study Society – a five-year research program aimed at addressing gaps in scientific and traditional knowledge baseline information about environmental and community health issues in the region (*www.wkss.nt.ca*). Other opportunities were defined in Environmental Agreements between mining companies, the Federal and Territorial governments, and Aboriginal organizations. In the case of the BHP Billiton Diamond Mine and the Diavik Diamond Mine at Lac de Gras, the Independent Environmental Monitoring Agency and the Environmental Monitoring Advisory Board are responsible for ensuring that traditional knowledge is meaningfully used (*www.monitoringagency.net*; *www.ainc-inac.gc.ca*).

The vision, design, and implementation of many of these TEK initiatives have been spearheaded by local communities. The leadership and commitment shown by Aboriginal organizations and their representatives ensured that, in every way possible, traditional knowledge was given equal consideration as science, in principle as well as in practice. In Lutsel K'e, the overall vision was to develop a community-based approach to address the environmental and community health issues of greatest concern to the community. In order to achieve this vision, a significant investment in building community capacity was required. Funding and support was sought and received from a variety of sources, including universities, non-governmental organizations, foundations, government departments, and industry. Under the direction of the Wildlife Lands and Environment Committee and an elders' committee, local youth were trained in basic research methods, the use of geographic information systems and database management, as well as in their native *Denesoline* (Chipewyan) language. By working with their elders on the land, youth were able to learn and practise Dene traditions and knowledge as well as to document and apply this knowledge to address their concerns about the impacts of mining in the region. Over a period of eight years, Lutsel K'e was able to achieve significant progress toward this vision, as described by Florence Catholique:

> We are trying to relay the traditional knowledge so that the Elders will be used, and maintain the youth in the school system. It's all very·complex. We've got to monitor, we've got to record traditional knowledge so the younger people can see it, so it's impressive to the young. The Wildlife Committee has more people employed than I have ever seen, it's an area where young people want to be, they're still interested in the land. But the next generation, if we don't keep our people out on the land, they're not going to [be interested in the land]. The community has to look at getting the people out on the land, to understand development. These things are being done with the Wildlife [Committee], that is their purpose. (Florence Catholique, Lutsel K'e, 2001)

Growing community capacity in Lutsel K'e paved the way for positive working relationships between the community, government, and industry. As a first step, the community sought to define its own indicators and baseline with respect to community and ecological health. Projects such as the *Traditional Knowledge Study on Community Health* and the *Community-Based Monitoring Pilot Project* provided a snapshot of health issues of concern to the community, including self-government, healing, and cultural preservation (Marlowe and Parlee 1998). A parallel project focused on *Denesoline* knowledge of ecosystem health resulted in the development of a range of indicators with respect to waterfowl, fish, caribou, and fur-bearing animals (Parlee and Marlowe 2001).

Baseline traditional knowledge studies have also been carried out on a site-specific basis. For example, a study in 1998 was carried out at *Gahcho Kue* (Kennady Lake), the site of a DeBeers Canada exploration project. In 2001, a similar study was carried out at *Na Yaghe Kue* (Snap Lake), where DeBeers Canada has proposed the development of Canada's third diamond mine. Both studies were aimed at documenting baseline information about plants, wildlife, and landscape features. Elders were also engaged in interpreting and predicting potential project impacts as well as developing recommendations for mitigation and management.

Monitoring based on traditional ecological knowledge, or "watching, listening, learning, understanding change in the environment," is a key theme in the work of the Wildlife, Lands and Environment Department (see the DVD in the back of the book). Between 1995 and 2003, the Wildlife, Lands and Environment Department established a monitoring system around a range of community-based socio-cultural and ecological indicators. These indicators aim at monitoring the cultural impact of mining activities on employees and their families, and on the quality of housing, health, and social service programs. Monitoring of caribou health and caribou movements based on scientific and traditional ecological

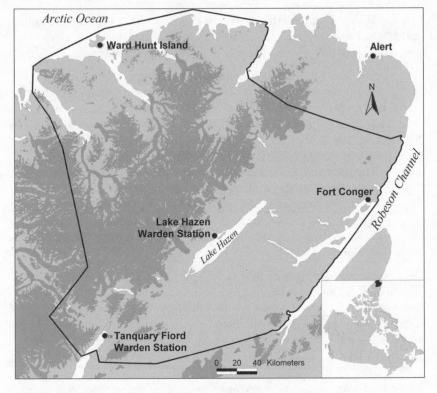

Figure 7.2 Quttinirpaaq National Park, northern Ellesmere Island.

knowledge has also been developed. It is hoped that the results of these projects can help the community better understand and deal with the effects of mineral resource development in their region. They are also using the information in the planning and management of their health and social service programs and self-government negotiations.

The work of the Wildlife, Lands and Environment Committee is an example of how local Aboriginal communities are building on the traditional knowledge and skills of their elders and developing a community-based resource management system.

Land management – Co-management of Quttinirpaaq National Park of Canada

'Quttinirpaaq' means 'top of the world' in Inuktitut, the Inuit language.

Quttinirpaaq National Park of Canada was established as a park reserve in 1986 to protect the integrity of an Arctic ecosystem. It is located at the northern end of Ellesmere Island in the Canadian Arctic Archipelago (Figure 7.2).

At 37,775 square kilometres, Quttinirpaaq is the second largest national park in Canada. The nearest communities are Grise Fiord, 640 kilometres to the south of the Park and Resolute Bay, 260 kilometres further south. Parks Canada has summer installations at Tanquary Fiord and Lake Hazen. The Department of National Defence has year-round military and research facilities to the south and north of the park, in Eureka and Alert. Although the park has been consulting the communities of Resolute Bay and Grise Fiord on an annual basis to discuss management and operational issues, it was not until the signing of the *Nunavut Land Claims Agreement* in 1993 that a joint Inuit/Federal Government management structure was established. The negotiations of an Inuit Impact and Benefit Agreement (IIBA) were completed in 1999 and provided detailed guidance on the co-operative management of Quttinirpaaq, Auyuittuq, and Sirmilik National Parks, all in Nunavut.

In order to help fulfill the purposes of the Agreement, a Joint Park Management Committee (JPMC) composed of three members appointed by the Qikiqtani Inuit Association and three members appointed by the Government of Canada was established for each park. As in most co-operative management arrangements in national parks under land claims, the JPMC of Quttinirpaaq advises the Minister of Parliament responsible for national parks on all matters related to park planning, management, and operation and on the means of accomplishing the park's goals. The JPMC oversees the development of park management plans, yearly work plans, budgets, research, and monitoring projects. It also helps the communities of Grise Fiord and Resolute Bay take advantage of the economic benefits associated with the park. A park management plan is required by the *Canada National Parks Act* and the IIBA. Although the processes for developing the management plan are similar in the Act and the IIBA, they are more extensive in the IIBA and require the additional involvement of organizations such as the Qikiqtani Inuit Association, the park's Joint Park Management Committee, and the Nunavut Wildlife Management Board. The Qikiqtani Inuit Association has official review functions, and the IIBA directs the association's president to write a foreword to the management plan. The management plans must be approved and recommended by the park's JPMC. The Nunavut Wildlife Management Board may also, at its discretion, approve any portion of the management plan that pertains to wildlife and wildlife habitat (F. Gertsch, pers. comm.).

The recent development of a park management plan highlights some of the "ways of working" adopted by the committee (Gertsch *et al.* 2004). Distance and means of communication between the communities hinder face-to-face meetings, so the work of the JPMC unfolds around one or two annual meetings and numerous teleconferences. Workshops are also organized to better comprehend the requirements of the IIBA, to learn about and further define the management planning process, and to understand some of the terminology used by Parks Canada (Gertsch *et al.* 2004). Time is spent on the land in the park to exchange views and observations about the area and to develop short and long-term conservation objectives. Although all meetings have simultaneous translation (English-Inuktitut) and all written material is available in both languages prior to the meetings, communication remains challenging. Time spent on the land helps build trust between members of the committee, staff, and advisors, and iterative meetings provide time and space for adequate consultation with community members. Issues are repeatedly discussed at meetings and in the communities and Inuit representatives only make decisions after they have reached a good understanding of the issues and after listening and learning from knowledgeable members of the communities. This requires time, but is a key function of the decision-making process.

Both Inuit *Qaujimajatuqangit* ("Inuit Knowledge of Old," see Box 7.1) and scientific knowledge relating to ecological structures and processes are presented and primarily by oral means. The Inuit representatives consult with community elders and park staff consult with scientists. Summaries of the information are shared at the meetings. Along with park rules and regulations, Inuit values and perspectives are communicated through oral histories, songs, and prayers. This plays a significant role in key decisions pertaining to, for example, the story lines selected and communicated to the public, decision rules followed to allocate research permits, define monitoring activities, and make decisions about tourism and other access/use of the park area. The "ways of knowing" clearly include many aspects of the knowledge system. It is more than a line in a database or a polygon on a map. It is more than factual information about the environment or knowledge about past and current use of the environment; culturally and spiritually based information also provide explanations and guidance (Usher 2000).

As expressed by the JPMC, Inuit culture and knowledge is strongly reflected by the strong desire to base their decisions on consensus, by comparing views and taking advice from the most experienced individuals. Management decisions also reflect a continuously changing environment, associated with dynamic and ongoing (on-site) learning. A decision is never final and is never set in stone. Keeping an eye on ongoing environmental changes and being able to adapt one's actions or decisions based on the observed outcome is perceived as critical to ensure the long-term viability of those ecological systems.

The Alaska Beluga Whale Committee of Alaska;
Wet'suwet'en First Nation of British Columbia; the Innu Nation of Labrador.

The resource management institutions represented below are further illustrations of co-management structures and local resource management initiatives. These examples will be referred to in the discussion section of this chapter along with references to the three cases described above.

The Alaska Beluga Whale Committee, Alaska, was established in 1988 by Alaska Native American beluga whale hunters, government biologists, and managers (Huntington 2000) to ensure that beluga whale stocks in Alaska remained viable and capable of supporting traditional subsistence harvests (Huntington 1992, Adams *et al.* 1993). The committee identifies data needs, establishes research priorities and methods, monitors hunter harvests of beluga, and makes management decisions as necessary. The committee meets annually; however, the research and monitoring activities continue throughout the year and often involve informal interaction between communities and members of the committee. At the annual meetings, hunters present the harvest reports and relevant observations about ice conditions and other environmental conditions. Researchers present the findings of previous years in language understood by all, yet do not avoid complexities and technicalities (Huntington *et al.* 2002). Interaction is encouraged and questions are asked by everyone to ensure understanding and generate new ideas. Traditional ecological knowledge is included through formal TEK studies and through interactions with hunters at annual meetings and when assisting researchers and managers in the field (Henry P. Huntington, pers. comm., June 2003).

The Wet'suwet'en First Nation of British Columbia (*www.wetsuweten.com*) has been using Geographic Information Systems (GIS) to map traditional land uses, wildlife habitat, and food sources. The Wet'suwet'en traditional social structure has five Clans, each of which is made up of thirteen Houses. Each House has a territory, usually based on watershed lines, which is the basis for mapping and managing resources. Using their knowledge, they are pursing economic opportunities for non-timber forest products. Black huckleberries will be the first product they attempt to manage and market. TEK has been used to identify areas of interest and potential management methods such as seven-to-ten-year rotational prescribed burning based on the House Territories. This project has gained the support of government, fire management agencies, potential buyers, and forest companies (Russell Collier, pers. comm., June 2003).

The Innu Nation, Labrador, in collaboration with the Government of New-foundland and Labrador, has recently completed an ecosystem-based forest management plan for 7.1 million hectares of boreal forest. A key feature of the plan is a protected area network that integrates conservation science, Innu traditional knowledge, and cultural land use data. The Innu Nation agreed that a minimum of 60 per cent of the land base needed to be set aside in order to adequately protect the ecological and cultural values of their homeland. An additional important aspect of this initiative was the establishment of an Innu forestry office and the hiring and training of Innu foresters and technicians. Their work consists of collecting field data, integrating Innu perspectives into forest operations, and ensuring on-going communication of their methods and observations to members of the community (Larry Innes, pers. comm., April 2003).

DISCUSSION

The initiatives described in this chapter provide insight into the current use of traditional ecological knowledge in northern Canada, the institutions that are involved, and how communities, governments, and others are working together to ensure that TEK is used in resource management decision making. The following discussion examines the role of legislation, management institutions, and communities in contributing to those changes.

Strong Legislative Guidelines, Policies and Acts
– Long-Term Commitment to Using TEK in Resource Management

The different northern initiatives presented in this chapter clearly illustrate that legislated guidelines, policies, and acts are an important foundation for the inclusion of TEK in the management of natural resources. Legislated agreements such as the *Inuvialuit Final Agreement* and the *Nunavut Land Claims Agreement* provide a general framework for ensuring that local communities and their knowledge are part of resource management institutions. Such agreements are useful in defining the relationship boundary and ensuring that time and effort spent on a given project actually lead to concrete results (R. Collier, pers. comm., 2003).

Opportunities for including TEK in resource management can also be defined in impact and benefit agreements. The *Inuit Impact and Benefit Agreement* (1999) for example, provides details related to the structure and responsibilities of the co-operative management of National Parks in Nunavut and the protection of Aboriginal and intellectual property rights. It also presents opportunities for including TEK in all aspects of the work.

In cases where there is no land claim or legislative basis, the requirements for including TEK have no firm legal foundation. In the case of the Lutsel K'e Dene First Nation, for example, the environmental assessment processes of the diamond mining projects have been an important impetus and source of funding

The implementation of the recently proclaimed Canadian Species at Risk Act (2002) provides an interesting example of these new governance rules and the use of tek in management decisions. The purpose of the Act is *to prevent endangered species or threatened wildlife from becoming extinct or lost from the wild, and to help in the recovery of these species. It is also intended to manage species of special concern and to prevent them from becoming endangered or threatened* (Environment Canada website; *www.ec.gc.ca/EnviroRegs*). It also recognizes that Aboriginal people have an important role in conserving species at risk and that their knowledge should be considered in the assessment of which species may be at risk and in developing and implementing recovery measures. To the extent possible, recovery strategies, action plans and management plans must be prepared in cooperation with provinces, territories, Aboriginal organizations, landowners, and other affected parties. Along with the other claims-based co-management boards (Nunavut Wildlife Management Board, Gwich'in Renewable Resources Board and Sahtu Renewable Resources Board) the Inuvialuit challenged their "advisory" designation in regard to the implementation of the Act. Supported by their land claims settlements, the co-management boards have gained decision-making power in the process operated by the Committee on the Status of Endangered Wildlife in Canada (COSEWIC). All relevant species status reports are to be reviewed by the relevant co-management boards before decisions are made. The process is so new that it is impossible to judge its success, but it is anticipated that it will be the co-management board's responsibility to ensure that tek is properly used, giving the communities that are home to tek control over the source and interpretation of the information and its relationship to other knowledge. Without these changes in the power structure, COSEWIC subcommittees would have consulted the northern co-management boards or involved them along with other stakeholders, but would not likely have recognized their legal responsibilities to ensure proper linkages with traditional knowledge-holding communities. With the legal agreements in place the use of tek is recognized as both a matter of ensuring that traditional knowledge *and* traditional management systems are part of the process.

for documenting and using traditional knowledge in many aspects of resource management. In the longer term, however, it is unclear what opportunities will exist for community members and their knowledge.

> There is a mistrust/distrust/lack of trust in whether anything will be done in providing information to the planning process. There is a lack of trust in participating unless one can demonstrate that governments are bound by some agreement to do something. (R. Collier, 2003)

In addition to the trust and commitment that comes with legislated arrangements, the new management arrangements present a redistribution in the balance of power (see Box 7.2). In some cases, there are changes of authority, from regional, provincial, territorial, or federal control to equal partnerships, institutionalized joint decision-making or legislated decision-making power at the local level. Such changes of final authority are recent. They give people a voice in decision making, they provide a mix of social, political, and economic benefits (Berkes 1997), and several writers have proposed that such changes are necessary for the use of TEK (Pinkerton 1989).

New Management Institutions – TEK and Decision Making

Accomplishments achieved by the different northern initiatives suggest that for TEK to be used in management decisions, respect for different knowledge systems must prevail: a respect for the knowledge and the ways of knowing. Although this may appear to be a given, it is not, and it requires continual efforts. In deciding on the beluga whale survey protocol in the Canadian Beaufort Sea (Inuvialuit Settlement Region) or the caribou survey of the Beverly and Qamanirjuaq herds (who range into the Lutsel K'e traditional territory), lengthy discussions arose on the value of the survey results and the "best" methodology; whether the survey accounted for all important areas; whether the surveys were conducted at the right time of year; whether the parameters measured should be quantitative (estimating population size) or qualitative (time of migration, movement patterns of the animals). In developing their beluga whale management plan, the Alaska Beluga Whale Committee realized that a statewide plan could not cover the range of environmental variations and associated hunting methods. In order to better reflect the varied conditions and practices, they opted for regional management plans. As a result, the conservation objectives and the strategies to deal with the harvest strike and loss rates were more easily defined and better adapted to the local biological and cultural situations (H. Huntington, pers. comm. 2003).

These discussions are critical to developing trust in the data gathered and "better" management strategies as well as trust in their validity and applicability. Most often, traditional knowledge holders have more confidence in a methodology that includes diverse types of information, a preference for observations and management options that are based on multiple ecological or socio-ecological variables and taken at fine-grain temporal and spatial scales, providing a better understanding of regional variations in individual species and within the system as a whole (Parlee and Marlowe 2001, Kendrick 2003). Traditional knowledge holders also often insist on consulting with elders in the community, to compare their observations to knowledge gathered by many generations of people living and working on the land of all living and non-living components of the environment.

Information is provided to the management boards in both formal and informal ways. The experience of the Alaska Beluga Whale Committee indicates that when results of a study are published in the scientific literature, the voice of the hunters has more weight with researchers and managers than it would if the hunters were speaking from only their collective wisdom. Similarly, the voice of the biologists has more credibility within the communities because community

West-Side Working Group members discussing Yukon North Slope fishing locations.
L-R: J. Archie (Aklavik Hunters and Trappers Committee (HTC)), D. Gordon (Aklavik HTC), B. Ayles (FJMC),
and C. Arey (Aklavik HTC). Photo by Ed McLean, Fisheries Joint Management Committee, 2001.

members know that it was their own hunters that collected the data and the samples. People conversant in both knowledge systems are instrumental to this work. They require good listening and communication skills, objectivity on the issue, and cultural sensitivity. They are community members or individuals that commit significant amounts of time to the community and develop long-term relationships with the elders. They travel on the land and develop a sense of local identity. Over time, networks of people develop and form the necessary basis to facilitate the inclusion of different knowledge systems in resource management (R. Collier, pers. comm. 2003).

In the case of Quttinirpaaq National Park and the Canada/Inuvialuit Joint FJMC, it appears that co-management plays a significant role in fostering interactions between TEK holders and scientists, between professionals, administrators, and politicians. Co-management institutions can create a place and a space where people interact, where cross-cultural learning occurs (Kendrick 2003). Significant efforts are made at communicating, listening, and learning from all people interested or affected by the issues. A common understanding of the management objectives develops, and the gathering of additional information creates a sense of shared community (Gray 1985, Stevens 1997, Singleton 1998, Wondolleck and Yaffee 2000). The presence of a local office and the hiring of community members often allows for critical ongoing exchanges of information (Kruse *et al.* 1998), a place for formal and informal discussions. Professionals trained in Western science work closely with traditional knowledge holders, and over time understanding, respect, and trust of the 'other' knowledge system takes place (Kendrick 2003). This presence of a space for mutual learning forces people to deal with language issues, to define their respective terminology, to interpret difficult concepts, and to understand the strengths and limitations of their knowledge system.

Communities Activities – TEK for the Purpose of Resource Management

The different initiatives introduced in this chapter exemplify an extraordinary level of creativity and adaptation within Aboriginal communities. In the case of co-management arrangements such as the Fisheries Joint Management Committee (FJMC) and Quttinirpaaq National Park, communities are using their knowledge to affect resource management decisions and policies at a regional, territorial, and potentially national level. In the case of the Lutsel K'e Dene First Nation and the Innu forestry Office in Sheshatshui, Labrador, communities are attempting to develop their own community-based approaches to resource management. These offices employ between five and ten people: an administrative assistant, biologists responsible for the research and monitoring programs, permit applications, and environmental impact assessment, GIS technicians responsible for all mapping, and data management and community researchers.

In all cases, Aboriginal peoples have been able to draw and build on their own community capacities. The documentation of traditional knowledge is one activity that enables communities to build on their existing capacities while at the same time learning new skills and approaches for sharing that knowledge in a resource management context. In Lutsel K'e, for example, students are audio and video recording the stories and experiences of their elders and community members and then transcribing those stories and experiences through the use of video, information databases, and GIS technology.

Over the last two decades, communities have been able to build on their existing resource management capacities and find new ways of bringing forward their traditional knowledge in a resource management context. In many areas of the North, communities have become key players in the design and implementation of resource management projects and the decision-making process; in the case of traditional knowledge research, they control what information is collected, how it is analyzed and interpreted, where it is stored, and how it is shared. These changes have resulted in additional community capacity and enhanced ownership and confidence in resource management institutions as well as the decisions made.

> It is tied to the aspirations of the community, self governance; it is part of the fabric of the community. It is bigger than we think. It is not a single issue. Any initiative here or at the political level matters the most in the community. There is so much dependence on a system that is foreign to us in aboriginal communities. We are so cut off from our traditional resources. (Traditional Knowledge Focus Group, Hudson Bay Ocean Working Group, June 2003)

Through the different initiatives, we have learned about the inherent differences between traditional and science-based knowledge systems. Significant time is spent listening and learning, defining one's terminology, exposing and communicating different worldviews. Significant efforts are made to adapt to the other culture, but making a place for TEK in resource and environmental management often remains a challenge. Where new institutions are created to meet new regulatory requirements as in the case of Lutsel K'e in the Northwest Territories, communities have to expend considerable effort on redefining the scope of the work, discussing the often predetermined biological indicators and monitoring protocol. In cases where we are experiencing a change in the management structure and strong linkages to a central authority (federal government), as with the Canada/Inuvialuit Fisheries Joint Management Committee and Quttinirpaaq National Park, there are comparable challenges. Communities have to invest significant amounts of time and effort at conveying their views of conservation (which often greatly differ from southern views), their approaches to resource management, and their overall understanding of the ecosystems, one of the main distinctions of which is having people as part of the system (Berkes and Folke 1998).

Aside from the redefinition of terminology and the discussion of resource management approaches and worldviews, questions are often asked about the validity of the information presented by traditional knowledge holders (a lack of trust in Western science is also true from the perspective of traditional knowledge holders). Western-trained biologists, scientists, and managers often have greater confidence in TEK results that are published or in TEK results that corroborate their own views and observations. The need for measurables, verification, and sound methodology is always sought even though traditional knowledge systems reside in very different rules and principles. There is more comfort when the information provided consists of empirical observations (or data; see Roots 1998). When the information provided consists of an assemblage of observations and some interpretation, confidence in the information is reduced and people search for a methodology to document the information as abstract data units. Efforts are continuously required to explain the fundamental differences between the knowledge systems, how information is obtained, retained, and validated. Respect usually prevails until a better understanding of the information presented is attained.

> When there is conflict between sets of information, it is accepted that there is a range of opinions, there is no confrontation and people move towards a conclusion without questioning one's credibility. (H. P. Huntington 2003)

CONCLUSION

The insights of this chapter are based on the work of many dedicated individuals, groups, communities, agencies, and governments across northern Canada. As conveyed by the different initiatives, here are some key elements to success:

A process that takes place over time. From a political, economic, and knowledge perspective, the use of TEK in resource management requires long-term commitments. Agreements are often put in place to define and secure new management structures and ensure that sufficient funding is given for the development of new relationships and the establishment of new institutions. A long-term perspective is critical in documenting the knowledge base and putting in place the necessary protocols for the protection of Aboriginal and intellectual property rights, clearly stating how the information will be collected, analyzed, interpreted, archived, and used. Time is spent on the land gathering and sharing knowledge and skills; time is spent meeting with elders, seeking guidance and wisdom. New roles and capacity develop, people get to know and respect each other.

A space needs to be created for the meaningful participation of traditional knowledge holders. They need to be part of the decision-making process; testimony and documentation can only contribute to decisions if they are properly interpreted and communicated with respect to the specific issue. Such space, achieved through working on common issues, spending time on the land, and developing a common vision, allows for the meaningful interchange of information and respect for different knowledge and management systems. This leads to people becoming conversant in both knowledge systems, people who can act as translators and who are able to oversee the implementation of jointly developed management goals and objectives.

Significant investment in developing the *capacity of local communities* is also essential for the successful inclusion of TEK in resource management decision making. More than a local benefit, capacity building is based on a commitment on the part of communities and other agencies to expanding and advancing the role of TEK in resource management. While community-based approaches to resource management were never replaced by the state system, knowledge, practices, and beliefs were significantly eroded as communities of subsistence harvesters conflicted with and were overruled by government regulations. This erosion of harvesting rights continued largely unchecked until 1982, when constitutional protection was finally afforded to Aboriginal and treaty rights by way of Section 35 of the *1982 Constitution Act* (Donihee 2002). As land claims are settled in different parts of the North, opportunities must arise for communities to further document their knowledge and knowledge systems and to further advance their vision of TEK in resource management.

ACKNOWLEDGMENTS

The authors wish to thank Katherine Cumming for conducting interviews; the Wildlife, Lands and Environment Department of Lutsel K'e First Nation, Henry P. Huntington, Russell Collier, Frances Gertsch, Graham Dodds, and Larry Innes for sharing experiences and insights and for providing a wealth of ideas on the use of traditional knowledge in resources management.

REFERENCES

Acheson, J.M., J.A. Wilson, and R.S. Steneck. 1998. "Managing chaotic fisheries." In *Linking social and ecological systems: management practices and social mechanisms for building resilience*, F. Berkes and C. Folke, eds. Cambridge: Cambridge University Press, 390–413.

Adams, M., K. Frost, and L. Harwood. 1993. "Alaska and Inuvialuit Beluga Whale Committee (ABWC) – an initiative in 'at home management.'" *Arctic* 46:134–37.

Ayles, G.B., and N.B. Snow. 2002. "Canadian Beaufort Sea 2000: The environmental and social setting." *Arctic* 55: 4–17.

Bailey, J.L., N.B. Snow, A. Carpenter and L. Carpenter. 1995. "Cooperative wildlife management under the Western Arctic Inuvialuit Land Claim." In *Integrating people and wildlife a sustainable future*, edited by J. A. Bissonette and K. R. Krausman. Proceedings of the First International Wildlife Management Congress. The Wildlife Society, 11–15.

Berkes, F. 1997. "New and not-so-new directions in the use of the commons: Co-Management." *The Common Property Resource Digest* 42: 5–7.

——. 1998. "Indigenous knowledge and resource management systems in the Canadian subarctic." In *Linking social and ecological systems: management practices and social mechanisms for building resilience*, edited by F. Berkes and C. Folke. Cambridge: Cambridge University Press, 98–128.

——, and C. Folke. 1998. "Linking social and ecological systems for resilience and sustainability." In *Linking social and ecological systems: management practices and social mechanisms for building resilience*, edited by F. Berkes and C. Folke. Cambridge: Cambridge University Press, 1–25.

——, P. George, and R.J. Preston. 1991. "Co-management." *Alternatives* 18: 12–18.

——, R. Mahon, P. McConney, R. Pollnac, and R. Pomeroy. 2001. *Managing small scale fisheries*. International Development Research Centre, Ottawa.

Bielawski, E., in Collaboration with Lutsel K'e Dene First Nation. 1992. *The Desecration of Nanula Tue: Impact of the Talston Hydro Electric Development on Dene Soline*. The Royal Commission on Aboriginal Peoples, Ottawa.

Canada. 1995. *Canadian Biodiversity Strategy. Canada's response to the Convention on Biological Diversity*. Minister of Supplies and Services, Hull.

CFFS (Conservation of Flora and Fauna Secretariat). 1997. *Recommendations on the integration of two ways of knowing: traditional indigenous knowledge and scientific knowledge. http://www/grida.no/caff/inuvTEK.htm*

Day, B. 2002. "Renewable resources of the Beaufort Sea for our children: Perspectives from and Inuvialuit Elder." *Arctic* 55 (Supp.): 1–3.

Donihee, J. 2002. Returning wildlife management to local control in the Northwest Territories. Master of Laws thesis, University of Calgary.

FJMC (Fisheries Joint Management Committee). 2001. *Beaufort Sea Beluga Management Plan. http://www.fjmc.ca*

———. 2003. *Fisheries Joint Management Committee Vision: 2000 to 2010. http://www.fjmc.ca*

F. Gertsch, G. Dodds, M. Manseau and J. Amagoalik. 2004. "Recent experiences in cooperative management in Canada's northern-most national park." In *Making Ecosystem-based Management Work, Proceedings of the Fifth International Conference on Science and Management of Protected Areas,* Victoria, BC, May 2003, edited by N.W.P. Munro, J.H.M. Willison, T.B. Herman, K. Beazley, and P. Dearden. Wolfville, Nova Scotia: Science and Management of Protected Areas Association.

Gray, B. 1985. "Conditions facilitating interorganizational collaboration." *Human Relations* 38: 911–36.

Green, N., and B. Binder. 1995. "Environmental impact under the Western Arctic (Inuvialuit) Land Claim." In *Integrating people and wildlife for a sustainable future. Proceedings of the First International Wildlife Management Congress,* edited by J. A. Bissonette and K. R. Krausman, 343–45. The Wildlife Society.

Harwood, L.A., S. Innes, P. Norton, and M.C. Kingsley. 1996. "Distribution and abundance of beluga whales in the Mackenzie Estuary, southeast Beaufort Sea, and west Amundsen Gulf during July 1992." *Canadian Journal of Fisheries and Aquatic Sciences* 53: 2262–73.

Harwood, L.A., P. Norton, B. Day, and P. Hall. 2002. "The harvest of beluga whales in Canada's Western Arctic: Hunter-based monitoring of the size and composition of the catch." *Arctic* 55: 10–20.

Harwood, L.A., and T.G. Smith. 2002. "Whales of the Inuvialuit Settlement Region in Canada's Western Arctic: An overview and outlook." *Arctic* 55: 77–93.

Huntington, H.P. 1992. "The Alaska Eskimo whaling commission and other cooperative marine mammal management organizations in Alaska." *Polar Record* 28: 119–26.

———. 2000. "Using traditional ecological knowledge in science: Methods and Applications." *Ecological Applications* 10: 1270–74.

———, P.K. Brown-Schwalenberg, K.J. Frost, M.E. Fernandez-Gimenez, D.W. Norton, and D.H. Rosenberg. 2002. "Observations on the workshop as a means of improving communication between holders of traditional and scientific knowledge." *Environmental Management* 30: 778–92.

Inuit Circumpolar Conference. 1992. *Principles and elements for a comprehensive Arctic Policy.* Centre for Northern Studies and Research, McGill University, Montreal.

Iwasaki-Goodman, M. 2004. "Resource management for the next generation: Co-management of fishery resources in Western Canadian Arctic region." In *Indigenous use and management of marine resources,* edited by J. Savelle, and N. Kishigami. Senri Ethnological Studies, National Museum of Ethnology, Osaka, Japan.

Kendrick, A. 2000. *Learning conceptual diversity through caribou co-management.* The Eighth Conference of the International Association for the Study of Common Property, Bloomington, Indiana.

————. 2003. *Caribou co-management and cross-cultural knowledge sharing.* Ph.D., University of Manitoba, Winnipeg, Manitoba.

Kruse, J., D. Klein, S. Braund, L. Moorehead, and B. Simeone. 1998. "Co-management of natural resources: A comparison of two caribou management systems." *Human Organization* 57: 447–58.

LaDuke, W. 1994. "Traditional ecological knowledge and environmental futures." In Colorado Journal of International Environmental Law and Policy, *Endangered peoples; Indigenous rights and the environment.* Boulder, CO: University Press of Colorado, 126–48.

Marlowe, E., and B. Parlee. 1998. *Traditional knowledge of community health.* West Kitikmeot Slave Study Society, Yellowknife. *http://www.wkss.nt.ca*

McCay, B.J., and J.M. Acheson. 1987. *The question of the commons – the culture and ecology of communal resources.* Tucson: University of Arizona Press.

McGhee, R. 1974. *Beluga hunters: An archaeological reconstruction of the history and culture of the Mackenzie Delta Kittegarymiut.* Toronto: Canadian Museum of Civilization and University of Toronto Press.

Nadasdy, P. 2003. "Reevaluating the co-management success story." *Arctic* 56: 367–80.

Norton, P., and L.A. Harwood. 1986. *Distribution, abundance and behaviour of white whales in the Mackenzie Estuary.* Environmental Studies Research Funds Report 035, Indian and Northern Affairs Canada, Ottawa.

O'Reilly, K., and E. Eacott. 1998. *Aboriginal Peoples and Impact and Benefit Agreements: Report of a National Workshop.* Northern Minerals Programme Working Paper 7.

Parlee, B. 1998. *Community-based monitoring: a model for northern communities.* Master in Environment and Resources Studies. University of Waterloo, Waterloo, Ontario.

————. and E. Marlowe. 2001. *Community-based monitoring project: Final report.* West Kitikmeot Slave Study Society, Yellowknife. *http://www.wkss.nt.ca*

Pinkerton, E. 1989. "Introduction: Attaining better fisheries management through co-management - Prospects, problems and propositions." In *Cooperative management of local fisheries: new directions for improved management and community development,* edited by E. Pinkerton. Vancouver: UBC Press, 3–33.

Pomeroy, R.S. and F. Berkes. 1997. "Two to tango: The role of government in fisheries co-management." *Marine Policy* 21: 465–80.

Posey, D.A. 1999. *Cultural and spiritual values of biodiversity. United Nations Environment Programme.* London, UK and Nairobi, Kenya: Intermediate Technology Publications.

Roots, F. 1998. "Inclusion of different knowledge systems in research." In *Terra Borealis: Proceedings of a workshop on traditional and western scientific environmental knowledge,* edited by M. Manseau, 42–49. Labrador: Institute for Environmental Monitoring and Research.

Ruddle, K. 1994. *A guide to the literature on traditional community-based fishery management in the Asia-Pacific tropics, Fisheries Circular No. 869.* Food and agriculture organization of the United Nations, Rome, Italy.

Singleton, S. 1998. *Constructing cooperation: the evolution of institutions of co-management.* Lansing: University of Michigan Press.

Stern, G.A., and M.G. Ikonomou. 2001. *Temporal trends of organohalogen compounds in Canadian Arctic beluga*. Synopsis of Research Conducted under the 2000–2001 Northern Contaminants Program, Indian Affairs and Northern Development, Ottawa.

Stevens, S. 1997. *Conservation through cultural survival: indigenous peoples and protected areas*. Washington: Island Press.

Tippett, F. 2000. *Towards a broad-based precautionary principle in law and policy: a functional role for indigenous knowledge systems (TEK) within decision-making structures*. Master of Laws thesis. Dalhousie University, Halifax, NS.

Usher, P. 2000. "Traditional ecological knowledge in environmental assessment and management." *Arctic* 53: 183–93.

Weaver, P.A. 1991. *The 1987 Beluga (Delphinapterus leucas) harvest in the Mackenzie River Estuary, NWT*. Canadian Manuscript Report of Fisheries and Aquatic Sciences No. 2097.

Wondolleck, J.M., and S.L. Yaffee. 2000. *Making collaboration work: lessons from innovation in natural resource management*. Washington, DC: Island Press.

CHAPTER 8

UNDERSTANDING & COMMUNICATING ABOUT ECOLOGICAL CHANGE:

DENESOLINE INDICATORS OF ECOSYSTEM HEALTH

Brenda Parlee (University of Manitoba)
Micheline Manseau (Parks Canada and
University of Manitoba)
and *Lutsel K'e Dene First Nation*

INTRODUCTION AND BACKGROUND

Ecological indicators are used by many indigenous peoples to understand and communicate about ecological change (Berkes 1999; Berkes *et al.* 2000a). "They have been used for centuries to guide environmental and livelihood planning and action, long before scientific knowledge attempted to understand the processes of environmental change and development" (Mwesigye 1996, 74). Among the Cree and Inuit of western Hudson Bay, indicators are the voices of the earth that are always talking to us (Tarkiasuk 1997). For many Aboriginal peoples, physical and spiritual signs and signals that the land is healthy are very important to their own feelings of health and well-being and that of their communities. As described by a Cree man from Chissasibi, "If the land is not healthy, how can we be?" (Adelson 2000: 6).

Recent work on traditional knowledge and ecological indicators has focused on specific resource management issues such as agricultural land management, desertification, sustainability in mountain forests, and climate change (Mwesigye 1996; Berkes *et al.* 2000; Kofinas *et al.* 2002). In some cases, the research has provided direct insight into the links between environmental and human health. An emerging body of literature on First Nations health in Canada, for example, reveals how indicators of environmental decline correspond directly with many social and human health problems (Hambly 1997). While the most meaningful indicators may be those that are developed on a site-specific basis (Berkes *et al.* 2000b, 388), there are commonalities in the way indigenous peoples interpret changes in the health of their environment. For example, the percentage of body fat of birds, caribou, and other animals at harvest is one ecological

Patty Lockhart classifying fish from Stark lake, near Lutsel K'e, Northwest Territories.
Photo by Jeanette Lockhart, Lutsel K'e Dene First Nation, 2002.

health indicator which appears to be common among many indigenous groups, including the Cree of northern Quebec (Berkes 1998, 8), the Gwich'in of Alaska (Kofinas *et al.* 2002), and the Maori of southern New Zealand (Lyver 2002). Many indigenous groups in circumpolar regions use similar indicators related to ice and weather conditions to communicate about complex changes associated with global warming (Riedlinger and Berkes 2001; Krupnik and Jolly 2001). As part of the key findings, these studies have provided insight into the sophisticated knowledge systems of local land-based cultures and their capacity to learn and adapt to ecological change.

This chapter focuses on ecological indicators developed by the *Denesoline* of Lutsel K'e Dene First Nation in the Northwest Territories to understand and communicate about changes in the health of their ecosystem or the "land"; this includes changes that have taken place over a long time period, at different geographic scales and change beyond natural variation. The study of these indicators also builds on previous arguments about the capacity of land-based cultures to learn and adapt to complex ecological change and the value of traditional knowledge in resource management (Berkes *et al.* 2000a).

THE *DENESOLINE* AND THEIR COMMUNITY

Lutsel K'e, formerly called Snowdrift, is a community of 377 Chipewyan Dene (*Denesoline*) located on the east arm of Great Slave Lake in the Northwest Territories. It is the most northerly Chipewyan-speaking Dene community, situated at the treeline (62° 24′ N / 110° 44′ w). Like many other northern Dene communities, Lutsel K'e has experienced significant social and economic change over the last fifty years. Traditionally the *Denesoline* were known as the most widely travelled of the Athapaskan peoples, inhabiting a vast area from Great Slave Lake east to Hudson Bay and from the mouth of the Coppermine River near the Arctic Circle to Wallaston Lake in present day Saskatchewan (Smith 1981, 271). Although the *Denesoline* now live in a more permanent settlement on Great Slave Lake, they still retain many aspects of their traditional harvesting economy, frequently travelling in an area of over 500 square kilometres from present day Yellowknife east to the Thelon River and from Aylmer Lake to Nanacho Lake in the south. Of particular significance is the Lockhart River; its headwaters flow southeast from MacKay Lake to Artillery Lake and then to Great Slave Lake (Figure 7.1). Straddling the border between the boreal forest and the tundra, the Lockhart River and Artillery Lake area is a rich ecosystem hosting a diversity of wildlife, vegetation, and landscape features representative of six different terrestrial eco-regions (Northwest Territories Centre For Remote Sensing 1998, March 13). Negotiations are underway between Lutsel K'e Dene First Nation and the federal government to protect this area as a National Park.

DEFINING ECOLOGICAL INDICATORS

Denesoline knowledge of this ecosystem was documented during the *Preliminary Traditional Knowledge Study in the Gahcho Kué Study Region* and *The Traditional Ecological Knowledge in the Kaché Kué Study Region* (Parlee *et al.* 2001). These projects were carried out in collaboration with Lutsel K'e Dene First Nation Chief and Council, the Wildlife, Lands and Environment Committee, and an Elders' Committee. *Denesoline* elders from Lutsel K'e defined the Artillery Lake and Lockhart River as the area of interest during project scoping in 1996 and again in 1999 (Parlee *et al.* 2001; Bielawski and Lutsel K'e Dene First Nation 1992). On the recommendation of the elders being interviewed, the identification of the indicators followed the *Denesoline* harvest calendar, beginning with waterfowl in early spring (May) followed by fish (June-August), caribou (August-October), and fur-bearing animals (December-February).

Community researchers were the primary information gatherers for both projects. Additional support was attained from an academic advisory committee. The community-based research effort for these projects was involved and substantial. Data collection occurred through individual and small group semi-directed interviews with twenty-seven to fifty *Denesoline* elders and harvesters. Most interviews were audio and/or video recorded by community researchers using translators during on-the-land workshops with elders and caribou harvesters. Data were also collected on 1:250,000 and 1:50,000 scale maps and integrated into the community geographic information system. Stories shared during small group interviews and elders' meetings were also recorded through minutes.

RESULTS

Over many generations, the *Denesoline* have developed a significant body of knowledge about the Lockhart River and Artillery Lake area. Much of this knowledge has accumulated through traditional harvesting practices, including hunting, trapping, and the gathering of berries and plants for food and medicine. Over 112 species of birds, wildlife, fish, and habitats were named and defined in Chipewyan, and ecological indicators were documented for those species most commonly harvested.

Barren ground caribou (*Rangifer tarandus groenlandicus*) is the most important source of traditional food for the Lutsel K'e Dene; the movements of the Bathurst and Beverly caribou herd have been recognized as a key driver of their traditional land use patterns and social organization (Irioto 1981; Jarvenpa and Brumbach 1988). In spring and fall, the *Denesoline* also include several species of geese and ducks as part of their diet; northern pintail (*Anas acuta*), scaup (*Aythya* spp.), and white winged scoter (*Melanitta fusca*). Lake trout (*Salvelinus namaycush*), lake whitefish (*Coregonus clupeaformis*), round whitefish (*Prosppium cylindraceum*), and lake herring (*Coregonus artedi*) are also an important part of the diet in summer months, as are many berries and plants, including cranberries (*Vaccinium vitis-idaea*), blueberries (*Vaccinium uliginosum*), Labrador tea (*Ledum groenlandicum*), and spruce gum (*Picea glauca, P. mariana*). During the winter trapping season, wolverine, wolf, and fox are also harvested in the region. The indicators, or signs and signals, used by the *Denesoline* to understand and communicate about change in the health of these species revolve around four major themes: body condition (Table 8.1), species abundance and distribution (Table 8.2), quality of land and water (Table 8.3), and *Denesoline* cultural landscapes and land features (Table 8.4).

Indicators can be defined and presented in many different ways (Meadows 1998); the indicators presented here are purposely framed as questions in terms that have meaning in the community of Lutsel K'e. Framed as questions, they become more than tools for describing ecological change; they become tools for ongoing learning and communication with the elders and harvesters that hold and have ownership of this knowledge.

Body Condition

The percentage of body fat is an indicator commonly used by *Denesoline* to interpret and communicate about the health of waterfowl, fish, caribou, and fur-bearing animals; if the animal is fat then the hunter is happy (JBR 10 15 98). In a caribou workshop in 1999, elder J.B. Rabesca described how a fat caribou could be identified by a wide chest, tail hidden in hindquarter, busy set of antlers, and a well-developed coat (JBR 10 15 98). Wildlife behaviour can also be an important sign of good body condition; if the caribou is jittery it is a signal to hunters that the animal is young and the meat more tender (JBR 10 15 98). Hunters can tell if the birds are fat by their behaviour during flight; fatter birds will fly lower over the water and are slower and clumsier when taking off or landing. Harvesters evaluate the length/weight ratio of fish to determine if they are fat; if the fish is expected to be fat but is found skinny, it is considered "sick" (Parlee *et al.* 2001).

Table 8.1
DENESOLINE INDICATORS OF BODY CONDITION

Size / shape	• Is the animal of normal size and shape? • Is the weight in proportion with the length of the fish? • Are there any deformities?
Fat	• Is the animal fat? Are they skinny? • Is there some fat around their organs?
Clean Organs	• Are there cuts, marks or parasites (white spots, dark spots) in their stomach, on their liver or other organs?
Colour / Texture and Taste of Fish Flesh	• Is the flesh firm or soft? • Is the flesh tasty? Does it taste like stagnant water? • Is the trout flesh red? • Is the whitefish flesh a good white or is it brownish / greyish?

Table 8.2
DENESOLINE INDICATORS OF SPECIES ABUNDANCE, DISTRIBUTION AND DIVERSITY

Animal Population	• Are there abundant fish and wildlife of all kinds? Are there abundant fish and wildlife valued as traditional food? Have the population of these species changed from the past? Have people seen some species of fish or wildlife that are uncommon or have never been seen before? Are there some fish that you don't see anymore?

Table 8.3
DENESOLINE INDICATORS OF LAND AND WATER QUALITY

Land	• Does the land (in this place) look and smell clean? • Has there been garbage left? • Is the ground or vegetation disturbed? • Have there been any machines or vehicles there? Were there any spills or leaks of fuel or other dangerous material?
Water Levels	• Does the water look higher or lower than normal? • Are there any small streams, creeks that have dried up? • Is travel more difficult in some areas as a result of lower water levels? • Has there been damage to boats and motors as a result of hitting the bottom? • Are there portages that were very good in the past but are now too wide or long?
Water Quality	• Are there some areas where the water is no longer good to drink and the fish good to eat? • Is your drinking water tasty? Does it turn black in tea? • Are you worried about contaminants in the water? • Are you worried about the chlorine in the water?

Table 8.4

DENESOLINE INDICATORS RELATED TO CULTURAL LANDSCAPES AND LAND FEATURES

Dechen Nene *forested areas south of the* *treeline dry flatland/wet* *Marshy Land*	• Is the animal of normal size and shape? • Is the weight in proportion with the length of the fish? • Are there any deformities?
Hazu Kampa *at the treeline*	• Are the animals fat? Are they skinny? • Is there some fat around their organs?
Hazu Nene *Barrenlands*	• Are there cuts, marks or parasites (white spots, dark spots) in their stomach, on their liver or other organs?
Eda *Caribou Crossing*	• Is the flesh firm or soft? • Is the flesh tasty? Does it taste like stagnant water? • Is the trout flesh red? • Is the whitefish flesh a good white or is it brownish / greyish?
Ts'u dzaii / Ts'u dza aze *Small Stands of Trees at* *the Treeline and in the* *Barrenlands*	• Were there ever or are there now hunting or trapping camps in this place? • Was this place ever used for shelter? • Is there drywood available for fuel? • Is there clean water nearby? • Is there green wood available for tent poles?
K'a *Heights of Land with* *Erratics*	• Was this area ever used as a hunting blind for caribou?
Thai t'ath *Eskers*	• Were there ever or are there now hunting or trapping camps in this place? • Was this area ever used for shelter? • Are there any wolf, fox or bear dens?
Nikele *Dry Flatland*	• Are there cranberries growing in this area? • Are there other berries or plants growing here that might be used for traditional medicine?
Ni horelghas nene *Wet Hummocky Land*	• Are there blueberries, cloudberries or cranberries growing in this area? • Are there other berries or plants growing here that might be used for traditional medicine?

The outward appearance, including well-developed plumage, scales or coat, also indicate whether the bird, fish, or animal is healthy. Any internal injury or disease such as broken limbs, lesions, parasites, poor colour, or smell is often a sign to hunters that the animal is unhealthy. The texture as well as the colour of the fish flesh is also important; if the flesh is too soft, for example, the fish are described as "spoiled." In some cases injuries or diseases are also signals that something is wrong locally or in the broader ecosystem as in the case fish in Nanacho Lake and Stark Lake (PM 04 20 00). For example, caribou arriving from their fall migration with shorn or broken legs are signs to elders that development activities in the region, including roads and other structures, may be negatively affecting caribou.

Wildlife Abundance, Distribution and Diversity

Indicators of species abundance, distribution, and diversity are also used by the *Denesoline* to understand and communicate about ecological health (Parlee *et al.* 2001). The abundance of caribou is particularly significant. Elders nostalgically describe periods when caribou were more abundant – "there were so many caribou, it would just feel like the ground was moving" (HC 02 02 00). They also describe periods when there were very few caribou and people were very hungry (AM 06 11 97). Today, there is still tremendous joy associated with the return of the caribou and fear associated with a population decline (JBR 09 15 99). The abundance of wolves and foxes is also a sign that the land is healthy and can also signal hunters of caribou in the area. Trappers are particularly happy when animals are abundant for social and economic reasons; as with other harvesting activities, species abundance increases opportunities and success of harvesting.

Species abundance is an indicator of ecological health strongly associated with respect. If people do not respect the animals then they will not come back, give themselves, or return to the people. For example, chasing caribou is forbidden. Hunters are also careful not to be arrogant toward the ducks and geese or play with (catch and release) the fish (JBR 09 15 99; EC 06 29 99).

The *Denesoline* associate the abundance of each species with different places in the Artillery Lake/Lockhart River area. If large numbers of birds are using the same staging areas and migration routes each year, it is a sign that the birds and the land are healthy. Changes in the range and habitat of different species are often signs to elders that something has changed in the region. For example, recent increases in the number of bears around the community and moose along the treeline in the Artillery Lake area have caused anxiety and confusion. To some elders it is a sign of habitat disturbance or loss in their region and to the south (JC 01 15 01; PC 01 15 01).

Characteristics and Quality of the Land and Water

Indicators related to wildlife habitat largely revolve around the cleanliness of the land and water as the base of the food chain. The cleanliness of the water is of particular concern (PC 01 29 01; ML 09 15 99).

That the land and water is free of visible signs of waste is particularly important; leaving garbage is a significant sign of disrespect in *Denesoline* culture. Many *Denesoline* are even more concerned about the waste on the land and in the water that they cannot see, including long-range pollutants (POPs), leaks and spills from vehicles and equipment, and bacteria and disease originating from remote locations. While most people feel that their land and water is generally very clean, there is concern about certain areas where development has taken place in the past or may occur in the future. Of particular concern is the quality of drinking water from lakes and rivers. Elders often highlight the increased levels of mercury in the Talston River and Nanacho Lake caused by a 1960s hydroelectric project and the perceived contamination of Stark Lake as a result of uranium exploration in the 1950s. The current and potential effects of diamond mining activity on the health of the land and water are also a concern. One elder talking about mining in the region remarked, "soon we won't even be able to drink our water from our own lake" (PM 04 02 00 in Parlee *et al.* 2001).

Denesoline Cultural Landscapes and Land Features

Signs and signals of ecological changes can also revolve around specific places or areas of the landscape commonly used by the *Denesoline*. Places such as *eda cho* "big caribou crossing," "*desnethch'e* "where the water flows out" and *des delghai* "white water river" also refer to specific areas of ecological as well as social significance. In addition to the information they provide about the biophysical landscape, they inform us of the ecosystem as a whole and the role of the *Denesoline* within the system.

For example, the *Denesoline* associate the fall migration of the caribou with key water crossings or bifurcation points on the caribou range. Caribou movements through these crossings are signals of where and when to look for caribou during the winter months. A large number of caribou crossing at *eda cho* at Artillery Lake, for example, is a signal that the herd is likely to over-winter in the eastern part of their range. Their use of crossings from McKay to Benjamin Lake indicate the winter grounds may be further to the west. Another useful land feature are the *ts'u aze di a si* or the small stands of black spruce (krummolhz) and willow found in the valleys and along rivers on the barrens. Given the scarcity of shelter and firewood on those open landscapes, these clumps of dwarf trees are valuable, particularly during winter. Hunters also use them as campsites and meeting places to exchange information about caribou movements, numbers, and behaviour (Parlee *et al.* 2001).

Some place names reflect on the *Denesoline* culture and spirituality. One of the most important cultural and spiritual sites is *tsankui theda*, or the "old lady of the falls," located on the Lockhart River; the *Denesoline* visit the site every year to seek spiritual guidance and direction. Other sacred sites in the Artillery Lake area include Beaver Dam and Hachoghe's Shovel; the significance and origins of the landscape features are also explained in *Denesoline* legends that have been passed on through oral histories.

These place names reflect these many different social, cultural, spiritual, and ecological values as an integrated whole. *Kahdele*, for example, is more than a physical descriptor for "areas of open water in winter"; the *Denesoline* have named, used, and recognized these places for thousands of years as critical for their own well-being as well as that of many wildlife species (Parlee *et al.* 2001). Early or permanent open water on rivers, lakes, and estuaries or *askui* is valued similarly among the Innu of Labrador (Innu Nation 2001). Birds depend on those areas to feed in spring, when returning from migration. Fish benefit from the high primary productivity of these areas. Fur-bearing animals depend on the abundance of food around those areas of open water at key times of the year when prey become scarce. The potential loss of those areas of open water means more than a change in the ice or freezing pattern; it relates to changes of an entire ecosystem.

DISCUSSION

The indicators developed by the *Denesoline* have enabled them to understand and communicate about complex changes in their environment for many generations. They reflect or capture different aspects of ecological health, and they provide us with insight into the quality and condition of species of key importance to the *Denesoline*. Furthermore, the indicators also reflect on the interconnections between individual species and the "land." Similar to the concept of ecosystem, the "land," or *nene* in the Chipewyan language, reflects on all aspects of the physical as an integrated whole; the Chipewyan concept also perceives a spiritual dimension. The *Denesoline* conceptualization of the land is also based on the understanding that human beings and the environment are interconnected. An undisturbed and productive tundra landscape lends itself to a stronger and healthier caribou population; clean water is critical for healthy populations of whitefish and trout and sustainable harvesting of these species is the foundation of sustainable and healthy communities.

The indicators themselves are not necessarily different from those already in use by NGOs and in government programs such as the Arctic Borderlands Knowledge Coop, EMAN North and the Department of Fisheries and Oceans (see the DVD at the back of this volume). More important, perhaps, than their technical character, these indicators are cultural symbols that reflect how the *Denesoline* see, hear, and feel about change in their environment. In addition to marking and measuring ecological change as part of their oral history, the *Denesoline*, like other land-based peoples, have experienced those changes, their sensitivity heightened by their dependence on resources for survival. As explained by one *Denesoline* elder: "Some people who don't care so much won't notice the changes" (ML 05 11 00).

Diachronic Indicators: Reflecting Change over a Long Time Period

Denesoline legends, as well as archaeological evidence, provide clues as to the longevity of their knowledge system, including their indicators of ecosystem health. For example, *Denesoline* knowledge of caribou movements around Artillery Lake is likely five thousand years old. Elders say this area has always been good for caribou; stone lanceolates (arrowheads used for killing caribou) found in that area have been dated back to 3000 BCE (Macneish 1951; Noble 1981). Some *Denesoline* legends, including "the Old Lady of the Falls" and "How the Bear who Stole the Sun," suggest that *Denesoline* knowledge of this area may date back to the post-glacial period.

After the world was created, things were not always the same. There were ups and downs. One time, the sun disappeared. After the sun was gone, it was only winter and there was lots of snow falling. There was no sun and that is how people stayed. (Excerpt from ZC in Parlee *et al.* 2001)

All of the other Dene people followed Hachoghe who was chasing another beaver down the river. They were heading toward the east arm of Tue Nedhe. After a while, the people noticed that the woman was still back at the falls. So Hachoghe picked two healthy people to go back and look for her. They went all the way back up the Lockhart River and they found her sitting at the falls. She had been sitting there a long time and so she was stuck in the earth. The two people told her that Hachoghe was asking for her to return to Tue Nedhe. She said, "I cannot return with you. I have been sitting here too long and now I will be here for all eternity." (Excerpt from ZC in Parlee *et al.* 2001)

The exact time period in which these legends originated is not clear; the connection between such narratives and signs and signals used today to understand ecological change is not always obvious. Both legends describe significant ecological events: glaciation and changing patterns of water drainage. Other Dene legends with similar geomorphological references have been dated to about 8000 BCE (Hanks 1997, 182).

Stories about the importance of respecting animals and about the behaviour of men and women are told more as cautionary tales with very human characters. This might suggest that these stories originated more recently, or within the past several generations. Other knowledge and experience with ecological change, such as the changes that occurred as a result of the Talston hydroelectric project and the Stark Lake uranium mine (1950s–60s) developed in the very recent past.

While the oral history about events that occurred a thousand years ago are clearly less detailed than information generated in the recent past, it is useful to consider how information about critical events has been retained through time and how this information is integrated as a whole over time. The strength of *Denesoline* traditional knowledge is not in accumulating objective empirical observations or "data" about isolated events that can be compared a thousand years from now. The strength is arguably in the capacity of the *Denesoline* to interpret and use their empirical observations day after day, year after year, and decade after decade. The test, of course, is survival; for without accurate knowledge of their environment, they would have succumbed to the harsh Subarctic environment.

Scaling-Up of Denesoline Knowledge

Indicators presented here reflect an understanding of ecosystem health around the Lockhart River and Artillery Lake. However, *Denesoline* knowledge was not limited to this geographic area; the large-scale movements of the Bathurst caribou herd meant that the *Denesoline* travelled, observed, and communicated observations over large areas.

Most *Denesoline* knowledge of caribou and caribou movements reflect their vantage point on the fall and winter range of the herd. The elders' characterization of the migration cycle begins when the caribou return to the Lockhart River/Artillery Lake region in the fall and ends when the caribou leave the area in March. In contrast, Inuit elders from the Bathurst Inlet area describe the migration from the spring and summer range of the herd (Thorpe *et al.* 2001).

Effective harvesting of the caribou required an understanding of caribou movements beyond the Lockhart River/Artillery Lake area. As a result, the *Denesoline* hunting parties were known to share information about caribou movements with one another to maximize the opportunities for harvesting. Such extensive social networking was made possible in part because of the *Denesoline* predilection for widespread travel; they are recognized as the most well travelled of all the Athapaskan peoples (Smith 1981). Successful interpretation and communication about ecological events or processes that would affect their movement on the land, such as changes in water levels, ice conditions, weather patterns, or grizzly sightings would also been key to successful hunts.

Traditionally, this scaling up of knowledge was important for successful harvesting; increasingly, there are other issues that make knowledge networking important.

In the western Hudson Bay region, for example, Inuit and Cree observations of weather and sea ice conditions were linked together to provide a regional picture of climate change (McDonald *et al.* 1997). The Arctic Borderlands Knowledge Coop provides a forum for communities in the Porcupine Caribou range to share their observations and experiences around such issues as non-renewable resource development and climate change (*www.taiga.net*). A circumpolar project, 'Rangifer,' aims to pull together local knowledge with respect to caribou (*www.rangifer.com*). In all these examples, a composite picture of regional ecological change is drawn from the local observations and the knowledge of local communities.

Recognizing Change beyond Natural Variation

While these indicators of health provide a general picture of how the *Denesoline* understand and communicate about the land, they are by no means employed uniformly; they are applied using traditional knowledge of natural variation. Based on continued interactions with the land and communication over the generations, the *Denesoline* are in a favourable position to determine whether changes are related to natural variation or anthropogenic activities (McDonald *et al.* 1997). For example, female caribou arriving at the treeline in early fall are much skinnier and rougher in appearance than later in the fall because they have been nursing their calves; harvesters do not consider these animals to be unhealthy. Fish in some barren land lakes are softer and skinnier than in lakes along the treeline; however, harvesters interpret this as "normal." Other examples of this natural variation relate to the abundance and diversity of waterfowl and fish. The population of fish in the east arm of Great Slave Lake is perceived as good or greater than in the past (EC 06 29 99). However, according to the elders, the abundance and diversity of waterfowl has declined. There used to be many more ducks and geese in the past compared to today (AM 04 20 00). They suggest that the population of black ducks or white-winged scoter (*Melanitta fusca*) is much lower today than it was in the past.

The capacity to understand and communicate about change beyond natural variation is expressed in the following way: elders distinguish between natural change as *edo* and change that is perceived as unnatural – *edo aja* – which translates directly as "something has happened to it"; what is considered un-natural disturbance is generally a disturbance that the community perceives as interrupting or interfering with recognized ecological patterns relationships or cycles. Many of the interferences described as *edo aja* are anthropogenic; the environmental effects of mining, hydroelectric development and long-range contaminants are all perceived as unnatural. Ecological events or changes that have not been documented within the social memory of the community are also described in terms of *edo aja*. For example, decreasing water levels in the region are described here by elder Maurice Lockhart:

> We have been losing water but I don't know why. All the small lakes [ponds] on the barrenlands are disappearing as well as the small streams and creeks that flow between them. That is why the water is no longer healthy to drink. (ML 08 28 00)

Other elders observing erratic weather events, including unseasonably warm weather and unpredictable winds and storms, attribute the change to global warming.

> The climate is changing. The wind blows harder than it did in the past. Its different – the wind picks up quickly and changes quickly. Now I don't know what has happened.... A long time ago my sister and I traveled on the Snowdrift River to Siltaza Lake. We never saw any rocks along that river but today you can see lots of rocks [the river is shallow]. (ND 05 11 00)

Of particular concern is the increased incidence of lightning storms and forest fires in the region. Elders have said that until recently (the last five years) they had never seen a forest fire caused by lightning. (PM 11 06 00)

> Regarding the forest fires – some scientists say it's good for new growth. But do you know what the caribou eat? If the lichen burns,

it will take over 100 years for the plants to grow back. Some scientists
say the forest fires are good, but it's not like that for us. We look after
the land and we respect the land and the animals. (PM 11 06 00)

This capacity to differentiate between natural and unnatural change in their local
environment is key to understanding a variety of resource management problems.
One major problem exists in the Nanula Tue area. In the 1960s, a hydroelectric
dam developed on the Talston River, flooded Nanula Tue, which was once an
important fishing and trapping area as well important habitat for overwintering
for caribou. As a result of this activity, the Lutsel K'e Dene are no longer able to
fish, trap, or hunt in that area, and winter travel through that area has become
dangerous (Bielawski and Lutsel K'e Dene First Nation 1992). Some of the problems that are now visible are described here by elder Pierre Marlowe.

Long ago at Nanula Tue, before they built the dam there were good
fish – just like Great Slave Lake fish. Now they have a dam on the
Talston River and the fish are different. I remember before they built
the dam, I trapped around there.... When the dam was built there
– there were lots of changes. You can't eat the fish now because it's
soft and skinny. (PM 1999)

Another such problem exists in a nearby lake, once a key fishing area for *Denesoline* hunters. In 1952, however, exploration for uranium in the area resulted
in the development of a small mining operation on a peninsula of land in the
lake. Today the elders recount their concerns about the water and the fish being
spoiled as a result of this uranium exploration.

The fish in Stark Lake are a problem. Since the mine [uranium
exploration site] was put there ... the fish are different – the water
too. In another ten years, maybe we won't be able to drink the water
from our own lake. There are lots of elders who have passed away
from cancer already because of it. (PM 04 20 00)

These changes not only have implications for the long-term health of the biophysical environment; they also have profound effects on the health of the
community. People worry about what will happen to the land and their children
in the future. As in other communities that depend significantly on the land
and resources for their livelihood, these unnatural changes are the cause of
significant anxiety.

People living directly from the land and water around them are
acutely aware of indications that things are right or wrong with
the natural world.... Unnatural disruptions – for example river
impoundment and regulation, or environmental contamination
– are profoundly disturbing and give rise to deep seated anxiety and
insecurity. (Usher *et al.* 1992, 114)

Jonas Catholique and J.B. Rabesca share their knowledge on the north shore of Great Slave Lake, NWT. Photo by Jeanette Lockhart, Lutsel K'e Dene First Nation, 2002.

As Usher *et al.* (1992) point out, the traditional economy is grounded in peoples' sense of security about their ability to access an abundant natural resource base. If the security of that resource base or their access to it is compromised, or is threatened, so too is the community. Bielawski (1992), in her work on the impacts of the Talston River hydroelectric development, suggested that the greatest impact was the frustration over their inability to prevent the damage that occurred, as much as it was the impact of the damage itself.

Communicating about Ecological Change

Where indicators have meaning within a community, they can also be vehicles of cultural continuity. Such symbolic indicators are sometimes described as "community indicators" because of their meaning to a specific community or people, or "beloved indicators" (Meadows 1998). At a very basic level these indicators are cultural symbols that help convey or tell others about a given experience or observation.

The symbolic value associated with the indicators developed by the *Denesoline* is visible in their cultural narratives. For example, the importance of the Artillery Lake as a caribou crossing is well defined in stories passed on by *Denesoline* elders. Other stories describe changing water levels in different lakes or rivers (ND 05 11 00), common migration routes for ducks and geese, and dangerous areas for travel in winter. Stories about the impacts of the Talston River hydroelectric project or the Stark Lake uranium exploration site are also told and retold to ensure that current and future generations are aware of the dangers of harvesting in those areas. In some cases, stories or words are not necessary to share information. For example, information about the fatness of ducks can be conveyed through the smells and sounds of meat cooking over an open camp fire. Traditionally, hunters could gain insight about the movements of caribou across their fall range by the numbers of animals harvested at different fall camps. Insights about the abundance of fish in a given location could be gained by observing family stores of dry-fish.

Learning and Adapting to Ecological Change

Historically, the capacity of the *Denesoline* to use these indicators to learn and adapt to their changing environment has been critical to their survival. Empirical observation over a long time period is the foundation of that capacity to learn and adapt. Such observations revolve around a diversity of indicators and measures as described in this chapter. Some indicators may be quantitative, as in the abundance of caribou or whitefish, or based on qualitative perception. In Lutsel K'e, *Denesoline* hunters used the information about movements and abundance of caribou and other wildlife to make decisions about where and when to hunt in order to feed their families. They watched signs of changing weather and ice conditions to ensure safe travel while trapping for furs in the barrens. Careful inspection of the condition of animals being harvested was important in preventing illness. However, empirical observation is only the first stage of knowledge generation (Berkes 1999; Roots 1998). Critical to a discussion on the role of TK in resource management is the recognition of how observation becomes knowledge and wisdom in Aboriginal culture. Observations of one hunter or elder in a community cannot necessarily be construed as traditional knowledge; it is only after these observations are verified and interpreted along with other observations from the past and present that it may be considered to be knowledge. Traditionally, this verification and interpretation would have occurred informally through family groups. Today, elders' committees and harvester councils often fulfil that role.

The capacity of the *Denesoline* to successfully adapt to their changing environment may be based on the horizontal or non-hierarchical nature of their traditional social order (Smith 1981). Although there were some important and wise elders who exerted influence over large numbers of people from time to time, decisions about how to work together, where and when to hunt, trap, and fish were fundamentally made by individuals within small family groups. The size of camps would increase or decrease depending on the size of the family and social need as well as on the work involved in harvesting. For example, the groups associated with caribou harvesting were traditionally larger than those associated with duck hunting or fishing because of the uncertainty associated with finding caribou in the vast geography of the fall and winter range.

This non-hierarchical social order still influences how decisions are made today, including how the *Denesoline* deal with ecological change. In the case of diamond mining activity, for example, individuals representing different family groups seek to be involved at all levels of planning and management of these projects, from the act of observation and monitoring to data interpretation and analysis, site management, and policy making. Although these roles and tasks are framed very hierarchically in a government or industry setting, for the *Denesoline*, they cannot be separated from one another. This is illustrated in the following quote from J.B. Rabseca, who, in one short statement, shares his empirical observations, hypotheses about potential effects, and recommendations for managing and mitigating those effects

> I have seen the caribou around that place [the mine]. I am concerned that if the caribou start eating the food around the mine area. Anything that spills on the ground is taken up by the plants. There is muskeg in that area too. The spills will stay in that area. Someone said that they would put up a fence in that area but they haven't done anything yet. If they put a fence in that area – we wouldn't worry about the caribou. It's not good to have caribou in the mining area.
> (JBR 02 14 01)

This integrated approach demonstrated by the *Denesoline* can be a guide to building an integrated resource management approach in which land users play a fundamental role, not simply as technical assistants or stakeholders, but as decision makers with a well developed understanding of complex ecological change.

CONCLUSION

> The report that has been put together is about our culture and our way of life. The documents show how we see things.... It tells what we understand about the animals and how they behave and how we live on the land.... We are not playing around. It is not a game. What we are talking about it is very serious.... (ZC 28 06 00 MT)

The health of northern ecosystems is changing at an alarming rate; "the earth is moving faster now" (Krupnik and Jolly 2001). The current and potential effects of non-renewable resource development, the presence of POPs and other contaminants in the food chain, and the impact of climate change are causes of significant anxiety for the *Denesoline* and others who have lived off the land for many generations. Addressing these issues of ecosystem health is complex; "environmental change does not lend itself to analysis by conventional approaches" (Berkes and Folke 2002, 336). In addition to addressing tough biophysical questions, there are many complex social, economic, and cultural implications to consider. This human dimension of ecosystem change is often overlooked; the debate over climate change is one example (Riedlinger and Berkes 2001).

The indicators presented here provide useful insight into some of the complex ecological changes being observed and experienced by northern communities, however, indicators are not an end result; they are only a window into what remains a relatively untapped system of local and traditional knowledge about our changing environment.

ACKNOWLEDGMENTS

This chapter is dedicated to the community members of Lustel K'e Dene First Nation whose knowledge, generosity, and commitment have made this work possible. Particular thanks to the Band Council, Wildlife Lands and Environment Committee, Florence Catholique, Evelyn Marlowe, Wally Desjarlais, Marcel Basil, Terri Enzoe, Nancy Drybones, Bertha Catholique, Archie Catholique, Lawrence Catholique, Charlie Catholique, August Enzoe, JB Rabesca, Albert Boucher, Jeanette Lockhart, Shawn Catholique, Stan Desjarlais, and James Marlowe. Funding for the research was provided by the West Kitikmeot Slave Study Society. Additional thanks to Dr. Fikret Berkes and the University of Manitoba for their support.

REFERENCES

Adelson, N. 2000. *Being Alive Well: Health and Politics of Cree Well-being.* Toronto: University of Toronto Press.

Berkes, F. 1998. Do Resource Users Learn from Management Disasters? Indigenous Management and Social Learning in James Bay. Paper read at Crossing Boundaries – 7th Annual Conference of the International Association for the Study of Common Property, at Vancouver, British Columbia.

———. 1999. *Sacred Ecology: Traditional Ecological Knowledge and Resource Management.* Philadelphia: Taylor and Francis.

———, and C. Folke. 2002. "Back to the future: Ecosystem dynamics and local knowledge." In *Panarchy: Understanding Transformations in Human and Natural Systems,* edited by C. S. Holling. Washington, DC: Island Press.

———, J. Colding, and C. Folke. 2000a. "Rediscovery of traditional ecological knowledge as adaptive management." *Ecological Applications* 10: 1251–62.

———, J. S. Gardner, and J. Sinclair. 2000b. "Comparative aspects of mountain land resources management and sustainability: Case studies from India and Canada." *International Journal of Sustainable Development and World Ecology* 7: 375–90.

Bielawski, E., and Lutsel K'e Dene First Nation. 1992. The Desecration of Nanula Tue: Impact of the Talston Hydro Electric Project on the Dene Soline. Ottawa: Royal Commission on Aboriginal Peoples.

Hambly, H. 1997. *Grassroots Indicators for Sustainable Development.* Vol. 23:1, *Grassroots Indicator Project*: International Development Research Centre. *http://www.idrc.ca/books/reports/V231/susdev.html*

Hanks, C.C. 1997. "Ancient knowledge of ancient sites: Tracing Dene identity from the late Pleistocene and Holocene." In *At a crossroads: Archaeology and First Peoples in Canada,* edited by T. D. Andrews. Burnaby, BC: Archaeology Press.

Innu Nation. 2001. The Askui Project Symposium. Paper read at Knowledge, Culture and the Innu Landscape: A multidisciplinary symposium, at Halifax, NS.

Irioto, T. 1981. "The Chipewyan caribou hunting system." *Arctic Anthropology* 18(1): 44–56.

Jarvenpa, R., and H. J. Brumbach. 1988. "Socio-spatial organization and decision-making processes, observations from the Chipewyan." *American Anthropology* 90(3): 598–615.

Kofinas, G., Aklavik, Arctic Village, Old Crow, and Fort McPherson. 2002. "Community Contributions to Ecological Monitoring." In *The Earth is Faster Now: Indigenous Observations of Arctic Environmental Change*, edited by I. Krupnik and D. Jolly. Fairbanks, AK: ARCUS.

Krupnik, I., and D. Jolly. 2001. *The earth is moving faster now: indigenous observations of arctic environmental change.* Fairbanks, AK: ARCUS.

Lyver, P. 2002. "Use of traditional knowledge by Rakiura Maori to guide sooty shearwater harvests." *Wildlife Society Bulletin* 30(1): 29–40.

Macneish, R. S. 1951. An Archaeological Reconnaissance in the Northwest Territories. *Annual Report for 1949–50. National Museum of Canada Bulletin* 123: 24–41.

McDonald, M., L. Arragutainaq, and Z. Novalinga. 1997. "Voices from the Bay: Traditional Ecological Knowledge of Inuit and Cree in the Hudson Bay Bioregion." *Northern Perspectives* 25(1).

Meadows, D. 1998. Indicators and Information Systems for Sustainable Development: A Report to the Balton Group. Hartland: Four Corners: The Sustainability Institute.

Mwesigye, F. 1996. "Indigenous language use in grassroots environment indicators." In *Grassroots indicators for desertification: experience and perspectives from Eastern and Southern Africa*, edited by H. V. Hambly and T. Onweng Angura. Ottawa: IDRC.

Noble, W. C. 1981. "Prehistory of the Great Slave Lake and Great Bear Lake region." In *Handbook of the North American Indians: Subarctic*, edited by H. Helm. Washington, DC: Smithsonian Institute.

Northwest Territories Centre For Remote Sensing. 1998, 13 March. Terrestrial Ecoregions of the Northwest Territories. Yellowknife: Government of the Northwest Territories. *http://www.gov.nt.ca/RWED/pas/images/ecoreg1.jpg*

Parlee, B., M. Basil, and N. Drybones. 2001. Traditional Ecological Knowledge in the Kache Kue Study Region: Final Report. Yellowknife NT: West Kitikmeot Slave Study Society. *http://www.wkss.nt.ca*

———, M. Manseau, and Lutsel K'e Dene First Nation. 2005. "Using Traditional Knowledge to Adapt to Ecological Change: *Denesoline* Monitoring of Caribou Movements." *Arctic* 53: 1.

Riedlinger, D., and F. Berkes. 2001. "Contributions of traditional knowledge to understanding climate change in the Canadian Arctic." *Polar Record* 37: 315–28.

Roots, F. 1998. "Inclusion of different knowledge systems in research." In *Terra Borealis: Proceedings of a workshop on traditional and western scientific environmental knowledge*, edited by M. Manseau, 42–49. Labrador: Institute for Environmental Monitoring and Research.

Smith, J. G. 1981. "Chipewyan." In *Handbook of North America Indians – Subarctic*, edited by J. Helm. Washington: Smithsonian Institute.

Tarkiasuk, Q. 1997. "Voices from the Bay: Traditional Ecological Knowledge of Inuit and Cree in the Hudson Bay Bioregion." *Northern Perspectives* 25(1).

Thorpe, N., N. Hakongak, and S. Eyegetok. 2001. Tuktu and Nogak Project: A Caribou Chronicle. Yellowknife: West Kitikmeot Slave Study Society. *http://www.wkss.nt.ca*

Usher, P.J., P. Cobb, M. Loney, and G. Spafford. 1992. Hydro-Electric Development and the English River Anishanabe: Ontario Hydro's Past Record and Present Approaches to Treaty and Aboriginal Rights, Social Impact Assessment and Mitigation and Compensation. In *Report for Nishanawbe Aksi Nation: Grand Council Treaty #3 and Tema-Augama Anishanabai.* Ottawa: PJ Usher Consulting.

CHAPTER 9

WILDLIFE TOURISM AT THE EDGE OF CHAOS:

COMPLEX INTERACTIONS BETWEEN HUMANS AND POLAR BEARS IN CHURCHILL, MANITOBA

R. Harvey Lemelin (Lakehead University)

INTRODUCTION

Whether wildlife tourists are attracted to Point Pelee National Park to view the annual bird migrations or to the Subarctic environment of Churchill, Manitoba to witness the large, predictable aggregation of polar bears, tourist behaviours and ensuing management strategies in these instances have generally been viewed in terms of the rationalist/functionalist paradigms of scientific inquiry. Through this approach, variances deviating from the norm have been frequently dismissed as exceptions or "noise" (McKercher 1999). Consequently, researchers in the tourism field have tended to overlook key events and individuals that are often implicated in triggering major shifts in the configuration of tourism developments in an area. As a result, "our understanding of the dynamics of change in tourism has suffered" (Russell and Faulkner 2003, 220).

Complexity provides an alternative perspective enabling researchers to acquire a greater understanding of the change process (Urry 2003). In contrast to the rationalist/functionalist paradigm mentioned earlier, complexity interprets systems as being inherently unstable and dynamic (Kellert 1993). Hence, individual differences and random externalities are, from this perspective, recognized as having the potential to precipitate major realignments in systems through disequilibrium and positive feedback processes. Thus, from a complexity perspective, change is the only constant. Since tourism developments along the Hudson Bay coastline in northeastern Manitoba can be associated with the actions of individual actors and key events, complexity can, therefore, provide a useful framework for understanding the integrated management of coastal zones and analyzing the creation and growth of wildlife tourism in this area.

A historical case analysis composed of a literature review, oral histories, on-site observations, and interviews was used to understand the events and patterns shaping the development of the wildlife tourism industry in Churchill, Manitoba. The selection of interview candidates associated with or involved with Churchill's polar bear viewing industry was based on a literature review and the author's extensive knowledge of the wildlife tourism industry in this area. In total, seven in-depth interviews were conducted with stakeholders involved in Churchill's polar bear industry. This included individuals associated with management agencies (federal and provincial), people currently employed or previously associated with the polar bear viewing industry, and long-term members of the community.

The chapter begins with an overview of the prominent research approach in tourism. Concepts of complexity are then outlined to provide a framework for the analysis that follows. The Churchill case study, particularly the role of key events (*i.e.,* polar bear alert programme) and stakeholders (*e.g.,* local entrepreneurs, non-local proponents) in the development of this tourism destination is then examined for illustrative purposes. Following is a re-examination of Russell and Faulkner's (2003) model of "contrasting inclinations of chaos-makers and regulators" and the relevance of this model to the Churchill context.

COMPLEXITY IN TOURISM

Past multidisciplinary approaches to tourism have meant that research in this area has been enriched by a variety of theoretical perspectives (Echtner and Jamal 1997, Hall 1995); however, it has also resulted in the transmission of a functionalist approach. An approach largely founded upon reductionism, and positivism or post-positivism, which has dominated most of the social science discourse in the twentieth century (Byrne 1998; Scoones 1999). On the other hand, models or theories describing the fragmentary and complex nature of tourism have been relatively ignored until the recent appearance of crisis management in tourism (Beirman 2003) and complexity in tourism (Faulkner and Russell 1997; Hall 1995; McKercher 1999; Urry 2003). The functionalist approach has served the tourism field relatively well in terms of advancing knowledge, in particular, conventional approaches to tourism research attuned to the analysis of relatively stable systems. However, it has also created omissions that have resulted in large gaps of understanding, *i.e.,* the dynamics of change in tourism (Russell and Faulkner 2003).

Appreciating that systems are inherently non-linear, dynamic, and inherently unstable, the study of complexity yields qualitative predictions where detailed quantitative predictions are impossible (Kellert 1993). Instead of discarding "anomalies" such as "noise" and/or "external stressors," complexity seeks to provide a framework from which human interactions with the environment can be best understood (Waldrop 1992). Of particular relevance to this investigation is the proposition that individual differences and random externalities are the catalysts for variety and adaptation (Russell and Faulkner 2003). Several concepts associated with complexity theory that are germane to this examination of the development of a tourism industry in Churchill, Manitoba are described next.

Polar bears, near Churchill, Manitoba. Photo by Doug Clark.

In the steady state perspective of functionalism, small system changes in the initial state are reflected in equally small shifts overall as any disturbances are improved by a negative feedback process that restores equilibrium within the system (Byrne 1998). In systems where positive feedback and non-linear relationships are more prevalent:

> [T]he effects of even minor random disturbances are accentuated and magnified by mutually reinforcing positive feedback loops. In tourism destinations, therefore, it is feasible for a single initiative, be it a single entrepreneur, a new technological innovation, or an event, [to] create a major shift in the evolution of a tourist destination. The changes they introduce can reverberate throughout the destination in question as other enterprises within the system adjust and find new niches, resulting in a new configuration of linkages. (Russell and Faulkner 2003, 226)

This phenomenon, also known as the 'butterfly effect' (Gleick 1987), occurs when small and apparently insignificant changes or perturbations precipitate a chain reaction culminating in a fundamental shift within system structures (Byrne 1998). This often results in a system on the virtual edge of chaos, perpetually changing and adapting to new stimulus.

Often described as a delicate balance between order and chaos, the edge of chaos is essentially a place "where the components of a system never quite lock into place, and yet never quite dissolve into turbulence either" (Waldrop 1992, 12), a place where new ideas are forever nibbling away at the edges of the status quo (Gell-Mann 1994). The challenge in healthy, complex adaptive systems

(CAS) is to keep order and chaos in balance by regulating themselves through a web of feedback and regulation, while at the same time leaving "plenty of room for creativity, change, and response to new conditions" (Waldrop 1992, 294). Hence, change and adaptation are essential components of CAS. Indeed, it is essentially meaningless to talk about the equilibrium of CAS since these systems are in perpetual motion, always unfolding and always in transition (Waldrop 1992). In fact, if a system does reach equilibrium, it isn't just stable, it's dead! (Waldrop 1992)

> And by the same token, there's no point in imagining that the agents in the system can ever optimize their fitness, or their utility, or whatever. The space of possibilities is too vast; they have no practical way of finding the optimum. The most they can ever do is to change and improve themselves relative to what the other agents are doing. In short, CAS are characterized by perpetual novelty. (Waldrop 1992, 147)

Paradoxically, while some systems may appear in a constant state of flux, feedback combinations can also be responsible for creating inertia, or what is also known as temporary system equilibrium (McKercher 1999). Temporary equilibrium in a tourism destination's evolution is a period of relative stability, where systems remain relatively unchanged for long periods of time. This is often reflected in a corresponding stability of visitor profiles. Changes are introduced in the system through internal changes (*e.g.*, competition) and through such 'non-tourism related externalities' (NTRE). NTREs are random external stressors such as fluctuating international market demands, social crises (*e.g.*, terrorism, war, epidemics), gradual or sudden changes in the natural environment compromising the sustainability of tourism development, non-local technological innovations creating new opportunities in the marketplace, and initiatives elsewhere affecting the competitiveness of the destination (Faulkner 2003). NTRE may occur without warning and plunge previously stable tourism destinations into a state of rapid change or decline (McKercher 1999).

According to Byrne (1998) and Scoones (1999), punctuated equilibria and temporary equilibria or system changes have not been well understood by most social scientists, managers, and policy makers espousing the virtues of static, prescriptive models of sustainable and economic growth, carrying capacity, and other social constructions founded upon equilibrium and stability. Indeed, a manager's responsibility is to minimize change, or at least provide the appropriate parameters to understand this change, in order to produce a degree of continuity that is consistent with existing management plans or policies and (in some cases) community consensus. On the other hand, entrepreneurs seek the flexibility that is required to respond to new threats and opportunities within their environment (*i.e.*, NTRE). Thus, managers/regulators seek to preserve a "functional" regime of equilibrium and linear change, whereas the

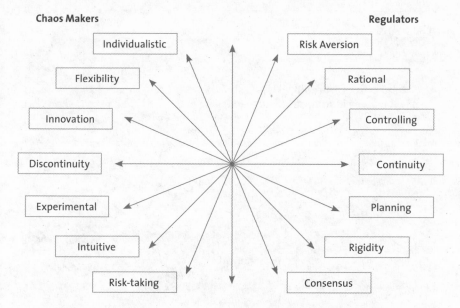

Chaos Makers

Regulators

Individualistic

Risk Aversion

Flexibility

Rational

Innovation

Controlling

Discontinuity

Continuity

Experimental

Planning

Intuitive

Rigidity

Risk-taking

Consensus

Figure 9.1 Contrasting inclinations of chaos-makers and regulators. (Russell and Faulkner 2003)

entrepreneurs/chaos-makers are associated with the disequilibrium, non-linearity, and spontaneity which are more symptomatic of complexity theory (Faulkner and Russell, 1997). Tourism evolution in a specific destination can then be construed in terms of ongoing tensions between entrepreneurs, often described as agents of change, and managers whose role is to moderate change or at least control these changes within certain acceptable parameters. The contrasting styles of entrepreneur (chaos-makers) and planners (regulators) are summarized in Figure 9.1. While the importance of the entrepreneurial and managerial relationship in tourism cannot be overstated, other factors such as non-local actors and social-ecological transformations should not be ignored. These factors are integrated into this analysis.

Entrepreneurial Activities and Management Strategies in the Cape Churchill Wildlife Management Area and Wapusk National Park

The local economy of Churchill, Manitoba (Figure 9.2) has been historically dependent on its strategic location in Canada's North and on wildlife tourism. As one of the earliest European settlements in North America, Churchill has witnessed dramatic and disruptive economic fluctuations over the course of its history as explosive boom-or-bust cycles repeatedly swept through the region. By the late 1970s, Fort Churchill, a military base located near the community of Churchill, had closed its doors. By the late 1990s, the economy of this small, Subarctic community of a thousand residents was well diversified, ranging from transportation services (Port of Churchill, Northern Transport Canada

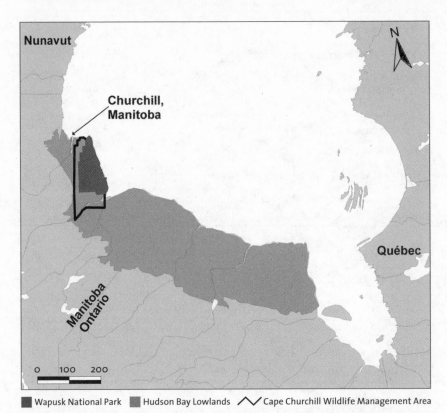

■ Wapusk National Park   ■ Hudson Bay Lowlands   ∧∨ Cape Churchill Wildlife Management Area

Figure 9.2 Location of Churchill, the cwma and Wapusk National Park in Northeastern Manitoba.

Limited, Omnitrax, and Calm Air) to health services provided by the Regional Health Authority and the Northern First Nations Transient Centre and various tourism initiatives.

The abundance of lakes, rivers, forests, and tundra, coupled with the long-standing tradition of wilderness outfitters, lodges, and other leisure facilities, has provided a firm foundation for Churchill's tourism industry. By the 1960s, Churchill was becoming a popular birding destination; however, it was not until a decade later that some small-scale polar bear outings were offered in the region (Interview with management stakeholder 2003). Through the help of various existing industries, such as hunting, fishing, birding, whalewatching, aurora borealis gazing, and polar bear viewing activities, Churchill's tourism industry continued to grow and diversify throughout the late twentieth century. The economic impact of nature tourism in 2002 was estimated at well over $3 million (Interview with community stakeholder 2003). One of the most important components of Churchill's wildlife tourism industry is polar bear viewing.

Churchill became an international tourism destination because polar bears congregate along the shores of the Hudson Bay to await the formation of sea ice in early to mid-November. It is this unique natural phenomenon that attracts thousands of wildlife tourists each year to this region. The ability of tundra vehicles to traverse the Subarctic environment provides the ideal mode of transportation to see and photograph this large, attractive, and relatively predictable predator, in relative safety and comfort. The viewing of polar bears in this area, however, is further facilitated by the habituation and tolerance of these animals to human presence, their curiosity, and their propensity to entertain wildlife viewers (Interview with community representative 2003). Without the tourists' demand to see polar bears in their natural environment, and subsequently, the polar bear's predictability, there would be no polar bear viewing industry in this community. It thus serves us well to remember that it is our responsibility to protect the polar bears and their environment (Interview with community representative 2003).

A majority of the bear viewing in Churchill actually occurs in two protected areas located twenty-one kilometres east of the community. Created in 1978 and divided into eight zones totalling over 848,813 hectares, the Churchill Wildlife Management Area (CWMA) (formerly known as the Cape Churchill Wildlife Management Area) was established to protect polar bear staging and denning areas, nesting grounds for geese, and habitat for caribou (Teillet 1988). Of particular interest to this study is Gordon Point (115 square kilometres), where a majority of the polar bear viewing occurs within this particular protected area. Even with the creation of Wapusk National Park (WNP) in 1997, and the transfer of 1.14 million hectares of the CWMA from Manitoba to Canada, the CWMA remains, at 848,813 hectares, the largest and most northerly Wildlife Management Area (WMA) in Manitoba (Manitoba Conservation 1999). Due to its relatively late arrival on the management scene, Parks Canada was rarely mentioned during the interviews. This was primarily due to the fact that WNP, more specifically the Cape Churchill polar bear viewing area, was at one time part of the CWMA. Hence, the focus of this study pertains mostly to the CWMA.

THE STUDY

A literature review, combined with extensive discussions with stakeholders involved in Churchill's polar bear viewing industry, encouraged the development of the following research question: How do stakeholders involved in the polar bear viewing industry perceive the transformations of polar bear management strategies in the CWMA?

The relatively late arrival of other federal agencies: Parks Canada, the federal agency mandated with the management of WNP, and Fisheries and Oceans (the federal agency mandated with implementing the integrated coastal zone management plans of the *Oceans Act* (1997); see ss. 31 and 32) in this area minimizes, from the interviewee's perspective, the role of these agencies in northeastern Manitoba. Hence, the majority of the discussion will pertain to the roles of the Manitoba Conservation and the Canadian Wildlife Service (CWS). The latter is an agency which manages wildlife for the federal government through Environment Canada. Heuristic inquiry, described in the following paragraphs, was used to explore the research question.

With strong roots in humanistic psychology (Maslow 1954; Rogers 1967), heuristic inquiry epitomizes the phenomenological emphasis on meanings and knowing through personal experiences; it exemplifies and places at the forefront the way in which the researcher is the primary conduit in qualitative inquiry: "... it develops a sense of connectedness between the research and research participants in the mutual efforts to elucidate the nature, meaning, and essence of a significant human experience; and, it challenges the extreme traditional scientific concerns about researcher objectivity and detachment" (Patton 1990, 73). Simply defined, heuristic inquiry is the combination of personal experi-

Polar bear gazing into tundra vehicle, Churchill, Manitoba. Photo by Harvey Lemelin.

ence and intensity that yields an understanding of the essence of the phenomenon (Patton 1990). From this perspective, interpretive sciences or, more specifically, heuristic inquiry can be used to understand the relationships between operators and management agencies. However, in order to understand the stakeholders' perceptions of the development of this industry:

> [One] must understand people-environment relationships (*i.e.*, the exchange of information, energy, goods and services between people, land management organisations and the environment. One must also understand the particular historical political and social context of these exchanges and relationships. (Stynes and Stokowski 1999, 443)

The process and the selection of the stakeholders associated to the bear viewing industry was based on the researcher's knowledge and contacts with the polar bear viewing industry. Stakeholders involved in the polar bear viewing industry were divided into three main categories:

1 *Management agencies*: Manitoba Conservation (formerly known as the Manitoba Department of Natural Resources), a provincial agency mandated to protect and manage wildlife resources in the CWMA; the Canadian Wildlife Service, a federal research agency conducting research on the Churchill polar bear population; and Parks Canada, the federal agency

responsible for the management of Wapusk National Park (WNP) and several national historic sites around the Churchill region.

II *Industry representatives*: Local entrepreneurs directly involved with the industry (*e.g.,* Great White Bear Tours, and Tundra Buggies), tour operators (Natural Habitat, Frontiers North, Travel Wild, Churchill Nature Tours), helicopter companies, and transportation services.

III *Other representatives*: Accommodations (hotels, bed and breakfast), restaurants, shops, the town of Churchill, Churchill Chamber of Commerce, not-for-profit environmental organizations (*e.g.,* Polar Bears International, Born Free), independent researchers, photographers, film crews, and the media (news crews), directly or indirectly involved in the industry.

Analyzing the roles of all these stakeholders in Churchill's polar bear viewing industry would be truly representative of complexity research. Unfortunately, such research is beyond the scope of the present study. Therefore, selected candidates for this study represented one or two affiliations with the aforementioned groups. In general terms, two interviewees represented management agencies, three interviewees represented industry interest, and two interviewees represented the other category. The focus of this study was on the roles and relations between the Manitoba Conservation and the Canadian Wildlife Service, the local entrepreneurs directly involved in the tundra vehicle industry (Tundra Buggies, and Great White Bear Tours), and the role of nature/wildlife photographers, film crews, and the media during the embryonic stages of this industry.

Data collection and analysis were based on the following: (1) seven in-depth interviews with stakeholders, representing a broad array of experiences; (2) attendance of public meetings related to the tourism industry in Churchill, Manitoba, and (3) examination of documents related to the wildlife tourism industry. Interviews were primarily conducted over the telephone. In addition, follow-up interviews for further clarification were also conducted. Participants had full access to the taped interviews throughout the editing process, and any concern, input, or retraction that the interviewees deemed necessary was respected and complied with by the researcher. Silverman (1993) indicated that this 'respondent validation' is particularly appropriate within community research. The intention of information accessibility was to promote a dialogue between the participants and the researcher. Following its completion, all records, along with a copy of the article, were returned to the participants. The methodology was conducted within the framework of an interpretive paradigm (Denzin and Lincoln, 1998). The focus on an interpretive approach in this research incorporated the everyday, lived experience of participants in polar bear viewing, as well as perceptions of prominent stakeholders involved in the industry. These experiences are discussed next.

The closure of York Factory and the military's (American and Canadian armed forces) withdrawal from the Churchill area in the mid-to-late twentieth century decreased human activity along the west coast of the Hudson Bay, resulting in declining human disturbance of polar bears in the area, and indirectly contributing to the growth of the polar bear population in the region (Interview with community representative 2003). As was the case in other locations (*e.g.,* Yellowstone National Park), bear population growth often resulted in increased wildlife-human encounters, which in turn stimulated wildlife management needs (Schullery 1992). "Wildlife management" in the region that had been previously overseen by the military, at times with dreadful impacts on the polar bear population (Interview with community representative 2003), was taken over by the Manitoba Conservation and assisted by the Canadian Wildlife Service. Ironically, the effort of these two agencies soon garnered attention from such environmental organizations as the *International Federation for Animal Welfare* (IFAW) and the *Smithsonian Institute* (Interview with management representative 2003). As explained below, the reverberations from these two organizations, in combination with other factors, would be felt for years to come in both polar bear management sectors and in the subsequent emergence of a new wildlife tourism industry.

In 1971, IFAW decided to support the "Operation Bear-Lift" program, which removed "problem" bears from the region by transporting them further up the Hudson Bay coastline away from populated areas (IFAW website 2003). While IFAW is no longer associated with the polar bear removal program, the success of this initiative is evident as the province of Manitoba continues to use non-lethal bear control policies to help minimize human-polar bear interactions in the region, under the *Polar Bear Alert* program (Interview with management representative 2003). In the mid-1970s, increasing human-polar bear interactions near the community, especially at the Churchill dump, resulted in the Smithsonian Institute sending one of its reporters/photographers (Dan Guravich) to cover the issue (IWA website 2003). The significance of this event cannot be understated. Although it was the Churchill dump which brought Guravich to the area, it was the unplanned polar bear 'outing' provided by a local entrepreneur by the name of Len Smith, and the eventual friendship between the two, that resulted in "the dramatic transformation and subsequent growth of wildlife tourism in the area as we know it today" (Interview with industry representative 2003). The eventual friendship between Len Smith and Dan Guravich resulted in "the dramatic transformation and subsequent growth of wildlife tourism in the area as we know it today" (Interview with industry representative 2003). This growth was related to two factors, the "creation" of tundra vehicles, and the awareness of Churchill and its polar bears brought about through media coverage.

The brainchild of the Smith and Guravich partnership was the construction of tundra vehicles, a technological innovation that replaced the smaller and crowded track vehicles used at the time to take wildlife tourists out along the

Hudson Bay coast with larger, safer, much more comfortable and mobile modes of transportation (Interview with industry representative 2003). This mode of transportation would also facilitate the creation of the first "Muktuk Salon," a semi-permanent lodge located in the CWMA, providing wildlife tourists with twenty-four-hour access to polar bears (Interview with industry representative 2003). While these vehicles were crucial to providing access to view the bears, all interviewees agreed that the prominence of Churchill as an international tourism destination was primarily due to extensive media coverage that the polar bear viewing industry received during its embryonic stage.

Examples of this early media coverage included articles appearing in such prominent magazines as: *Smithsonian Magazine*, 1978 (see David 1978); *Canadian Geographic*, 1984 (see Braummer 1984); and various documentaries: *24 Hours: Polar Bears*, 1980 (see CBC 1980); the *Polar Bear Alert*, 1982 (see National Geographic 1982); and *Heartland* 1987 (see Imax Corporation 1987). International media attention, combined with contributions from various professional photographers (*e.g.*, Robert Taylor, Joseph Van Os, and Dan Guravich), generated a tremendous amount of early interest in the location. The interest subsequently grew into inquiries as to the possibilities of personally witnessing this spectacular and unique natural phenomenon. It was another photographer by the name of Joseph Van Os who capitalized on this growing demand by non-photographers by creating one of the first tour packages to the area (Interview with industry representative 2003).

The earliest beginnings of the tourism industry in Churchill were also supported by the province of Manitoba through various operation grants provided by *Tourism Manitoba* for the construction of some tundra vehicles by Len Smith (Interview with management representative 2003). The creation of the Cape Churchill Wildlife Management Area was yet another illustration of this mutual relationship. By selecting a provincially designated protected area such as a wildlife management area over other provincial and federal protected area designations (*e.g.*, provincial or national parks), the province of Manitoba could continue to promote the industry and accommodate the research being conducted by the CWS, while also curtailing some of the possible negative impacts of the industry on the ecosystem (*e.g.*, number of tundra vehicles) (Interview with management representative 2003). This was accomplished through the creation of the Cape Churchill Wildlife Management, the establishment of a permit system for tundra vehicles (set at 18 vehicles in 2003), and various other guidelines pertaining to the industry (*e.g.*, use of existing trails by tundra vehicles) (Interview with management representative 2003). Lastly, the most revealing aspect of this collaboration was the process of identifying bears by Manitoba Conservation and the CWS. As the photography of polar bears became more frequent, so did the displeasure over the practice of painting large numbers on polar bears for identification by the management agencies. The Canadian Wildlife Service responded by painting smaller numbers on the upper shoulders of the bears, and eventually ceased the practice by turning to cream-coloured ear tags, which were nearly invisible to most polar bear viewers (Interview with management representative 2003).

The co-operative relationship between entrepreneurs and the management agencies was evident through various initiatives, including an environmental rehabilitation project conducted by one operator in the summer of 2000 (Interview with community representative 2003), proposed regulations changes to the CWMA itself (see Manitoba Conservation 1999 – Draft Wildlife Management Plan for the CCMWA), operator certification (see Manitoba Government2002 – the Manitoba Resources Tourism Operators Act), and increased polar bear protection in the province (see Manitoba Government 2003 – amendment to the Polar Bear Protection Act). These recent initiatives indicate that the co-operative relationship between entrepreneurs and the province of Manitoba's management agencies, in this case Manitoba Conservation, is ongoing. Parks Canada, the agency currently responsible for the management of Wapusk National Park, recently released a management plan which will address anthropogenic use in the park (Interview with community representative 2003).

While disagreements between managers and entrepreneurs have flared throughout the past four decades, these have been relatively minor, and usually quickly addressed (Interview with industry representative 2003). However, such a co-operative relationship was not so evident between some wildlife photographers and the scientific community, with the former seeking greater access into the area, criticizing polar bear research, and often promoting such controversial tactics as the baiting of polar bears. The scientific community, on the other hand, often resisted the increasing number of non-researchers into the area for fears of impacts on research sites and on the animals (Interview with community representative 2003). In many cases, it was Manitoba Conservation that mediated and defused the tension between these two groups (Interview with management representative 2003). Today, much of the tension between researchers and photographers has dissipated. In fact, in order to subsidize research costs, some wildlife photographers and media crews have accompanied CWS researchers during their polar bear research.

Through innovations that have precipitated fundamental shifts, the path of tourism development in the Churchill region has been linked with key events and individuals who have had a significant impact. In the beginning, it was Alan and Bonny Chartier (Churchill Wilderness Encounters), who conducted the first birding tours and other forms of wildlife tourism in the area (Interview with community representative 2003). However, the region's coming-of-age as a tourist destination was in the 1970s and 1980s, when a procession of remarkable individuals including Dwight Allan and Louise Foubert, Len and Beverly Smith, and Donny and Marilyn Walkoski focused their attention largely on the polar bear viewing aspect of the industry (Interview with community representative 2003).

Prior to the development of tundra vehicles and seasonal camps set up in the coastal area, wildlife tourists visiting Churchill had relatively little access to remote areas where polar bear aggregation occurred. At the dawn of the twenty-first century, over three thousand polar bear enthusiasts can now view bears from one of eighteen tundra vehicles, two tundra vehicle camps (operated by Tundra Buggies and Great White Bear Tours) and two helicopter companies permitted into the CWMA and WNP. Furthermore, extensive scientific studies (Ramsay and Stirling 1982, 1990; Stirling *et al.* 1993), and media coverage from magazines (*Canadian Geographic*, 1997 [see Comeau 1997]; *National Geographic*, 1998 [see Eliot 1998]), to documentaries (*Bears*, 2001 [see National Wildlife Federation 2001]) and videos (*Polar Bear*, 1997 [see BBC 1997]) continue to promote the region as a popular destination for researchers, photographers, media crews, and wildlife tourists.

The consolidation of the industry in the mid-1990s by two companies (Len and Beverly Smith, owner and operator of Tundra Buggies and Donny and Marilyn Walkoski, owners and operators of Great White Bear Tours) did not diminish the growth of wildlife tourism in the area. In fact, the consolidation era led to larger vehicles accommodating up to forty-five people, smaller vehicles accommodating photographers, film crews, and researchers in the CWMA and WNP, and various other bear viewing enterprises occurring outside of both protected areas (*e.g.*, White Whale Lodge, GreatWatch'ee Lodge) (Interview with management representative 2003).

From a complexity perspective, it would appear that tourism in this area has therefore reached a temporary equilibrium point whereby shifts in consumer demand require only minor adjustment in marketing strategies in order to maintain levels of business. All of these developments can be, to varying degrees, attributable to the butterfly effect of a diminishing human presence along the Hudson Bay coastline in the mid-to-late twentieth century, the growth and diversification of global tourism opportunities, the introduction of one technological innovation – the tundra vehicle – and the role of chaos makers.

CHAOS MAKERS IN CHURCHILL

Much of Russell and Faulkner's (2003) analysis of entrepreneur and manager relationships in the Australian gold coast area were based on a polarized, dichotomous relationship, with the former, according to the authors, acting as "new guards" or agents of change and the latter representing "old guards" or defenders of the status quo. This basic assumption, in addition to two oversights – the attraction (be it a landscape, flora and fauna, or a combination of all three) and the role of external agents in Churchill – are discussed next.

Notwithstanding the contributions of various other stakeholders in Churchill's tourism industry, including the hospitality and transportation sectors and local government, which were not mentioned in this study, the focus of this article was dedicated to highlighting the relationship between entrepreneurs and managers (*i.e.*, Manitoba Conservation). While it is certainly true that tensions and discords were detected between the two, especially between wildlife photographers associated with wildlife tourism and the scientific community associated with management agencies, it was Manitoba Conservation that often mediated and diffused these situations. Thus, the rapport between the management and entrepreneur stakeholders was at best co-operative (*i.e.*, through Travel Manitoba grants), and at worse ambivalent (*e.g.*, regarding the relatively benign presence of Manitoba Conservation in the Churchill Wildlife Management Area) (Interview with industry representative 2003).

Churchill's success was due to the aforementioned "co-operative tension" between entrepreneurs and managers in combination with external proponents involved in the polar bear viewing industry in Churchill. The role of external chaos makers was not discussed in Russell and Faulkner's (1999 and 2003) analysis. Yet, without the influence of early nature photographers such as Dan Guravich, Joseph Van Os, and Robert Taylor, and the international recognition derived from their photography of the area, one wonders if the polar bear viewing industry in Churchill would have developed as quickly as it did. (Interview with industry representative 2003). Furthermore, the role of these tourism ambassadors to Churchill cannot be overstated, since one individual in particular, Dan Guravich, encouraged the development of two technological innovations – the construction of tundra vehicles and the first tundra lodge – while another, Joseph Van Os, created the first tour packages to the area (Interview with industry representative 2003). Therefore, it appears that in the context of Churchill, both entrepreneurs and management agencies embodied both functionalist and complex characteristics, and benefited from the changes that occurred (*i.e.*, the appearance of new tourism industry) and the relative stability or punctuated equilibrium of this industry throughout the past twenty-five years. The only true agents of change, or "chaos makers," in this scenario, using the metaphor of Russell and Faulkner (2003), were reporters, photographers, and film crews. Bound by neither policy mandates nor profit motives that often directed the activities of managers and entrepreneurs, this communication medium, so beneficial to Churchill's wildlife tourism industry during its embryonic stages, is now contributing to the appearance of NTRE, as in, for example, its coverage of the role of climate change and its potential impacts on polar bears (see CBC 1999; Linden 2000; Hawaleshka 2003). The challenge created by this coverage is that some individuals have associated such negative impacts as declining weight and falling birth rate in the Churchill polar bear population to the industry (Interview with community representative 2003). What these reporters and journalists fail to realize is that many of the challenges facing the polar bears are not of local origin; they are, in fact, international issues (*i.e.*, global warming, the effects of biomagnification or toxin accumulation in polar bears) (Interview with community representative 2003).

Often sensationalizing or contriving polar bear images or activities, these agents of chaos have also, at times, created unrealistic expectations in polar bear viewers. This "sensationalistic approach to wildlife" is due, in part, to photographers and film crews who use or hire animals in enclosures or on wildlife ranches to create an illusion of real "wildlife" in its natural habitat (Schullery 1992; Wilson 1991). Many of these types of images have made Churchill famous, for example the aggressive and predatory nature of polar bears, the cute and adorable "teddy bear," or the supposedly "natural" interactions between polar bear and husky dogs (see Comeau 1997). What wasn't disclosed in the production of some of these pictures and film footages is how they were manufactured, or that they violated existing management guidelines and industry protocols (Interview with community representative 2003). Nor is there ever any mention of how long these productions took, or how much editing went into making that one-hour montage (Interview with community representative 2003). Much like Schullery's (1992) observations of visitors to Yellowstone, casual discussions with wildlife tourists in Churchill reveal that media influences were deeply ingrained:

> I think that a lot of these programs reinforce in us some old tendencies. They make us feel that we don't have a good photograph unless we "fill the frame" – get close enough so that the whole picture is loaded with an animal. They make use expect more or less constant action out there, when in reality most animals spend almost all their time doing stuff that, when we do it, we consider fairly uninteresting. (Schullery 1992, 249)

It should be noted, however, that many of Churchill's polar bear photographs have been produced by wildlife photographers who have used ethical and sound practices, and the same can be said for some media news groups. Also, while it is true that some individuals travel to Churchill solely to capture a polar bear on film, there are thousands of others who travel to the region to learn about and witness the majesty of polar bears in their natural environment.

Ironically, neither climate change nor contrived polar bear footage have had as much impact on the industry (so far) as the media coverage of terrorism attacks of September 11, 2001 in New York City and Washington, DC, the U.S.-led coalition's war on Iraq in 2003, and the recent outbreak of Severe Acute Respiratory Syndrome (SARS) in Toronto, Ontario (2003). The coverage of these NTRE has occurred rapidly throughout the past three years and with little warning, and while the industry has proven somewhat resilient to these stressors, there has still been nevertheless a decrease in visitation numbers in both Canada (see Canadian Tourism Commission 2003) and Churchill in the last few years (Interview with community representative 2003). What we could be witnessing in this particular instance is Gleick's (1987) "butterfly effect," where remote and apparently unrelated changes in one medium, in this instance, the realm of information dissemination, is beginning to reverberate deep within Canada's national and regional tourism industry, possibly shifting tourism one step closer to the edge of chaos.

It appears that many of the attributes associated with chaos makers and regulators in Russell and Faulkner's (2003) model are attributable to both entrepreneurs and managers in Churchill, while most of the characteristics associated with chaos makers are specifically relevant to non-local agents

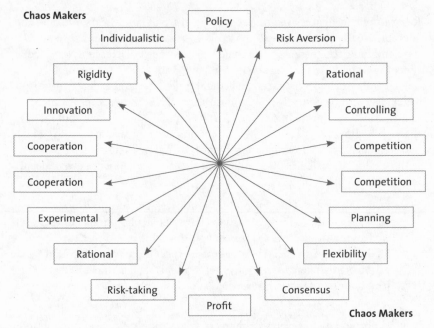

Chaos Makers

Policy

Individualistic

Risk Aversion

Rigidity

Rational

Innovation

Controlling

Cooperation

Competition

Cooperation

Competition

Experimental

Planning

Rational

Flexibility

Risk-taking

Consensus

Profit

Chaos Makers

Figure 9.3 Chaos-makers and analogous characteristics of entrepreneurs and management agencies in Churchill, Manitoba.

(*e.g.*, photographers, media). Therefore, a refinement of the model for the Churchill context has been developed for this study (see Figure 9.3 above). At the centre of the diagram are the attractions – polar bears and the natural environment. Flowing outward, yet still linked to the bears and the environment, is a large circle representing the complex-adaptive-tourism-system (CATS) surrounding the polar bears and the principal stakeholders involved in the wildlife tourism industry, in this particular context, the managers and the entrepreneurs and associated partners. Motivated by profit, and by policy mandates, both groups contribute in varying degrees to the stability and evolution of the industry that are represented by various arrows, also symbolic of the characteristics of chaos makers and regulators. Around, yet detached from the circle, are the chaos makers (*e.g.*, media). Devoid of policy, profit, or responsibility motives, these agents are free to move through time and space promoting stability and well-being in one location, while destabilizing and punctuating equilibria in others.

DISCUSSION AND CONCLUSIONS

The application of complexity and the conceptual approach developed by Russell and Faulkner (2003) was in this particular scenario an attempt to understand the dynamic relationships that existed between various stakeholders (*i.e.*, the entrepreneurs, managers and chaos makers) who were involved in the integrated use of protected areas (*i.e.*, the CWMA and WNP) along the northeastern coast of Manitoba. By focusing on stakeholder interactions, "one is forced to acknowledge contemporary as well as historical process which affects the outcomes and impacts of tourism development strategies" (Hall 1995, 102). Thus, rather than assuming that alternative conceptual approaches (*e.g.*, complexity) are mutually exclusive from traditional approaches (functionalism), each approach should be applied to certain domains or phenomenon within a field, depending on where they prove to be the most useful. In this particular instance the use of complexity proved more useful.

Ongoing research in the CWMA and WNP provides the necessary technological innovations and expertise for the development of bear viewing recreation opportunities. Because of this long tradition, Churchill possesses both the local and scientific expertise and the underlying social infrastructure needed to develop new wildlife recreation opportunities while protecting the attraction. Through innovative social arrangements, local entrepreneurs working in cooperation with managers have opened a new niche in the pre-existent tourist economy and moved to capture the benefits associated with a new breed of tourist. However, it is important to recognize that "no matter how robust the rest of a regional tourism sector may be, if its primary attractions lose favour with the traveling public or if the transport sector abandons it, the entire destination suffers" (McKercher 1999, 427).

Indeed, should the turbulence of the last three years persist, a fundamental realignment of product and re-focusing of marketing activities may be necessary if the viability of the destination is to be maintained. In addition, traditional management frameworks based on "co-operative tensions" between operators and managers, a process viewed quite favourably by the interviewees, may have to be re-examined. Various NTRE, such as the 2003 SARS outbreak in Toronto, and the 9/11 attacks in New York, prove that unpredictable external events can have drastic impacts on ecotourism and the economy of Churchill. As well, there have been several internal changes such as the increasing presence in the region of Parks Canada, Environment Canada through the Canadian Wildlife Service and the *Species at Risk Act*, Fisheries and Oceans Canada through the *Oceans Act*, and the appearance of new stakeholders, NGOs such as Born Free.

Canada's Oceans Strategy provides a mechanism for integrated coastal zone planning. But these various external and internal changes bring uncertainties and challenges for the implementation of integrated management. In Churchill, these changes are somewhat mitigated by the stability of the polar bear viewing industry and the recognition of new tourism opportunities (*e.g.*, viewing the aurora borealis in the winter time, or witnessing the annual migration of thousands of beluga whales in the summer). The future vitality of the community will depend on the complex-adaptive strategies of these systems, and the extent to which the entrepreneurs, managers, and chaos makers have the creativity and energy to identify and pursue these opportunities while maintaining the integrity and survival of its principal attraction – the polar bears.

REFERENCES

Bierman, D. 2003. *Restoring tourism destinations in crisis: A strategic marketing approach*. London: CABI Publishing.

BBC (British Broadcasting Corporation). 1997. *Wildlife Special: Polar Bear*. Produced by Martha Holmes. 60 min. BBC Worldwide Publishing. Videocassette.

Braummer, Fred. 1984. "Churchill: Polar bear capital of the world." *Canadian Geographic*. 20–27.

Byrne, David. 1998. *Complexity theory and the social sciences – an introduction*. New York: Routledge.

CBC (Canadian Broadcasting Corporation). 1976. *The nature of things: Polar bears*. Produced by the CBC. 60 min. Television Series.

———. 1980. *24 Hours: Polar bears*. Produced by the CBC. 95 min. Television Series.

———. 1999. *The National: The shrinking polar bears*. Produced by the CBC. 15 min. Television Series.

Canadian Tourism Commission (CTC). 2003. Canadian tourism commission releases results of SARS and Iraqi war economic impact research. *http://www.canadatourism.com/en/ctc/aboutctc/articledetails.cfm?articleid=46842* (31 July 2003).

Comeau, P. Dangerous liaison. 1997. *Canadian Geographic* (September/October): 57–60.

David, Richard, C. 1978. "The Hudson Bay polar bear – dignified and sly, cheeky and fierce – still needs friends and protectors." *Smithsonian Magazine* 8(4): 70–81.

Denzin, Norman, and Y.S. Lincoln, eds. 1998. *The landscape of qualitative research: theories and issues*. Thousand Oaks, CA: Sage.

Echtner, C.M., and T.B. Jamal. 1997. "The disciplinary dilemma of tourism studies." *Annals of tourism research* 24(1): 868–83.

Eliot, J.L. 1998. "Polar bears: Stalkers of the high arctic." *National Geographic* 193: 52–71.

Environment Canada. 2003. Species at Risk Act. *http://www.ec.gc.ca/press/2003/030605-2_b_e.htm* (31 January 2004).

Faulkner, Bill. 2003. "Towards a framework for tourism disaster management." In *Aspect of Tourism 9: Progressing tourism research*, edited by Chris Cooper, Michael Hall, and Dallen Timothy, 244–68. Toronto: Channel View Publications.

Faulkner, B., and R. Russell. 1997. "Chaos and complexity in tourism: In search of a new perspective." *Pacific Tourism Review* 1: 93–107.

Fisheries and Oceans Canada. 1997. *An approach to the establishment and management of marine protected areas under the Oceans Act*. Ottawa: DFO.

Gell-Mann, M. 1994. *The quark and the jaguar*. New York: W.H. Freeman and Company.

Gleick, J. 1987. *Chaos*. New York: Penguin Books.

Hall, C. M. 1995. "In search of common ground: Reflections on sustainability, complexity and process in the tourism system – a discussion between C. Michael Hall and Richard W. Butler." *Journal of Sustainable Tourism* 3(2): 99–105.

Hawaleshka, Danylo. 2003. "Nature under siege." *Maclean's* 116(22): 24–34.

Herrero, J., and S. Herrero. 1997. "Visitor safety in polar bear viewing activities in the Churchill region of Manitoba, Canada." *Report for Manitoba Natural Resources and Parks Canada*. Calgary.

Imax Corporation. 1987. *Heartland.* Produced and directed by Norma Bailey and Richard Condie. 40 min. Imax Corporation.

IWA (International Wildlife Adventures). 2003. Main webpage. *http://www.wildlifeadventures.com/2002/North/churchil3.htm* (12 July 2003).

IFAW (International Funds for Animal Welfare). 2003. Main webpage. *http://www.ifaw.org/page.asp?id=625* (3 August 2003).

Jamal, T.B., and D. Getz. 1994. "Collaboration theory and community tourism planning." *Annals of Tourism Research* 22(1): 186–204.

Kellert, S.H. 1993. *In the Wake of Chaos.* Chicago: University of Chicago Press.

Linden, E. 2000. "The Big Meltdown." *Times* 156(1): 24–31.

Manitoba Conservation. 1999. *Management Plan Draft: Churchill Wildlife Management Area.* Unpublished Report.

Manitoba Government. Legislative Electronic Publications. 2002. *Province promotes polar bear protection and sustainable eco-tourism: Ashton.* *http://www.gov.mb.ca/chc/press/top/2002/11/2002-11-04-01.html* (4 November 2002).

———. 2003. *Proposed legislation would protect Manitoba polar bears.* *http://www.govmb.ca/chc/press/top/2002/07/2002-07-03-02.html* (1 September 2003).

Maslow, A. 1954. *Motivation and personality.* New York: Harper.

McKercher, B. 1999. "A chaos approach to tourism." *Tourism Management* 20: 425–34.

National Geographic. 1982. *Polar bear alert.* Produced and directed by James Lipscomb. 60 min. National Geographic Videos. Videocassette.

National Wildlife Federation. 2001. *Bears.* Directed by David Lickley and produced by Goulam Amarsy and James Marchbank. 40 min. National Wildlife Federation. Documentary.

Patton, M.Q. 1990. *Qualitative evaluation and research methods.* Newbury Park, CA: Sage Publications.

Ramsay, M., and I. Stirling. 1982. "Reproductive biology and ecology of female polar bears in western Hudson Bay." *Le naturaliste canadien* 109: 941–46.

Ramsay, M., and I. Stirling. 1990. "Fidelity of female polar bears to winter-den sites." *Journal of Mammalogy* 71: 233–36.

Rogers, C.R. 1967. *Person to person: the problem of being human; a new trend in psychology.* Walnut Creek, CA: Real People Press.

Russell, R., and B. Faulkner. 1999. "Moves and shakers: Chaos makers in tourism development." *Tourism Management* 20: 411–23.

———. 2003. "Movers and shakers: Chaos makers in tourism development." In *Aspect of Tourism 9: Progressing tourism research,* edited by Chris Cooper, Michael Hall, and Dallen Timothy, 220–43. Toronto: Channel View Publications.

Scoones, I. 1999. "New ecology and the social sciences: What prospects for a fruitful engagement?" *Annual Review of Anthropology* 28: 479–507.

Schullery, P. 1992. *The bears of Yellowstone.* Worland, WY: High Plains Publishing.

Stirling, I., N. Lunn, and A. Derocher. 1993. "Possible impacts of climatic warming on polar bears." *Arctic* 46: 240–45.

Stynes, D., and P. Stokowski. 1999. "Alternative research approaches for studying hard-to-define nature-based human values." In *Nature and the human spirit,* edited by B. Driver, D. Dustin, T. Baltic, E. Gary, and G. Peterson, 421–54. State College, PA: Venture.

Teillet. D. 1988. The Cape Churchill Wildlife Management Area: Management guidelines. *Report produced for the Manitoba Department of Natural Resources.* Unpublished document.

Urry, J. 2003. *Global Complexity.* Cambridge: Blackwell.

Waldrop, M.M. 1992. *Complexity.* Toronto: Touchstone Book.

Wilson, A. 1991. *The culture of nature.* Toronto: Between the Lines.

CHAPTER 10

ECONOMIC DEVELOPMENT BASED ON LOCAL RESOURCES:
COMMERCIAL HARVESTING OF CARIBOU ON SOUTHAMPTON ISLAND

Brock Junkin (Government of Nunavut)

Northern economic development based on local resources has been discussed at least since the Mackenzie Pipeline Inquiry of the 1970s, but there has been relatively little documentation or evaluation. Under land claims agreements, northern communities are turning to community economic development based on local wildlife, marine mammals, and fishery resources. This chapter is about one such experience. What can we learn from the case of commercial harvesting of caribou in the Nunavut community of Coral Harbour on Southampton Island (see map on inside cover of book), northern Hudson Bay? How can commercial harvests be organized? How can such economic enterprises be organized in a way that is consistent with Inuit cultural values? Before getting to the Southampton Island caribou case and these questions, some background is needed about land claims agreements and wildlife co-management.

WILDLIFE MANAGEMENT IN NUNAVUT

The Roman principle of *res nullius* has typically governed wildlife management in Canada (Hughes and Roe 2000). Under this principle wildlife is managed by the Crown on behalf of its citizens. No one owns the wildlife until it is harvested. Evolution of the relationship between Canada and her Aboriginal people over the past several decades has led to the inclusion of Aboriginal rights in the constitution. One outcome of the changing attitudes and rights associated with this evolving relationship is that responsibility for wildlife management has become a co-management issue. As a result, Canada's indigenous peoples now participate in the management of both the land and its resources. This sharing of power and responsibility for wildlife management has resulted in a new outlook of protecting the land "for use rather than from use" (Pimbert and Pretty 1997).

Table 10.1

SOUTHAMPTON ISLAND CARIBOU HERD POPULATION ESTIMATES FROM 1967 TO 2003

Year	Caribou Population
1967	40
1978	1,200 ± 340
1987	5,400 ± 1,130
1990	9,000 ± 3,200
1991	13,700 ± 1,600
1997	30,381 ± 3,982
2003	17,891 ± 2,128

Source: Pers. Comm. 2003 Wildlife Division of Department of Sustainable Development, Government of Nunavut.

The Nunavut Territory has been created under the *Nunavut Land Claims Agreement*, which is sometimes referred to as the "Inuit Final Agreement" (Canada 1993). This agreement created the Nunavut Wildlife Management Board (NWMB) to function as wildlife co-management body for Nunavut. This board is tasked as follows under the Article 5.2.33 of the agreement: *Recognizing that [the] Government [of Nunavut] retains ultimate responsibility for wildlife management, the NWMB shall be the main instrument of wildlife management in the Nunavut Settlement Area and the main regulator of access to wildlife and have the primary responsibility in relation thereto in the manner described in the Agreement* (Canada 1993). Although the Inuit have been welcomed to the table to take part in the co-management of these resources, final authority for wildlife management remains with the Minister of the Nunavut Department of Sustainable Development as specified in the statement of principles in Article 5.1.2, which states that: *This Article recognizes and reflects the following principles: ... [the] Government [of Nunavut] retains the ultimate responsibility for Wildlife Management....* (Canada 1993)

On a day-to-day basis the Inuit gain access to wildlife resources by making application for a quota, otherwise known as a "total allowable harvest" (TAH) allocation. Local Hunters and Trappers Organizations make application to the NWMB to be allocated a certain level of harvest. As noted above, the NWMB, subject to ministerial authority, has the discretion to recommend the level of harvest that will be allowed. These recommendations are based on the scientific understanding and traditional knowledge of that resource. The recommendation is then sent to the Minister of Sustainable Development, Government of Nunavut for approval. If the minister is not satisfied with the recommendation, he/she can send it back to the NWMB for further review. Once it has ministerial approval, the NWMB advises the local Hunters and Trappers Committees of the decision made concerning their application for harvesting rights. Responsibility for administering these rights is passed to the HTO, which makes allocates the TAH allocation according to locally perceived needs.

Table 10.2

SOUTHAMPTON ISLAND CARIBOU HERD HUNTING QUOTAS FROM 1978 TO 2004

Year	Hunting Quota
1978	25 subsistence harvest only
1990	400 subsistence harvest only
1992	400 subsistence and 250 commercial
1993	1,000 commercial and unlimited subsistence
1994	5,000 commercial and 1,000 subsistence
1997	6,000 commercial and unlimited subsistence
2004	3,000 commercial and unlimited subsistence

Source: GNWT 1994a & 1994b Wildlife Division of Department of Sustainable Development, Government of Nunavut.

HISTORICAL CONTEXT OF THE CORAL HARBOUR, SOUTHAMPTON ISLAND CARIBOU HERD

Southampton Island is located in Hudson Bay, in Canada's northern Territory of Nunavut. Europeans arriving on this island during the 1800s and early 1900s viewed the island's indigenous caribou herd as a ready supply of fresh meat. By the early 1950s, this relatively new hunting pressure on the caribou herd was being further increased by the introduction of improved and more effective hunting technologies and a buoyant fox pelt industry. Local hunters and trappers involved in the fox pelt industry increased the number of dog teams being kept, thereby necessitating the provision of more caribou to feed the dogs. Ultimately the herd was decimated. Some ten years later the then Government of the Northwest Territories (GNT) took steps to restore the island's caribou herd. In 1967 forty caribou were transferred to Southampton Island from neighbouring Coates Island.

From the modest beginning of forty animals in 1967 (MacPherson and Manning 1967) the population had grown to the point that by the early 1990s there were fears of population crashes brought on by over-grazing of the habitat (Ouellet *et al.* 1993; Ouellet and Heard 1994). The absence of predators such as wolves, and the rich foliage of lichens preferred by caribou, contributed to near geometric growth (Ouellet *et al.* 1993). A brief review of caribou population estimates from 1967 to 2003 gathered through aerial and other surveys is provided in Table 10.1. The data document rapid and dramatic population increases. As the increasing size of the caribou population became more evident some cautious hunting quotas were established. Changes to the hunting quotas from 1978 to 2004 are provided in Table 10.2. In 1978 a subsistence harvest of twenty-five head was allowed. By 1993 a thousand commercial and unlimited subsistence harvesting was allowed.[1] There was considerable discussion around the quota for 1994, when it rose by four thousand animals from one thousand to five thousand. It appears that the intention behind this increase was to stabilize the herd so that no further increase would occur.

Meanwhile, this rapid growth of the Southampton Island herd was a matter of increasing concern to the local Hunters and Trappers Organization. The Aiviit HTO subsequently requested that the wildlife management office of the new Nunavut Territory's Department of Sustainable Development work with them to develop a herd management strategy. Wildlife management staff agreed that if the herd was left unchecked it would outstrip its range and eventually crash. They supported the HTO's request, and forwarded it to the Regional Kivalliq Wildlife Board (KWB) (then the Keewatin Wildlife Board) for consideration. The KWB forwarded it to the Nunavut Wildlife Management Board (NWMB) with the recommendation that the NWMB approve the initiation of a herd management strategy for the Southampton Island caribou herd.

In 1995 a commercial harvest was introduced as a herd management strategy. Over the nine years which followed the harvest was managed under three different organizational structures. The first was a corporation owned by the local Aiviit Hunters and Trappers Committee called Tunniq Harvest. The second involved a subcontract to a private Inuit-owned company called the Southampton Meat Company. The third was a community development corporation of the Coral Harbour community.

Herd Management Constraints

All three management structures have been faced with operating constraints. In a general sense major challenges include maintaining a consistent supply of product, high processing costs, lack of infrastructure, difficulties accessing working capital, and modest economies of scale. As well, however, there are other constraints unique to this particular enterprise. These include the following:

- The harvest of caribou occurs on the land, and the Canadian Food Inspection Agency (CFIA) and European Union (EU) regulations require that the ambient temperature cannot rise above $-18°$ C. This in effect means that the harvest can only occur in the winter. By early spring, the temperatures are already too warm to continue with harvesting activity.
- Days on Southampton Island are as little as three hours long at the beginning of winter, not long enough to efficiently pursue a harvest. It is not until the middle of February that there is sufficient daylight to undertake any harvesting activity.
- Infrastructure at the site can accommodate a limited number of people and caribou. Consequently, even when weather and logistics co-operate, the most that harvest operators have been able to harvest after five years of operation is five thousand animals out of a quota of six thousand.
- The weather itself can be an unpredictable constraint. A week of bad weather in the available six-week window can have a significant impact on the efficacy of the harvest.

Table 10.3

SOUTHAMPTON ISLAND CARIBOU HERD HARVEST ACTIVITY UNDER TUNNIQ HARVEST FROM 1995 THROUGH 1998

Year	Number of caribou harvested	Gender breakdown
1995	2,356	gender breakdown unknown
1996	1,839	gender breakdown unknown
1997	3,365	gender breakdown unknown
1998	2,956	70% female

Source: Wildlife Division of Department of Sustainable Development, Government of Nunavut.

- The streamlined carcasses are shipped to Kivalliq Arctic Foods, which operates an EU and federally certified plant in Rankin Inlet. At current yield levels of three to five thousand animals this plant is at maximum capacity. While there are modest plans to expand the Rankin plant, it is important to resist tendencies to over-capitalize which may lead to skewed decision making on future quotas.
- Access to organizational skills has been a constraint in the past. The ability to manage the harvest profitably and meet the requirements of the CFIA and the EU has been problematic in part because of a lack of expertise. As the harvest has evolved, however, a cadré of individuals has developed around the enterprise.

MANAGING THE CARIBOU HARVEST 1995–98:
TUNNIQ HARVEST CORPORATION

Tunniq Harvest was incorporated as a wholly owned subsidiary of the Aiviit HTO under the Companies Act of the Northwest Territories, on 2 March 1995. A total of two class "A" common shares were issued, and that was the extent of the capital stock for the four years the company was in operation. In that same year the Aiviit HTO allocated its existing total allowable harvest (TAH) or quota to Tunniq Harvest. There was no requirement for a formal agreement, and no need to put out a call for proposals. The sole shareholder of the caribou harvest operator during the period of 1995 through 1998 was the Aiviit HTO.

In the interests of good resource management the commercial harvesters were encouraged to take females. The rationale was that females bore the calves and the more females there were the more calves would be born. Reducing the number of males would not affect reproduction because the remaining males could still impregnate whatever number of females remained. Commercial harvesters, however, preferred to take males, as they were bulkier and thus would yield greater profits. Consequently, some friction developed from time to time between the conflicting goals of the harvesters and those responsible for management of the resource. The level of harvest activity during this time is summarized in Table 10.3. A gender breakdown of the animals taken was only available for 1998.

Table 10.4
FINANCIAL RESULTS OF TUNNIQ HARVEST CORPORATION FROM 1995 TO 1998

Financial Results	1995	1996	1997	1998
Sales	$366,635	$326,210	$587,947	$608,510
Expenses:				
Accounting and legal		18,451	9,300	14,500
Administration fees		25,764	9,603	8,500
Bad debts	2,400	3,609		
Consulting			41,465	10,000
Food	19,203	20,148	17,079	24,000
Freight	175,833	111,732	183,468	284,346
Fuel	17,354	25,267	39,514	45,000
Handling and storage	8,600			
Inspections and permits	5,988	4,811	14,253	8,000
Insurance		2,508	1,870	2,500
Interest and bank charges		8,044	17,874	16,000
Marketing	32,170			
Materials, equip. and supplies		20,700	29,213	32,658
Miscellaneous	3,469	1,250	2,906	7,500
Office supplies		573	531	500
Packaging	12,517	12,781	14,923	10,392
Processing	3,599			
Rental	12,913	40,850	35,286	44,800
Repairs and maintenance	2,799	1,937	44,106	20,500
Telephone		103	2,595	
Travel		31,44z1	11,866	30,000
Wages and Benefits	154,119	197,127	316,835	255,830
Total expenses	450,964	527,096	792,687	815,026
Net income or (loss)	(84,329)	(200,886)	(204,740)	(206,516)
Other income				
Government funding		150,254	155,660	217,109
Net income or (loss)	($84,329)	($50,632)	($49,080)	$10,593

Source: Financial Statements of Tunniq Harvest Corporation.

The initial solution for marketing the 1995 harvest was to contract Arctic Foods Limited (AFL) to handle the whole marketing effort on a commission basis. AFL undertook the activity and successfully marketed all 120,000 pounds of stream-lined carcasses. Of these, 15,000 pounds went to the meat plant in Rankin Inlet, and the rest went to Tricon Commodities International in Edmonton, Alberta. Management did not think that it was getting fair value for the services of AFL and so they decided to market the following year's harvest themselves. They secured a contract with Grandview Farms in Ontario, an abattoir specializing in game meats. This new marketing arrangement had the following advantages as noted in the 1996 harvest report: additional revenues of $0.20 per pound; additional revenues available from the sale of remnant parts; significant savings on the cost of management services; a commitment from Grandview Farms to train local personnel; and no hidden charges such as storage, check-trim (an adjustment for wastage at the abattoir), etc. as had been required by AFL.

This remained a satisfactory arrangement with the harvest operator right through its 1998 season, the final year of its operation. The Rankin Inlet plant received as much meat as it needed during this period, with the balance going to Grandview. The financial results from 1995 to 1998 are provided in Table 10.4. The results for 1995 through 1998 included losses of $84,000 in 1995, $50,000 in each of 1996 and 1997, and then a profit of $11,000 in 1998. The level of government subsidy received in a given year was based on a budget and business plan prepared prior to the harvest.

General Analysis and Assessment

The following analysis of Tunniq Harvest's management of the caribou harvest operation is based on the three objectives cited in the initial proposal for the caribou harvest: herd management, job creation, and social support for the community of Coral Harbour.

In terms of herd management the harvest could be deemed a medium success in that on average three thousand caribou were taken in each of its four years of operation. However, this level of harvest would have resulted in the herd increasing to the Island's estimated carrying capacity of forty thousand or less in a short period of time (Parker 1975). Since no statistics were kept to track the relative female/male ratio of animals taken, it was not possible to estimate with confidence how the potential for herd growth would be affected by this factor. The herd management objective, while helped by this four-year initiative, would not be met, and the expected crash only postponed a few years. Additional action was needed in order to improve the herd's long-term outcomes.

In terms of job creation, on average thirty-five part-time positions were created for six weeks, or the equivalent of approximately four person-years. Given that the government's average annual subsidy was about $131,000, that would mean each person year cost $33,000 in direct government resources to create. Compared to the annual cost of social assistance which would include a housing subsidy

The caribou harvest portable camp situated on a frozen lake. Photo by Richard Connelly, 2003.

of $24,000 and direct payments of $12,000 annually, the saving was marginal but significant enough to allow the experiment to continue.

MANAGING THE CARIBOU HARVEST 1999–2002: SOUTHAMPTON ISLAND MEAT COMPANY – PRIVATE OPERATOR

The series of losses experienced by Tunniq Harvest were a matter of mounting concern to the Aiviit HTO. The non-profit board of volunteers put in place to oversee the Tunniq Harvest operation had been ill-prepared to provide guidance to what proved to be a complex business enterprise. By mutual consent and pursuant to a motion to privatize the operation in the fall of 1998, subject to the terms and conditions of their Memorandum of Agreement responsibility for the operation was transferred. The new harvest operator, the Southampton Meat Company, was incorporated in March 1998, under the Companies Act of the Northwest Territories. Five shares of capital stock were issued and there were no changes throughout the four years under review. There were five equal shareholders, all of them Inuit and residents of Coral Harbour. While there were five equal shareholders, there was a clear delineation of authority and responsibility among the five.

Unlike the Tunniq Harvest arrangement, under which there was an informal granting of the TAH to the harvest operator, in this case there was no longer a close relationship in terms of ownership. The proponent of the harvest was a private firm, and this called for a more formal arrangement in which the Aiviit HTO formally granted access to the resource in return for some compensation. On the one hand, this new arrangement represented a chance for the HTO to make a profit from the compensation arrangement. On the other hand it

Table 10.5
SOUTHAMPTON ISLAND CARIBOU HERD HARVEST ACTIVITY UNDER PRIVATE OPERATOR FROM 1999 TO 2002

Year	Number of caribou harvested	Gender breakdown
1999	1,094	47% female
2000	2,166	54% female
2001	3,696	57% female
2002	3,834	25% female

Source: Wildlife Division of Department of Sustainable Development, Government of Nunavut.

provided the HTO the chance to be released from the burden of managing the harvest operation.

The 1999 harvest got off to an inauspicious beginning. The new proponent made marketing arrangements similar to those that had been made in the past. Unfortunately that customer, Grandview Farms, ran into financial difficulties and could not complete its obligations. For a period of time it appeared that the harvest would have to be called off. However, discussions were initiated with a former minority customer, Keewatin Meat and Fish Plant, and that group agreed to increase its purchase of streamlined carcasses in order to save the harvest, and to augment its own opportunities. A deal was struck and the harvest was salvaged – though just barely. Harvest statistics for the four years beginning 1999 to 2002 are provided in Table 10.5, and the financial results from this arrangement are detailed in Table 10.6. The first year produced a net loss of $14,000, but in the following three years profits of $9,000, $41,000 and $64,000 were realized.

As can be seen by comparing the financial information from Tunniq Harvest to that of the Southampton Island Meat Company, the need for government involvement in subsidizing the harvest had declined markedly with the private operator. In fact, for the final year of operation under the Southampton Meat Company in 2002, it could be argued that no subsidy was necessary at all. However, officials from the Department of Sustainable Development did not consider this a viable scenario. They felt that the harvest should continue to receive support in order to allow government some ability to influence the ongoing herd management.

General Analysis and Assessment

As for the Tunniq Harvest analysis, this review of the Southampton Island Meat Company is based on the three objectives cited in the initial reasons for the harvest: herd management, job creation and social support. With an average of 2,700 caribou taken in each of the four years the harvest could be deemed only a modest success in terms of herd management. However, since a decline in the average number of animals taken was occasioned by one unfortunate event in

Table 10.6
**FINANCIAL RESULTS OF THE SOUTHAMPTON ISLAND MEAT COMPANY
FROM 1999 TO 2002**

Financial Results	1999	2000	2001	2002[*]
Sales	$213,128	$303,985	$525,209	$561,000
Expenses:				
Accounting and legal	5,105	4,161	4,774	4,500
Administration fees	11,000	22,826	29,721	28,000
Amortization	0	0	300	0
Bad debts	100	0	0	0
Camp mobilization/ demobilization	19,329	600	0	1,000
Consulting	11,000	14,700	0	15,000
Food	105,369	142,529	176,435	160,000
Freight	1,860	5,966	5,395	6,000
Fuel	2,418	9,949	20,709	20,000
Inspections and permits	250	1,100	12,996	10,000
Interest and bank charges	928	1,124	499	1,200
Materials, equipment and supplies	8,822	4,948	4,516	6,000
Repairs and maintenance	0	0	3,783	12,000
Telephone	835	2,382	2,461	3,000
Vehicle	11,321	1,676	668	2,000
Wages and Benefits	48,304	186,906	288,724	290,000
Total expenses	226,641	398,867	550,981	558,700
Net income or (loss)	(13,513)	(94,882)	(25,772)	2,300
Other income				
Government funding	0	103,700	75,000	75,000
Net income or (loss) before taxes	(13,513)	8,818	49,228	77,300
Income tax	0	0	8,076	13,141
Net income or (loss) after taxes	($13,513)	$8,818	$41,152	$64,159

Source: Financial statements of Southampton Island Meat Company.
[*]Final audited results could not be obtained for this year. These figures are pro forma and based on third party inquiry.

1999, we may get a more accurate evaluation if we treat 1999 as an aberration and instead average the number of animals harvested over the other three years of operation – 2000, 2001, and 2002. In this case, we find an improved average of 3,200 animals harvested per year. While this number is encouraging, it should be borne in mind that a similar or different circumstance such as bad or unpredicted weather could arise in the future and once again throw the harvest statistics off course. So, in terms of analyzing the success of the harvest as a herd management technique, it would be imprudent to count on the higher average. Thus, the lesson to be learned from both this experience and those previous is that current herd management practices will only rarely meet the TAH of six thousand. The reality, then, is that while the caribou harvest was managed as a private operation the herd continued to move steadily, though less quickly than under the Aiviit Corporation, toward a population crash.

In terms of job creation, the same crew of approximately thirty-five part-time workers was required to operate the hunt. If more caribou were harvested this crew worked longer and earned more money. The employment of four person-years noted previously continued to apply here, give or take 20 per cent. What is noteworthy, however, is that the average annual subsidy declined to $16,000 under the Southampton Meat Company. This was considerably less than had been required during the first four years. When the lower subsidy was taken into account the harvest looked more productive in terms of job creation. Discussions with Department of Sustainable Development staff indicated that subsidies at the level of $75,000 would continue in order to allow departmental influence on herd management. A more important element of the private sector experience was that the streamlined carcasses were being processed in Rankin Inlet through Keewatin Meat and Fish. Thus fourteen full-time jobs were being created in another Nunavut community. Under Tunniq Harvest these jobs had been transferred south.

MANAGING THE CARIBOU HARVEST 2003
COMMUNITY DEVELOPMENT CORPORATION (CDC)

The contract for the Southampton Island caribou harvest is let by the Aiviit HTO in two-year increments. As described earlier, the first two increments of the contract had been let to the Aiviit HTO's wholly owned subsidiary. In 1998 the harvest was privatized and let to the Southampton Island Meat Company for two more increments. In 2002 the HTO compiled new terms of reference for the harvest. Under the new terms of reference the evaluation process for contract bids put greater emphasis on protecting the environment, and in particular, remediation of the abattoir site. Southampton Meat Company chose not to bid. The Coral Harbour Community Development Corporation (CHCDC) did place a bid, and was awarded the tender for the two-year period ending 30 April 2004.

The CHCDC was incorporated in June 2001 for the purpose of engaging in enterprise within the community, nurturing businesses to viability and then divesting them to the private sector. It was incorporated under the Canada Business Corporations Act as a corporation without share capital. The venue of legal construct is not dissimilar from a society in that the votaries of the organizations are members and do not own shares, as would be the case in a corporation with share capital. Hence there is a one-member, one-vote effect. Anyone who is of age, shares and supports the objectives of the corporation, and is also a resident of the Hamlet of Coral Harbour, may apply for membership to the corporation. Membership is subject to the approval of the Board of Directors. Membership of the Board of Directors is largely based on ex officio appointments from community organizations such as the HTO, the District Education Authority, the Hamlet Council, and the Tourism Association, together with three elected members-at-large. The board was structured in this manner to guard against stacking the board with special interest groups that may not be at one with the corporation's objectives.

The 2003 harvest was the most successful to date in meeting its primary goal of herd management. At conclusion of its first year of operation it had exceeded the performance of all previous harvests by harvesting over 5,000 animals. It also set a half-day record of 120 animals harvested and a full day record of 226 animals harvested. Most importantly, the ratio of female to male animals taken improved dramatically, with 60 per cent of the harvested animals being female. In the previous year the number of females taken had been only 25 per cent. Disturbance to the environment was minimized as well, since the CHCDC did not have to move the camp to achieve this level of harvest. The stock of immediately available caribou seemed not to be disturbed by six weeks of hunting effort.

The CHCDC used the same meat processing plant that had been used by its predecessor. To allow for ease of administration rather than increased capacity, Kivalliq Arctic Foods (KAF) (formerly Keewatin Meat & Fish Plant) marginally increased its processing capacity. KAF also began to explore new markets so that the company would have a buffer should one of its existing markets collapse. It made initial forays into the European Union once it received its EU certification. While the EU offers the potential for a long-term export market, the KAF recognized that serious challenges existed in attempting to enter that market. Europe is a well established competitive venue for game meats. Scandinavian countries supply reindeer, for example, and New Zealand is a source of red deer. Nonetheless, the product of both these competitors stems from game ranching, which may be disadvantageous relative to free-range Southampton caribou. The caribou have not received hormones or other growth enhancers, they have not been injected with antibiotics or been genetically modified. In a marketplace fed on fears – founded or unfounded – most of the foregoing qualities could increase the appeal of Southampton caribou and the willingness to pay a premium for it. Despite the logic inherent in this analysis, the unexpected can and will occur. In the case of the 2003 harvest, KAF was able to buy the entire product only to find its own marketing stymied by a Mad Cow scare in an unrelated market. It is uncertain early in 2004 what the outcome will be for KAF.

Product ready for shipping. Photo by Department of Economic Development and Transportation, Government of Nunavut.

Reproduced in Table 10.7 below are the CHCDC's interim financial results as of the end of 30 April 2003. These financial statements have yet to be audited, but any changes that may come about as a result of an audit are not likely to change significantly the conclusion as to this year's success or failure. The level of operating subsidy remained the same as in previous years at $75,000. However, given the level of the operating loss, an additional subsidy of $60,000 was negotiated with the Government of Nunavut to allow for the harvest de-mobilization and clean up. A new capital subsidy of $338,000 was introduced to allow for recapitalization in 2003. Such a capital subsidy had not previously been required. When the harvest was first proposed it had been projected that the harvest would be at least marginally profitable and in this way provide for its own capital replacement needs in the future. Unfortunately this did not turn out to be the case, and so the need for a continuing capital subsidy will have to be revisited.

General Analysis and Assessment

Evaluating the CHCDC's performance relative to that of the two earlier opera-tors, and relative to the objectives of the harvest, is not a simple task. Regard-ing herd management, it is clear that the CHCDC substantially out-performed both the HTO's Corporation and the private sector agency. Not only were the overall harvest numbers up significantly, but the ratio of females to males was dramatically increased.

Table 10.7

INTERIM FINANCIAL RESULTS OF THE
CORAL HARBOUR COMMUNITY DEVELOPMENT CORPORATION FOR 2003

Revenue:	
Meat	$700,017
Country food	3,288
Total Revenue	703,305
Operating costs:	
Ammunition	17,168
Casual labour	585
Food	84,721
Freight	38,194
Fuel	72,057
Inspections	19,654
Labour	646,291
Licences	500
Miscellaneous	362
Mobilization	5,768
Office expense	214
Overseer	1,284
Packaging	23,675
Propane	863
Rental	84,720
Repairs: Buildings	2,002
Repairs: Equipment	5,985
Small tools	12,844
Supplies	24,805
Telephone	4,685
Total operating costs	1,046,378
Net operating income	(343,073)
Subsidy	135,000
Net income	($208,073)

Source: Financial statement of Coral Harbour Community Development.

From an economic development perspective, the CHCDC harvest was clearly the most successful of the three organizations. Gross sales were up substantially. Antlers, tail cartilages, and penises were put into storage to be sold to customers in the Far East pharmaceutical industry. If these can be sold as expected they will generate additional revenues of about $100,000. CHCDC has initiated a pilot project to make use of the fur on the hides. It is hoped that this caribou felt project will make use of an otherwise discarded by-product. Hides at this time of year are not good for use in the leather industry because of the damage done by a caribou parasite, which has the effect of peppering the hide with holes. Employment rose from thirty-five to fifty-nine persons during the height of the harvest, and a felt project of the kind described would employ a substantial number of people on a more permanent basis.

Other obvious community benefits resulted under this organizational model. For example, harvest assets now belong to the community and are controlled by an organization with economic development as its purpose. The innovative and more efficient building techniques used by CHCDC when building housing for the crew will be applied elsewhere in the community. The CHCDC also took care to ensure that all parts of the animals were used. Carcasses that had been rejected by the inspectors for gut or shoulder shots, for example, were salvaged for local use when possible. Delicacies such as the tongues and tunniq were brought into the community for general distribution for the first time in the history of the herd harvest. No useable country foods were left on the land. Better land management practices included the daily removal of all harvest debris to the dump. When summer arrived it was buried. Previously this material had been left on the land. Finally, there has been a growth in understanding and participation. Former operators tended to withhold information from the community. The open and transparent structure of the CHCDC means more people know what is going on. As a result they can do their jobs better and are better able to step into the breach should there be staff shortfalls in critical areas.

Despite the innovations, the increase in gross profits and the many community benefits realized under the CHCDC's first year of operation, in terms of economic viability the CHCDC performance was a disappointment, showing a net loss of $208,000. A number of factors contributed to the 2003 loss. The CHCDC assumed that the previous operator would leave the basic infrastructure and some supplies. Instead, these were removed and the CHCDC incurred unexpected start-up costs and delays, including last-minute construction of new facilities to house and feed the crew. Since they did not find out that they would receive nothing from the previous harvester until after the time of year when supplies could have been transported by ship, they had to rely on air transport, and so incurred extraordinary air freight costs. CHCDC operations were further hampered by the lack of adequate mobilization equipment such as bulldozers and Bombardiers. The bulldozers, Bombardiers, snow machines, generators, generator engines, and abattoir equipment they did have broke down frequently. Operations were further constrained by management inexperience in running a commercial abattoir and harvest on the land, and new Canada Food Inspection Agency (CFIA) regulations and requirements added to considerably to the operator's workload. Finally, though concentrating the harvest on females was desirable from a herd management perspective, it led to an increased payroll burden since females yield significantly less meat per carcass. To survive in the longer run the CHCDC will need to produce a higher net profit. The CHCDC harvest suggests that the corporation has both a capable workforce and a viable product. The question is how to make it profitable. Strategies leading to a more profitable outcome have emerged through discussions with harvest staff and others involved.

The CHCDC's second year of operations should demonstrate a better financial return. It will not be necessary to re-capitalize the harvest infrastructure in 2004. The company now has the gear it needs to conduct the harvest and will not have to engage in an expensive scramble at the last moment to ready operations. Air freight costs will be dramatically reduced as substantial pre-planning was able to be done for the coming harvest. A small crew was maintained to the end of July 2003 to ensure that the harvest infrastructure would be ready for the 2004 harvest and equipment breakdowns should be minimized in 2004. Overall, the expertise gained from the 2003 harvest should result in more efficient operations. The CHCDC has complied with the new CFIA guidelines, and the CFIA has signed off on the harvest plan. This time-consuming activity will not be repeated in 2004. Finally, the wage structure will be changed to a per pound basis rather than the current per head basis to offset the decline in the average carcass weight due to concentrating the harvest on females. In 2003 the remuneration rate for line workers was $1.60 per head. When changed to a per pound basis the rate will translate into 2.5 cents per pound. This will equate to a reduction in wages of approximately 17 per cent. This reduction in wages will likely be accepted by harvest staff, albeit grudgingly.

Transportation will remain a problem in for the CHCDC. The product will continue to have to be transported across the land via Cat train. Once it reaches Coral Harbour it has to be transported via air to Rankin Inlet for final processing. Both modes of transport are very expensive when compared to the costs of moving freight in the South.

CONCLUDING OBSERVATIONS

The initial reason for establishing a caribou harvest was not one of economic development but rather wildlife management. The Southampton Island caribou herd was growing so quickly that left unchecked the herd was expected to collapse as food sources were exhausted. Since this was a herd which supplied much of the country food for the people of Coral Harbour, there was some urgency to see the herd managed, particularly given the history of the extinction of the former herd in the 1950s. Culling the herd, while at the same time providing some economic benefit via a commercial harvest, appeared to be a viable solution. This chapter describes the three organizational models which have been used to manage the Southampton caribou harvest, and evaluates the outcomes against two major criteria: herd management and economic development for the broader community.

The project has passed through two organizational frameworks and entered a third. The first included a wholly owned corporation of the local Hunters and Trappers Organization and subsequently an additional four years with a private corporation. The project is now in its third operational structure – a community development corporation model. The experience gained has allowed us a clear view of relative performance. In terms of bottom-line efficacy, the private operator substantially outperformed the HTO Corporation and the CDC. Initially this might suggest that a private operator is the management model of choice for this enterprise. However, it should be remembered that the first objective of the HTO was herd management. Further, the *raison d'être* for any Nunavut HTO is not making profit but rather wildlife management. That said, it might have been predicted that HTO and community efforts at managing an the enterprise for profit would pale against those of an organization whose focus is profitability.

The community development corporation structure has a humanistic orientation which has a focus on profit but for the broader good. This broader *raison d'être* is more consistent with the values of the Aboriginal people in the local communities. The anticipated broader distribution of net profits in the community by way of reinvestment in other enterprises, and not via profit sharing, will likely generate more local economic activity than if the profits had gone to shareholders of a private corporation. In this latter case one might have expected the money to be spent on imported goods, new snow machines, and trucks.

Given the tenuous nature of the caribou herd discussed in this chapter, it is incumbent upon those that exploit them to do so in such a way that they will continue to be available for future generations. The Southampton Meat Company had profit as its goal. The CHCDC adopted what can be described as a humanistic approach, using profits from the harvest to improve community well-being (Johnstone 1998). The experience highlighted in this chapter suggests that the Hamlet of Coral Harbour is better off with the broadly based humanistic model of a community development corporation. Not surprisingly, then, and in keeping with the Aboriginal cultural backdrop of sharing, honing traditional skills and knowledge, and working collaboratively, a communal approach to organizing this commercial harvest of country foods has been most acceptable to the community of Coral Harbour.

This case study has demonstrated that the choice of organizational model and its goals and orientation are important considerations in creating a development model for the North. This finding is consistent with results of The World Economic and Social Survey conducted by the United Nations and published in 2000. The survey reports that "institutions, including the political system, are crucial in influencing who appropriates an economic surplus and how it is spent. For example, a more equal distribution of ... income ... results in a larger share of income gains being received by peasants. This reduces capital flight and the import of luxury goods and has a positive effect on the private expenditures on education and on the demand for locally produced goods and services, thus stimulating development" (United Nations 2000, 124).

The Southampton Island caribou harvest experience has been successful from a herd management perspective, and despite the lack of profitability, the most recent harvest suggests a turn for the better in this regard. As well, the harvest has had indirect economic benefits, as it has provided from thirty-five to fifty-nine part-time jobs in Coral Harbour and fourteen full-time jobs in Rankin Inlet. The harvest builds on existing community strengths by using local resources to replace imports and to enhance the export of these resources.

In conclusion, then, it is likely that the community development corporation will not perform as well as the private operator in terms of profitability. Wages are likely to be somewhat more generous, and environmental issues will be dealt with more responsibly. There will be a greater reliance on local personnel, goods, and services than might otherwise occur. These behaviours will reduce short-term profitability. Nonetheless the long-term viability of this operation and its indirect effects is likely to be enhanced. The community development model will remain viable and have greater positive long-run impacts on the community than either the HTO corporation or the private corporation model.

REFERENCES

Canada. 1993. *Agreement between the Inuit of Nunavut Settlement Area and Her Majesty the Queen in right of Canada*. Referred to as the Inuit Final Agreement. Tunngavik and Indian Affairs and Northern Development, Ottawa.
Coral Harbour Community Development Corporation. 2004 Interim financial statements for 2003.
GNWT. 1994a. *Workshop on Harvesting and Management of Southampton Island Caribou*. May 5. Unpublished.
——. 1994b. *Minutes: Workshop of the Management of Southampton Island Caribou, May 18–19*. Unpublished
Hughes, R., and D. Roe, eds. 2000. *Northern Eden, Community based Wildlife Management in Canada*, Edmonton: Canadian Circumpolar Institute Press.
Johnstone, H. 1998. "Financing Ventures in a Depleted Community." In *Perspectives in Communities*, edited by Gertrude A. MacIntyre, 99. Sydney, Cape Breton: UCCB Press.
MacPherson, A.H., and T.H. Manning. 1967. "Mercy Mission: Caribou for the Hunters of Southampton Island." *Eskimo*. Vol. 78. Churchill. *Nunavut Land Claims Agreement*. 1999. Government of Canada. Ottawa: Queens Printer.
Ouellet, Jean-Pierre, and D.C. Heard. 1994. "Dynamics of an Introduced Caribou Population." *Arctic* 47(1) (March): 88–95.
——, D.C. Heard, and S. Boutin. 1993. "Range Impacts Following the Introduction of Caribou on Southampton island, NWT, Canada." *Arctic and Alpine Research* 25(2): 136–41.

Parker, G.R. 1975. "An Investigation of Caribou Range on Southampton Island, NWT."
Canadian Wildlife Service Report Series, 33. 82 pp.

Pimbert, M.P., and J.N. Pretty. 1997. *Diversity and Sustainablity in Community-based Conservation*. Paper for the UNESCO-IIPA regional workshop on Community-based Conservation, February 9–12, 1997, India.

Southampton Island Meat Company. 2003. Financial statements for 1999–2002.

Tunniq Harvest Corporation. 1999. Financial statements for 1995–1998.

United Nations. 2000. *World Economic and Social Survey: 2000*. New York.

NOTES

1   The numbers provided in this table represent the total number of animals that could be killed. During each harvest some of the animals harvested would be rejected for a variety of reasons including gut/shoulder shots and diseases. The number of carcasses rejected ranged from 2 per cent to 6 per cent, but averaged close to 3 per cent.

RESILIENCE *&* INSTITUTIONS

CHAPTER 11

CROSS-SCALE INSTITUTIONS & BUILDING RESILIENCE IN THE CANADIAN NORTH

Fikret Berkes (University of Manitoba)
Nigel Bankes (University of Calgary)
Melissa Marschke (University of Manitoba)
Derek Armitage (Wilfrid Laurier University)
Douglas Clark (Wilfrid Laurier University)

INTRODUCTION

In areas experiencing social and environmental transformations, such as the Canadian North, there is a need to develop the capacity to respond and adapt to change, and to explore policy directions that can help build resilience to deal with change. In the area of environmental management, Folke *et al.* (2002) have suggested that the creation of flexible multi-level governance systems that can learn from experience and generate knowledge to cope with change may be one such policy direction. The response of the community itself, through its own institutions, is key to effective adaptation to change, but support from regional and national governments is also important in the creation of multi-level governance.

Berkes and Jolly (2001) have argued that co-management institutions in the Canadian western Arctic under the *Inuvialuit Final Agreement* (IFA) have the potential to provide such multi-level governance. These institutions are instrumental in relaying local concerns across multiple levels of political organization. Participatory management has the potential to enhance local adaptation capabilities by shortening the links between different decision-making levels. Co-management mechanisms evolving in the Canadian North, especially when they take into account local and traditional environmental knowledge, speed up communication and bridge different systems of knowledge. Tightly coupled systems (*i.e.*, those involving close feedback relationships) reduce the response time to change, a necessary but not a sufficient condition for successful adaptation.

In most areas of social and environmental management, responding to change is rarely a one-step solution. More likely, it is an iterative process of learning-by-doing, or adaptive management (Holling 1978; Lee 1993; Kristofferson and Berkes, this volume). Since complex systems tend to be characterized by high degrees of uncertainty, policies and actions at any one time are necessarily based on incomplete information, and management is modified iteratively as understanding evolves.

The use of indigenous perspectives to guide such adaptive management is important in that northern peoples are often experts in learning-by-doing. Both scientific management and traditional management systems can learn from one another and from their joint experience with resource management issues. Learning across institutions is key (Diduck *et al.*, this volume); this is learning at the level of community institutions (such as hunter and trapper committees), regional organizations, national organizations, and international organizations such as the Arctic Council.

In this chapter, we use cross-scale analysis to deal with institutions at various levels – local, regional, national, international. There are two main ways in which these institutions may be linked across scale. Using terminology from Young (2002) and Ostrom *et al.* (2002), cross-scale linkages may be *horizontal* (across geographic space) or *vertical* (across levels of organization). It has been hypothesized that cross-scale linkages, both horizontal and vertical, may speed up learning and communication, thereby improving the ability of a society to buffer change, speed up self-organization, and increase capacity for learning and adaptation (Berkes 2002).

The use of adaptive management and the creation of multi-level governance (or co-management) systems is a shift from the usual top-down approach to management. It will not solve uncertainties inherent in change but will help deal with those uncertainties in an institutional context that encourages learning and adaptation. This is the essence of the resilience approach. Managing for resilience enhances the likelihood of sustaining nature and society in a changing environment in which the future is unpredictable. Resilient social-ecological systems are those that are able to absorb shocks without collapse. Building resilience means nurturing diversity, creating options, and increasing the capability of the system to cope with uncertainty and surprise (Berkes *et al.* 2003).

The concept of *resilience* provides a window for the study of change. The resilience of a systems is defined in terms of (1) the magnitude of shock that a system can absorb and still remain within a given state, or the ability to buffer disturbance; (2) the self-organization capability of that system, and (3) its capacity for learning and experimentation (Folke *et al.* 2002; Resilience Alliance 2004). The first attribute of resilience is difficult to study directly. Hence, the other two attributes may be used as rough measures of resilience.

The objective of this chapter is to explore the idea that cross-scale linkages help deal with change by building resilient systems. Using several resource management cases, we approach this objective by analyzing the mechanisms by which resource management systems build capacity for self-organization, learning, and adapting. The Canadian North is an appropriate setting in which to address the objective because of the existence of a number of experiments in cross-scale management through land claims agreements. We explore the objective with reference to the following questions:

1 Land claims based co-management connects local-level institutions and government agencies. What are the mechanisms by which co-management may contribute to learning and self-organization across levels of political organization?
2 Local experts (traditional knowledge holders) and scientists have been interacting in co-management committees, working groups, and conferences. What is the role of improved communication through sharing knowledge and views? How can traditional knowledge and science be combined toward more resilient systems?
3 A number of cross-scale linkages have been created through co-management arrangements. How do these horizontal and vertical linkages function in dealing with resource and environmental management problems?

The chapter explores these questions through a consideration of five cases or examples: co-management in the Inuvialuit region under the Fisheries Joint Management Committee (FJMC); the West Side Working Group (WSWG) fisheries traditional knowledge study; management of narwhal in Nunavut; management of polar bears across the Arctic; and the effort to deal with persistent organic pollutants (POPs) in the Arctic. Through these cases, we look for lessons and insights regarding building capacity for self-organization, learning, and adapting.

CO-MANAGEMENT IN THE INUVIALUIT REGION

In the Canadian western Arctic, co-management institutions, evolving since the signing of the *Inuvialuit Final Agreement* (IFA) of 1984, provide cross-scale linkages for feedback horizontally across the region and vertically across levels of organization from the local Hunter and Trappers Committees (HTCs) to regional agencies and beyond. These linkages have the potential to facilitate the transmission of community concerns, such as those about food chain contaminants and climate change, to the regional, national, and international levels (Berkes *et al.* 2001), and thereby help northern societies to respond to environmental problems.

Resource co-management, or the sharing of power and responsibility between the government and local resource users, emerged through the settlement of land claims in northern Canada. Under the *Inuvialuit Final Agreement*, a series of co-management boards were created in the Inuvialuit Region, Canadian western Arctic (see Figure 5.1). One of these is the FJMC. It consists of two Inuvialuit representatives, two government- appointed representatives, and a rotating chair (Fast *et al.* 2001). The FJMC is a consensus-based organization. Quarterly meetings and teleconferences help to ensure that information is exchanged among members of the board, engaging the FJMC in joint problem solving and adaptive learning.

The FJMC can address different concerns, from local fishing issues to regional oil and gas development policy. This is because the FJMC communicates with the HTCS in each of the seven IFA communities and directly advises the Minister of the Department of Fisheries and Oceans (DFO) on matters pertaining to fisheries and marine mammals in the region. An annual 'community tour' and meetings with the minister help the FJMC to facilitate sharing information and concerns among the various levels of governance. By dealing with a broad range of issues (*e.g.,* monitoring and harvest information for fish and marine mammals, cross-boundary issues, combining scientific and traditional knowledge), the FJMC is able to garner an in-depth perspective regarding fisheries management issues.

Incorporating local perspectives is an essential component of co-management, enabling local systems to be recognized and legitimized. Decisions requiring local input include data gathering, harvesting and allocation decisions, local knowledge, long-term planning, and inclusive decision-making (*e.g.,* Pinkerton 1989). Participatory approaches are essential for the consideration of multiple perspectives on management issues and for the inclusion of local and traditional knowledge in management. In participatory management, not only *what* information is included, but also *how* local perspectives are incorporated into the decision-making process become important. Communities need to have the capacity to set their objectives and know what work they want done at the local level.

Co-management needs to be experimental and flexible so that both local-level and government-level institutions can learn from their mistakes and gradually build capacity to deal with new circumstances and change in general. The Beaufort Sea Beluga Management Plan under the IFA illustrates how the adaptive management perspective of Holling (1978) may be combined with the idea of co-management, in what might be called an adaptive co-management approach. The FJMC, in co-operation with local HTCS and the DFO, developed the Beaufort Sea Beluga Management Plan that is widely supported throughout the region even though compliance is voluntary (FJMC 2001).

With the recent increase in oil and gas exploration in the Beaufort Sea region, the FJMC is searching for a 'legislative fit' to help with formal policy recognition for this plan. For example, under the 1997 *Oceans Act* there is the provision for taking a flexible planning approach in creating marine protected areas (MPAS) (Fast *et al.* 2001). Creating an MPA in the Beaufort Sea beluga areas is being considered as one way to guarantee protection for the beluga as interest in oil and gas development builds in those areas. This is an example of how management can be adaptive. The FJMC started with an informal management plan that is working and, in response to development pressures, moved to the use of new and existing legislation for beluga protection.

Table 11.1

MEMBERS OF THE WEST SIDE WORKING GROUP OF
THE FISHERIES JOINT MANAGEMENT COMMITTEE, THE INUVIALUIT REGION

Aklavik HTC:	local agency
Aklavik Elders Committee:	local agency
FJMC:	regional agency; coordinating role
DFO:	national agency
Parks Canada:	national agency

Co-management is meant to establish a dialectic process, functioning not only from the top down but also from the bottom up (McCay and Jentoft 1996). Co-management arrangements can take many forms, depending on the issues and context. The work of the FJMC demonstrates how an adaptive co-management approach enables changes in the locus, scale, and scope of decision making to be made appropriately, depending on the issue being addressed. The inclusion of local perspectives, which are often not heard, is an integral component of any co-management system. Indigenous forms of communication and organization are vital to decision-making processes that take into account local-level knowledge, as illustrated by the next example.

USING TRADITIONAL KNOWLEDGE FOR FISHERIES MANAGEMENT

The FJMC has been conducting traditional knowledge studies to feed into fisheries management plans that incorporate both traditional and scientific knowledge. The WSWG, with the facilitation of FJMC, was formed to initiate a traditional ecological knowledge (TEK) fishing study for the rivers west of the Mackenzie River to the Yukon-Alaska border (Figure 5.1). Representatives from the Aklavik HTC, the Aklavik Elders Committee, the FJMC, the DFO, and Parks Canada all sit on the board of this working group, chaired by a representative of the Aklavik HTC (Table 11.1).

The FJMC, knowing that there would be multiple objectives to be satisfied from a TEK study, facilitated a process to support both community and scientific priorities. The objectives of the TEK fishing study included (a) a traditional knowledge component related to the local context and based on Inuvialuit oral histories, and (b) a scientific component, using TEK related to fish biology and habitats, that could contribute to the fisheries management plan. The unwritten objective of this research was to facilitate a process among stakeholders (*e.g.*, community experts and fisheries scientists) that would enable learning. Because the FJMC coordinates multiple levels (local-regional-national) and is a respected co-management body, it was able to bring together people and institutions that normally do not collaborate.

The TEK Fishing Study, carried out during February and March 2002, enabled elders and others to share their knowledge of fish species, fishing methods, and changes in species over time. Since fishing is part of a series of land-based

Table 11.2

SOME CHANGES AFFECTING FISHING IN THE AREA FROM THE MACKENZIE RIVER TO THE YUKON-ALASKA BORDER (the "West Side")

1930s	RCMP posts/stores began closing; good muskrat trapping opportunities in the Delta; by the 1940's most people move towards Aklavik.
1960s	Introduction of the snowmobile; fewer and fewer dog teams (last in early 1970s); less people fishing (catching less fish).
1980s	Changes in water levels and fish migrations; closing of Fish Hole; after-effects of oil and gas development.
1990s	People returning to the coast, *e.g.,* Shingle Point; only some people are fishing now.

activities undertaken by the Inuvialuit, a story emerged blending the history of fish and fishing practices with the impacts of other influences in the region. A historical perspective sheds insight as to why fishers have changed their fishing locations and practices. For example, as more people moved toward Aklavik and the Mackenzie River Delta to pursue muskrat-trapping opportunities, the coastal areas became less frequently used. With the introduction of the snowmobile, it became less important to fish for dog food. Many changes have affected the Inuvialuit, and understanding changes in fish harvesting is only part of a much bigger story (Table 11.2).

The TEK Fishing Study generated a wealth of information. From an historical perspective, a better understanding emerged of how people fished in relation to other activities on the land. Changes such as market prices, introduction of the snowmobile, and more recently global warming have impacted where and how people fish. Although fishing intensity declined with the snowmobile and the closing of fishing on the Big Fish River for conservation reasons, families are returning to Shingle Point and Running River in recent years to fish during the summer months. Perhaps char and herring are less abundant (or less harvested) along the coast, but more freshwater species are being caught. Physical changes in the landscape such as erosion (due to permafrost thaw) are also affecting fishing. The water in some coastal areas is said to be far less salty than before. The results of this study illustrate how people remain in touch with their landscape, and just how much local knowledge exists about particular resources and places.

The synthesis workshop at the end of the first phase of the TEK Fishing Study enabled elders and members of the West Side Working Group to learn from each other. Elders from different communities who had once fished in the area were able to share stories (something that does not happen as often as one would think!) and to generate more information about their experiences on the land. When thinking about potential fisheries management, elders suggested measures be taken to (a) protect "fish holes," *i.e.,* stop development in intensive fishing areas; (b) leave the spawners; (c) remove diseased fish to prevent pollution of the system; and (d) use common sense. Elders wanted to ensure that

Table 11.3
PRINCIPLES IN ARTICLE 5 OF THE NUNAVUT LAND CLAIMS AGREEMENT OF 1993

1 Inuit are traditional and current users of wildlife;
2 The legal rights of Inuit to harvest wildlife flow from their traditional and current use;
3 The Inuit population is steadily increasing;
4 A long-term, healthy, renewable resource economy is both viable and desirable;
5 There is a need for an effective system of wildlife management that complements Inuit harvesting rights and priorities, and recognizes Inuit systems of wildlife management that contribute to the conservation of wildlife and protection of wildlife habitat;
6 There is a need for systems of wildlife management and land management that provide optimum protection to the renewable resource economy;
7 The wildlife management system and the exercise of Inuit harvesting rights are governed by and subject to the principles of conservation;
8 There is a need for an effective role for Inuit in all aspects of wildlife management, including research; and
9 Government retains the ultimate responsibility for wildlife management.

coastal areas are protected and managed so that future generations can continue to experience the landscape and fishing activities. This process enabled DFO scientists and others to better understand how local knowledge can contribute to resource management by providing contextual information, local detail on resources, and baseline data. The report produced from the TEK study is being used as the baseline to which the DFO will add scientific data.

The example illustrates how a co-management body can help to foster relationships across several levels. The FJMC, through its coordinating role, enables various players to communicate in order to learn from each other. Resilience is enhanced within this system through feedback mechanisms; as stakeholders begin to learn and to understand different worldviews, better management planning is possible. Prior to coordinating the WSWG, communication between scientists and community members was limited. The creation of the WSWG and the TEK study has enabled a dialogue to begin, an important step for mutual learning. Next we turn to the Nunavut region and a case on narwhal management.

INSTITUTIONAL DYNAMICS OF NARWHAL MANAGEMENT IN NUNAVUT

Historically, quotas for narwhal have been set by the DFO and have remained relatively static since 1977. This centralized management approach, and top-down quota allocation process, has typically not been responsive to Inuit desires for increased quotas, or assertions by hunters that narwhal populations have been increasing (Diduck *et al.*, this volume). Thus, until recently, key local and regional organizations have had a limited influence on the narwhal management process.

However, consistent with the principles articulated in the *Nunavut Land Claims Agreement* of 1993 (Table 11.3), an experimental community-based narwhal management process was established in five communities in the region in 1999. This process has been encouraging more effective cross-scale institutional and organizational linkages. In particular, the community-based narwhal

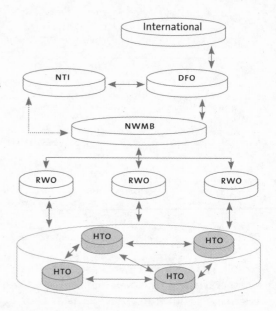

The DFO is linked to national and international narwhal issues, including coordination with Greenland.

The department of Fisheries and Oceans and the NWMB link federal and Inuit priorities. As the claims implementation organization, Nunavut Tunngavik Inc. plays a limited management function but is the 'voice' of the Inuit.

As a co-management entity created under a land claim, the Nunavut Wildlife Management Board plays a central role linking local, regional and national scale priorities.

Regional wildlife organization in Nunavut mediate concerns among HTOs in the Keewatin, Kitikmeot and Kivalliq (Baffin) regions.

Community-based hunters' and trappers' organizations reflect local values and goals, and can provide traditional knowledge that is transmitted to the NWMB.

The horizontal linkages among HTOs vis-a-vis narwhal management are less formalized.

Figure 11.1   Community-based narwhal management process in Nunavut.

management process is fostering greater participation among stakeholders, devolving decision making to community-level institutions, and encouraging the integration of Western science and traditional knowledge in decision making. Central to this resource management regime are local Hunters and Trappers Organizations (HTOs), Regional Wildlife Organizations (RWOs), the Nunavut Wildlife Management Board (NWMB), and the DFO, along with Nunavut Tunngavik Incorporated, a claims implementation organization (Figure 11.1).

Under the pilot community-based narwhal management process, DFO-set quotas have been lifted and harvest levels managed by HTOs in each of the five communities. In turn, communities are responsible for monitoring and regulating the harvest (Diduck et al., this volume).

However, there have been difficulties associated with the new management process. For example, the lifting of the DFO quota of 50 narwhal in the community of Qikiqtarjuaq in 1999 resulted in 127 narwhal landed, 40 struck and sunk, and another 79 wounded and escaped. The significant increase in harvest levels, and concerns with harvest methods and the subsequent wastage, led the DFO to temporarily close the Qikiqtarjuaq narwhal hunt in 1999–2000. Despite this apparent setback, the community-based narwhal management process may yet offer much in the way of developing new cross-scale institutional and organizational linkages. Such cross-scale linkages are more likely to transmit priority concerns, issues and knowledge from the bottom up, while linking the regional and national management institutions required to address problems of narwhal management that are not always bounded by convenient jurisdictional scales, geographic space or short time frames.

The community-based narwhal management process is a significant and important management innovation. However, the barriers and constraints to the development of a cross-scale institutional and organizational framework that fosters adaptation and resilience are not insignificant, and there are several interrelated issues that necessitate further analysis.

First, the current process is premised on a notion of 'community,' a notion whose geographic, political, and normative dimensions have not been adequately critiqued. As known from international experience, there are limitations created by class, stratification, conflict, representation, and division of resource use based on gender, ethnicity, and wealth (Agrawal and Gibson 1999; Li 2002). Exacerbating the problem, remote communities throughout the world have been integrated into increasingly privatized, individualized, and commoditized social and economic systems (Brosius *et al.* 1998). Do these concerns apply to Nunavut? The extent to which HTOs reflect or represent a diversity of local goals with respect to narwhal management is largely unexplored.

Second, the manner in which communities in the region are increasingly embedded in extra-local socio-political and economic structures is influencing, among many factors relevant to community-based narwhal management, property rights dynamics and formal governance institutions (*e.g.,* hamlet government). This continues to result in a shift from an historical emphasis on subsistence resource use toward economic growth opportunities, individualized income development opportunities, and the increased commercialization of resource appropriation strategies (*e.g.,* profiting from the sale of narwhal tusks). The point here is not to disparage the motives or rights of Inuit hunters to engage in diversified livelihood strategies, but to illustrate that such poorly understood dynamics influence the development of resilient, cross-scale institutional and organizational linkages.

Third, the current management framework seeks to foster cross-scale linkages by connecting local, regional, and national institutions and organizations. Yet the actual roles of the various management instruments still appear, somewhat counter-intuitively, hierarchical, positional, and competitive in orientation (McCay and Jentoft 1998). Unclear roles and responsibilities and inconsistent communication among individuals and organizations engender interactions among communities and the HTOs, the HTOs and the NWMB, and the NWMB and DFO, that are at times characterized by competition, tension, and conflict. Moreover, despite a policy shift toward devolution, partnership, the use of traditional knowledge, and community-based management arrangements in the Baffin Island region, there is still a tendency among some community members to accept the DFO as the primary source of information, knowledge, and authority regarding key resource stocks because of their historically dominant management role.

Finally, although a significant advance over top-down management, the community-based narwhal management process is still operationalized in a Euro-Canadian resource management framework (Rodon 1998; White 2001).

A Euro-Canadian management framework requires communities to develop formalized management bylaws, monitoring protocols, and reporting practices. To suggest these management requirements are inappropriate is disingenuous; yet, the requirements represent a further, largely unexplored dynamic associated with the development of resilient cross-scale institutional and organizational frameworks. It is also worth noting that the DFO remains the ultimate decision-making authority where narwhal is concerned. In practice, however, interventions by the DFO staff or the minister (*i.e.*, the 'negative option' approach) in the management process will become politically less palatable as decentralized management regimes evolve. Consequently, while the innovative framework for community-based narwhal management is seeking to build horizontal and vertical linkages among institutions and organizations necessary for resilient and adaptive decision making, there are several processes and dynamics that require further exploration and analysis.

CROSS-SCALE INTERACTIONS FOR POLAR BEAR MANAGEMENT

In 1973, Canada, the United States, the former Soviet Union, Norway, and Denmark signed the *International Agreement for the Conservation of Polar Bears and their Habitat* (the Agreement). This was spurred by international concern about rapidly increasing harvests of polar bears. The agreement is widely recognized as a success and is considered to have been instrumental in the establishment of effective polar bear conservation regimes throughout the Arctic (Fikkan *et al.* 1993; Prestrud and Stirling 1994; Ross 2000).

In Canada the linkages within and between different levels of the polar bear management system are short and tight, often dependent on a few individuals who work across several levels of the institutional scale, from local to international (Diduck *et al.*, this volume) (Figure 11.2). For example, the same provincial and territorial biologists may be involved in conducting population surveys and writing management agreements at the local level, drafting policies at the territorial level, consulting with the Federal-Provincial Technical and Administrative Committees for Polar Bear Research (PBTC and PBAC, respectively) at the national level, and serving on the IUCN's Polar Bear Specialist Group (PBSG) at the international level. Federal and academic biologists conduct similar tasks across the same span of the institutional scale. Continuity, shared experiences, and close relationships among this small number of peers have created high cohesion and consistent norms, goals, and standards in the group. The specific roles of key individuals are difficult to assess, but in such a small group they probably contribute substantially to these outcomes.

Despite those many positive attributes, or perhaps even because of them, real access to decision making in the network has generally been selective. Until recently, local stakeholders (mainly Aboriginal) were welcomed at PBTC and PBAC meetings only as observers. This is not to suggest that the dominant actors were unwilling to include them. Canadian polar bear managers have long been sensitive to Aboriginal rights and needs, and indeed were strong advocates for

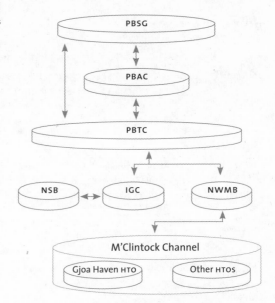

The Polar Bear Specialist Group coordinates polar bear conservation internationally. Members are government biologists of the five nations signed to the 1973 Agreement; academic biologists, managers, and co-management organizations attend as invited specialists.

The Polar Bear Administrative Committee sets policies and regulations; it consists of senior federal, provincial, and territorial managers.

The Polar Bear Technical Committee coordinates research and management activities and provides technical advice to the PBAC; it consists of government and academic biologists, field-level managers, and representatives of co-management organizations with settled land claims.

Horizontal linkages have emerged to address regionally-specific needs, such as the polar bear management agreement between the Inuvialuit Game Council and the North Slope Borough, and among communities affected by the M'Clintock Channel harvest moratorium.

Figure 11.2 Horizontal and vertical linkages among polar bear management institutions.

them in negotiating the agreement (Prestrud and Stirling 1994). Rather, the point is simply that existing network structures predate co-management regimes and tend to change slowly.

Both committees have formalized the participation of co-management bodies as members, but discourse – especially of the PBTC – remains overwhelmingly scientific. Likewise, representatives of Aboriginal co-management organizations attend PBSG meetings as invited specialists, but the group's outputs (*e.g.,* Wiig *et al.* 1995; Derocher *et al.* 1998; Lunn *et al.* 2002) clearly indicate a dominant paradigm of science-based conservation. This may be changing; the successful Inuvialuit-Inupiat co-management regime in the southern Beaufort Sea (see below) is being brought into the mainstream discourse (Brower *et al.* 2002; Johnson 2002).

In comparison to the vertical orientation of the dominant actors, horizontal linkages (Young 2002) have probably been important for local stakeholders for a long time. The 1988 Inuvialuit-Inupiat Polar Bear Management Agreement is a clear example of such a linkage, empowering both parties and integrating the management efforts of two groups of people interacting with a single, shared bear population. Interestingly, this agreement inverts the usual power relationship between government biologists and local stakeholders by establishing a Joint Commission consisting of two representatives designated by each of the Inuvialuit Game Council and the North Slope Borough Fish and Game Management Committee. The commission appoints a Technical Advisory Committee to review harvest data, research results, and management recommendations, placing biologists in an advisory role to the local actors in the Joint Commission.

National Linkages

Regional Linkages

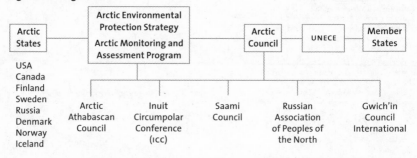

Figure 11.3 Management of POPs: national and regional linkages.

Another example of self-organizing horizontal linkage occurred in the wake of the apparent decline of the M'Clintock Channel polar bear population. In 2002, the communities that harvested that population were invited to hunt bears from an adjacent community whose quota came from a different, healthier population. Such sharing of hunting opportunities fits the pattern of intercommunity trade as an adaptive response identified by Berkes and Jolly (2001) in the western Arctic.

It appears that while the Canadian polar bear management regime was designed with vertical relationships in mind, it is certainly capable of accommodating self-organizing horizontal linkages where they meet local social or ecological needs. Such horizontal linkages seem to be becoming more common, and their development processes may learn from pre-existing linkage mechanisms. For example, the U.S.-Russia *Polar Bear Conservation Agreement* signed in 2000 enables the development of an indigenous peoples-to-indigenous peoples agreement for its implementation. The intent is to model this stakeholder-scale agreement after the existing Inuvialuit-Inupiat Agreement (Johnson 2002).

CROSS-SCALE DYNAMICS AND THE GLOBAL RESPONSE TO POPS

In the mid-1980s, Inuit in northern Canada and university and government scientists came to appreciate that Inuit country food, especially marine mammals, was contaminated by organochlorines (Dewailly and Furgal 2003). Over the next fifteen years, atmospheric chemists and others were able to describe a picture of long-range atmospheric transport of these substances

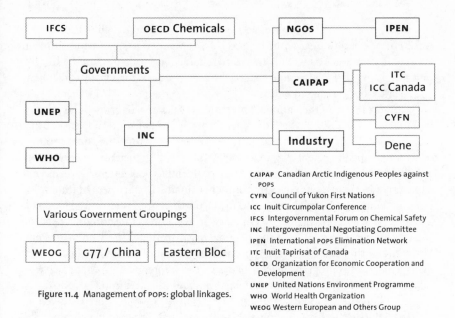

CAIPAP Canadian Arctic Indigenous Peoples against POPS
CYFN Council of Yukon First Nations
ICC Inuit Circumpolar Conference
IFCS Intergovernmental Forum on Chemical Safety
INC Intergovernmental Negotiating Committee
IPEN International POPS Elimination Network
ITC Inuit Tapirisat of Canada
OECD Organization for Economic Cooperation and Development
UNEP United Nations Environment Programme
WHO World Health Organization
WEOG Western European and Others Group

Figure 11.4 Management of POPs: global linkages.

from areas of use in the South to their subsequent deposition in high latitudes (Reiersen *et al.* 2003). Once made available to the northern environment, these persistent organic pollutants (POPs) bioaccumulate in fatty tissues and biomagnify as they move up the food chain.

On 23 May 2001, a diplomatic conference adopted the text of the *Stockholm Convention on Persistent Organic Pollutants* as a global companion to the regional *POPs Protocol* previously adopted in June 1998 under the auspices of the United Nations Economic Commission for Europe's (UNECE) *Convention on Long-range Transboundary Air Pollution* (LRTAP). Both instruments serve to prohibit the production and use of listed chemicals, and in the case of POPs that are an unintended by-product of other industrial purposes (*e.g.,* dioxins and furans), their continuing minimization and, where feasible, ultimate elimination. Both instruments acknowledge the special situation of Arctic ecosystems and of indigenous communities dependent on those ecosystems.

What were some of the horizontal and vertical cross-scale linkages that facilitated this outcome? Figure 11.3 sketches the national and regional linkages, and Figure 11.4 the global linkages. We discuss each in turn.

The identification of the POPs problem as a regional and global issue, rather than just a local issue, coincided with the emergence of a circumpolar Arctic consciousness (Rothwell 1996). Institutionally, this consciousness was reflected in the *Arctic Environmental Protection Strategy* (AEPS) of 1991 and subsequently in the formation of the Arctic Council in 1996. The AEPS laid a fundamental building block for a POPs agreement by establishing the Arctic Monitoring and Assessment Programme (AMAP). But the eight Arctic states recognized that the

UNECE offered a better forum for addressing the POPs issue. Why? The UNECE was better scaled to deal with the issue, as it included eastern and southern European countries as well as the Arctic states, and it also offered a legally binding framework convention that could form the basis for a POPs agreement. The Arctic Council offered neither.

Several features of the national and regional linkages diagram deserve comment (Figure 11.3). First, the Northern Contaminants Program (NCP) of the Government of Canada was instrumental in bringing together three groups: (1) scientists (health scientists, toxicologists, atmospheric scientists) concerned with identifying and describing the POPs problem; (2) key federal government departments (Health Canada, Environment Canada, Department of Indian Affairs and Northern Development, and the Department of Fisheries and Oceans), and territorial and provincial governments; and (3) northern Aboriginal peoples, at least through their national and regional organizations. These linkages were horizontal rather than vertical. Aboriginal organizations were involved in the management of the program, helping to ensure that a comprehensive approach was taken to both ecosystem and human health issues (Shearer and Han 2003).

Second, at the regional scale, Arctic indigenous peoples were involved at all levels. Both AEPS and AMAP were, and remain, unique among comparable international programs in encouraging participation by indigenous peoples. The Canadian NCP and the international fora also served to enhance and create new linkages *between* different indigenous peoples both nationally and, with the collapse of the former Soviet Union, across the circumpolar world.

Third, several factors made it possible to move the POPs issue from Arctic fora to the UNECE. Obviously there was significant overlapping membership between the two organizations at the state level, and Canada and Sweden assumed the task of persuading the ECE to accord a high priority to the POPs issue. But overlapping personal responsibilities were also important. For example, David Stone of DIAND was a key player in the NCP, served as chair of AMAP and became co-chair of the UNECE task force on POPs as part of the lead up to the Protocol. Indigenous people were far less well represented in the UNECE forum than they were in the Arctic fora (Fenge 2003) but nevertheless made a significant intervention (Bankes 1998; Selin 2003).

The scale limitations of a regional agreement were appreciated from the outset and the global POPs negotiations began as the regional POPs negotiations concluded. At the state level, many of the same key individuals (such as David Stone) continued to be involved. Indeed they were also involved in one other set of global chemicals negotiations that had recently concluded (the Rotterdam or PIC *Convention on Prior Informed Consent* (Buccini 2003)). Many of these individuals were also connected through other multilateral chemicals fora such as the OECD chemicals groups and the Intergovernmental Forum on Chemical Safety (IFCS). These fora helped to bridge developed and developing countries as well as NGO and industry chemicals experts, and helped parties form the common premises that allowed the global negotiations to begin (Buccini 2003).

Formally convened under the auspices of both the United Nations Environment Programme (UNEP) and the World Health Organization (WHO), the global POPs negotiations followed the model of most recent multilateral environmental agreements. They were structured around what is denominated an open-ended (meaning open to all states) Intergovernmental Negotiating Committee (INC). Although inclusive at the state level, such negotiations are still state-centred, and thus NGOs, indigenous peoples and industry organizations are relegated to observer, rather than participant, status.

Several features of the global diagram (Figure 11.4) deserve comment. First, NGOs were far more heavily involved (albeit in an observer capacity) in the global negotiations than they had been in the regional UNECE negotiations. A key organization that brought a coalition of NGOs together was IPEN (International POPs Elimination Network). So too were indigenous peoples, and there was extensive collaboration between these two communities. Second, Canadian indigenous peoples formalized their alliance on the POPs issue by forming CAIPAP (Canadian Arctic Indigenous Peoples Against POPs) but also actively engaged (as did industry and NGO representatives) the official Canadian delegation both before and during the various sessions of the INC negotiations (INC 1 to 5). At CAIPAP's request, the official Canadian delegation included a Dene.

Third, the imagery of the negotiations was important in depicting linkages. Presiding, conscience-like, over the negotiations for the entire two years was an Inuit carving of a woman holding a baby. The carving was presented to UNEP Executive Director Topfer by Sheila Watt-Cloutier, President of ICC, at the outset of the negotiations. Topfer in turn presented the carving to John Buccini, the chair of the negotiations. Fourth, while NGO and indigenous representatives served to publicize the negotiations and to keep up the political pressure for a successful outcome, we should not underestimate the critical linkages represented by traditional state groupings in international organizations, *e.g.,* the European Union, the Group of 77 and China. These linkages remain the crucial ones in diplomatic negotiations. But it is also clear that the linkages created by non-state actors can help bridge differences at critical points. One such issue was the elimination of DDT that, from time to time, threatened to scupper the negotiations (Watt-Cloutier 2003). The Inuit made it clear that they did not seek an agreement to protect Inuit mothers at the expense of those exposed to malaria.

DISCUSSION

We started the chapter to explore the idea that cross-scale linkages help deal with change by building flexible multi-level governance systems that can learn and generate knowledge to deal with change (Folke *et al.* 2002). We approached this objective by analyzing the mechanisms by which the capacity for self-organization, learning, and adapting can be built into resource management systems. In all five cases, there are elements of co-management connecting local-level institutions and government agencies, and improved communication through sharing knowledge and views between holders of traditional knowledge

Coastal inlet in James bay, near Chisasibi, Quebec. Photo by Fikret Berkes.

and scientific knowledge. In this section, we focus on cross-scale linkages that enable co-management and communication, first analyzing participatory aspects of management, second noting epistemic communities, and third identifying some shortcomings and constraints.

Aboriginal Participation

All five cases involve cross-scale linkages, but the relative importance of local Aboriginal participation varies across cases, roughly in line with the geographic and political scale of the case. The range is from the fisheries traditional knowledge study (the smallest scale and the most local) to the Arctic-wide POPs case. The FJMC has existed since 1984 and has a substantial track record of indigenous involvement (Fast *et al.* 2001). The WSWG and the TEK Fishing Study build on that track record. The particular study described in this chapter is one of several TEK fishing studies designed to provide indigenous input for fishery management plans in the region. There is emphasis on Aboriginal participation in the narwhal case as well. The DFO no longer sets the narwhal quota; under the *Nunavut Land Claims Agreement,* communities are responsible for monitoring and regulating the harvest. In the polar bear and the POPs cases, however, Aboriginal participation is less obvious or visible.

The management of polar bears follows the international agreement of 1973 that predates the Inuvialuit and Nunavut agreements. In any case, these agreements do not have specific provisions for polar bears but wildlife co-management in general. The provisions of these agreements have been implemented by formalizing the participation of land-claims-based co-management agencies in the two technical and administrative committees (PBTC and PBAC) for polar bear management. However, the discourse in these committees remains technical and scientific, not very inviting for Aboriginal knowledge and inputs.

In the POPs case, part of the impetus for international action came from the local level, especially following the Broughton Island study of 1989 that showed beyond doubt the extensive contamination of the Arctic with organic pollutants and the serious danger to Inuit health (Myers *et al.,* this volume). The key challenge for northerners was getting the POPs issue on the international agenda. They were able to do this because POPs could be characterized as a human health issue and because AMAP provided a credible information base. This effectively created important feedback linkages that triggered regional negotiations but at the same time confirmed that a regional agreement was by itself inadequate.

Both the POPs story (Downie and Fenge 2003) and the linkages (Figs. 11.3 and 11.4) are complex but they are also informative. The POPs case is a crisis-based issue (as opposed to ongoing management as in the other cases). This provided flexibility and fluidity, allowing the creation of cross-scale linkages. Most of those linkages were horizontal in nature not vertical – because the international system itself tends to be horizontal (Bankes, this volume). The POPs case provides insights regarding national and international politics, the timing of interventions, the strategic choice made by the eight Arctic states for UNECE over the Arctic Council (UNECE provided a better match in scale), the role of key groups (*e.g.,* CAIPAP) and key individuals, and the formation of epistemic communities.

Direct Aboriginal participation in the POPs international fora is not extensive, but it is probably fair to say that vertical linkages have been effective for the people of the North to convey their concerns up the scale. Thus, we see effective communication through vertical linkages and also through horizontal linkages. Some horizontal linkages were highlighted in the POPs story (*e.g.,* new linkages between indigenous groups nationally and internationally). There are likely to be many more examples of these, often self-organized, horizontal linkages in the POPs and other cases, but they have not been explored to any extent.

For example, in polar bear management, there usually are discussions between communities during the establishment of quotas and local management agreements for shared populations. How and where do these discussions take place? How are decisions made – negotiated or by consensus? Aside from the Inuvialuit-Inupiat Agreement, outcomes of horizontal processes have not been well documented or assessed. There is a marked contrast between the vertical orientation of mainstream institutional players (largely governmental) and the horizontal orientation of local-level stakeholders. Is this horizontality a response to power imbalances or is it merely reflective of a different, perhaps more pragmatic, indigenous way of doing business?

Key Players and Epistemic Communities

There is strong evidence in each of the cases (except perhaps the narwhal example) of the role of "key players" who have been instrumental in building cross-scale linkages and in the shared recognition of a problem across scale. Such shared problem identification is the hallmark of what Haas (1992) has termed epistemic communities. Members of such communities share principled beliefs, notions of validity, and policy goals that cut across political boundaries. Two of the cases identify epistemic communities. The first is the expert group identified in the polar bear case, characterized by continuity, shared experience, close relationships among a small number of peers, a high degree of cohesion, and consistent norms, goals, and standards. The polar bear case is significant in that it shows how cross-scale linkages may develop despite the initial vertical design, and how learning may take place.

The second is the grouping brought together by the Northern Contaminants Program of the Government of Canada: scientists (health scientists, toxicologists, atmospheric scientists) concerned with identifying and describing the POPs problem; key government departments; and representatives of northern Aboriginal peoples. Just how cohesive the group was with respect to shared beliefs, norms, and goals has not been explored. There may have been more than one epistemic community, for example, that of atmospheric scientists, health scientists, or circumpolar indigenous groups. An indirect evidence of the existence of epistemic communities, both nationally and internationally, is the apparent transfer of learning from one issue (POPs) to another (climate change).

Shortcomings and Constraints

In addition to providing insights about the significance of cross-scale linkages for participatory management and communication, the case studies also reveal some shortcomings and constraints. We discuss three: the use of traditional knowledge, the long lead times in co-management, and the continuing challenges of implementation.

The first concerns inadequacies in the use of local and traditional knowledge to transcend Western science-based, conservation-oriented, or harvest-oriented management. Only one case shows use of traditional knowledge by design: the

West Side Working Group illustrates how local and traditional knowledge can be used to elucidate the historical context. It highlights the historical dimensions of fishery west of the Mackenzie, land use changes, and how they can be factored into fishery management plans. The narwhal case highlights some of these contextual challenges as well, but there is no management mandate under the *Nunavut Land Claims Agreement* to conduct a narwhal historical study, as there is for the bowhead whale (Hay *et al.* 2000).

In the narwhal case, the use of local and traditional knowledge is limited, and in the polar bear and POPs cases, this seems to be an area of weakness. However, the potential does exist for these cases. Considerable polar bear knowledge exists among the Inuit, and in the case of POPs, local knowledge can complement science in a number of ways. Indigenous readings of signs of environmental quality indicate a range of possibilities, including the construction and use of environmental quality indicators (Cobb *et al.*, this volume) and community-based monitoring (Manseau *et al.*, this volume, and Parlee *et al.*, this volume). Only the narwhal case makes explicit mention of indigenous monitoring, but again the potentials are considerable for all of the cases.

A second constraint that emerges from the case studies concerns the systemic difficulties and long lead times in forging real partnerships of governments and local people. It is well known that it takes a long time to build co-management, some ten years in the case of Pacific Northwest salmon (Singleton 1998) and the Beverly-Qamanirjuaq caribou herd (Kendrick 2000). As these two studies illustrate, the time-consuming aspect of participatory management is building trust among the parties and the development of mutual respect for different ways of knowing. Hence, it is not surprising to see effective cross-scale linkages with the FJMC and the *Inuvialuit Final Agreement*, building on nearly twenty years of joint management experience, and initiatives such as the Beaufort Sea 2000 Conference that facilitated exchange between scientific and traditional knowledge holders and initiated discussion on societal goals (Anon. 2000).

However, even with the WSWG case, linkages between Inuvialuit communities and federal government departments were possible only because of the key facilitation role of the FJMC. The larger issue is the historically entrenched conventional, centralized, top-down, regulations-oriented management, based on expert knowledge. Hence the dominant discourse tends to be scientific, as in the polar bear case. Even though there is recognition and respect for native rights, there remain vestiges of a paternalistic approach in the way committees carry out their business and the way roles are defined. The inputs of Aboriginal parties are often hard to detect in the written outputs of the various committees. This is true in the polar bear case as well as in many others; indigenous rights are respected but indigenous voices and messages are not usually heard.

A third constraint is related to the continuing challenges of management and implementation. A key point to the entire discussion is that management must become more flexible and adaptive. Such flexibility is easier at the local level (*e.g.*, FJMC and WSWG cases) but more difficult as higher and higher organizational levels are considered. Making linkages and engaging in negotiations are one thing; regional and global implementation another. With respect to the POPs case, for example, this raises several issues: Can linkages created during the negotiation phase be strengthened? Will they prove to be effective? Can linkages between NGOs and between indigenous peoples be maintained as the issue moves from being an *ad hoc* problem to a continuing one?

CONCLUSIONS

We have emphasized the creation of multi-level governance systems that can learn and generate knowledge to deal with the problems of the North. Social learning typically involves an iterative process of learning-by-doing or adaptive management (Lee 1993). Because of uncertainties in the system, information will always be incomplete. Hence, it is more important for governance systems to be capable of *learning* in this adaptive sense, rather than *possessing* conventional management knowledge and skills to be applied top-down (Folke *et al.* 2002; Kristofferson and Berkes, this volume).

The response of local institutions is key to coping and adapting to change. Conventional resource and environmental management science does not have the methods in its tool kit to deal with complexities and uncertainties. What is needed is a different kind of management regime that goes beyond the received wisdom of centralized management, empowering local institutions for self-organization and adaptive management. The cross-scale linkages provide the connections to, and support from, higher levels of governance, so there is mutual learning and adapting.

Under the various land claims agreements, the Canadian North provides lessons, inviting us to start listening to practical insights and to review mechanisms that are already in place. The cases highlight different mechanisms that have been helpful in building cross-scale linkages that may have some influence on the resilience of linked systems of people and environment. The key is perhaps to identify and build on the appropriate mechanism in different contexts. For example, in the POPs and polar bear cases, there is an internationally sanctioned dimension to the linkages. In the narwhal case, the experimental management regime (if it survives) will end up institutionalizing a new cross-scale approach that promises to be more responsive and resilient to change. However, effective international linkages with Greenland for this shared resource is still lacking.

In the polar bear case, the existing sophisticated cross-scale approach effectively leaves out indigenous voices. Hence, the issue is how to insert Aboriginal co-management and build on the strengths of the mechanisms set up under the 1973 international agreement. In the WSWG case, the drivers or mechanisms appear more *ad hoc* than the others. The links between Aboriginal and government agencies exist in each of the cases. But additional efforts are probably needed to improve two-way communication throughout the network of linkages, especially with respect to communication to and from the communities.

REFERENCES

Agrawal, A., and C. Gibson. 1999. "Community and conservation: Beyond enchantment and disenchantment." *World Development* 27(4): 629–49.

Anon. 2000. Beaufort Sea 2000 Conference. Renewable Resources for Our Children. Conference Summary Report. Fisheries Joint Management Committee, Inuvik. *http://www.fjmc.ca/publications/Beaufort%20Sea%20Conf.pdf* (August 2003)

Bankes, N., 1998. "Steps towards the international regulation of POPs." *Northern Perspectives* 25(2): 18–21.

Berkes, F. 2002. "Cross-scale institutional linkages for commons management: Perspectives from the bottom up." In *The Drama of the Commons*, edited by E. Ostrom, T. Dietz, N. Dolsak, P.C. Stern, S. Stonich and E.U. Weber, 293–321. National Academy Press, Washington DC.

——, and D. Jolly. 2001. "Adapting to climate change: social-ecological resilience in a Canadian western Arctic community." *Conservation Ecology* 5:18. *http://www.consecol.org/vol5/iss2/art18*

——, J. Colding, and C. Folke, eds. 2003. *Navigating Social-Ecological Systems: Building Resilience for Complexity and Change.* Cambridge, UK: Cambridge University Press.

——, J. Mathias, M. Kislalioglu, and H. Fast. 2001. "The Canadian Arctic and the Oceans Act: the development of participatory environmental research and management." *Ocean & Coastal Management* 44: 451–69.

Brosius, P., A. Tsing and C. Zerner. 1998. "Representing communities: Histories and politics of community-based natural resource management." *Society and Natural Resources* 11: 157–68.

Brower, C.D., A. Carpenter, M.L. Branigan, W. Calvert, T. Evans, A.S. Fischbach, J.A. Nagy, S. Schliebe, and I. Stirling. 2002. "The Polar Bear Management Agreement for the Southern Beaufort Sea: An evaluation of the first ten years of a unique conservation agreement." *Arctic* 55: 362–71

Buccini, J., 2003. "The long and winding road to Stockholm: the view from the Chair." In *Northern Lights Against POPs: Combating Toxic Threats in the Arctic*, edited by D. Downie and T. Fenge, 224–55. Montreal: McGill-Queen's University Press.

Dewailly, E. and C. Furgal. 2003. "POPs, the environment and public health." In *Northern Lights Against POPs: Combating Toxic Threats in the Arctic*, edited by D. Downie and T. Fenge, 3–21. Montreal: McGill-Queen's University Press.

Derocher, A., G. Garner, N.J. Lunn, and Ø.Wiig, eds. 1998. Polar Bears. Occasional Paper of the IUCN Species Survival Commission (SSC) No. 19. IUCN, Gland, Switzerland.

Downie, D., and T. Fenge, eds. 2003. *Northern Lights Against POPs: Combating Toxic Threats in the Arctic.* Montreal: McGill-Queen's University Press.

FJMC. 2001. *Beaufort Sea Beluga Management Plan.* Fisheries Joint Management Committee, Inuvik.

Fast, H., J. Mathias, and O. Banias. 2001. "Directions toward marine conservation in Canada's Western Arctic." *Ocean & Coastal Management* 44: 183–205.

Fenge, T., 2003. "POPs and Inuit: influencing the global agenda." In *Northern Lights Against POPs: Combating Toxic Threats in the Arctic*, edited by D. Downie and T. Fenge, 192–213. Montreal: McGill-Queen's University Press.

Fikkan, A., Osherenko, G., and Arikainen, A. 1993. "Polar Bears: the Importance of Simplicity." In *Polar Politics: Creating International Environmental Regimes*, edited by O. Young and G. Osherenko, 96–151. Utica, NY: Cornell University Press.

Folke, C., S. Carpenter, T. Elmqvist *et al.* 2002. *Resilience for Sustainable Development: Building Adaptive Capacity in a World of Transformations*. International Council for Scientific Unions (ICSU), Rainbow Series No. 3. Paris. *http://www.sou.gov.se/mvb/pdf/resiliens.pdf* (August 2003).

Gunderson, L.H., and C.S. Holling, eds. 2002. *Panarchy: Understanding Transformations in Human and Natural Systems*. Washington, DC: Island Press.

Haas, P.M. 1992. "Introduction: epistemic communities and international policy coordination." *International Organization* 46: 1–35.

Hay, K., D. Aglukark, D. Igutsaq, J. Ikkidluaq, and M. Mike. 2000. Final Report of the Inuit Bowhead Knowledge Study. Nunavut Wildlife Management Board, Iqaluit.

Holling, C.S., ed. 1978. *Adaptive Environmental Assessment and Management*. New York: Wiley.

Johnson, C. 2002. *Polar bear co-management in Alaska: co-operative management between the US Fish and Wildlife Service and the native hunters of Alaska for the conservation of polar bears*. In *Polar Bears*, edited by N.J. Lunn, S. Schliebe, and E. Born, 139–41. Occasional Paper of the IUCN Species Survival Commission No. 26, IUCN, Gland, Switzerland.

Kendrick, A. 2000. "Community perceptions of the Beverly-Qamanirjuaq Caribou Management Board." *Canadian Journal of Native Studies* 20: 1–33.

Lee, K. 1993. *Compass and Gyroscope*. Island Press, Washington, DC.

Li, T. 2002. "Engaging simplifications: Community-based resource management, market processes and state agendas in upland Southeast Asia." *World Development* 30 (2): 265–83.

Lunn, N.J., S. Schliebe, and E. Born, eds. 2002. Polar Bears. Occasional Paper of the IUCN Species Survival Commission No. 26, IUCN, Gland, Switzerland.

McCay, B., and S. Jentoft. 1996. "From the bottom up: participatory issues in fisheries management." *Society and Natural Resources* 9: 237–50.

———. 1998. "Market or Community Failure? Critical Perspectives on Common Property Research." *Human Organization* 57: 21–29.

Ostrom, E., T. Dietz, N. Dolsak, P.C. Stern, S. Stonich and E.U. Weber, eds. 2002. *The Drama of the Commons*. Washington, DC: National Academy Press.

Pinkerton, E., ed. 1989. *Co-operative Management of Local Fisheries*. Vancouver: UBC Press.

Prestrud, P., and I. Stirling. 1994. "The International Polar Bear Agreement and the current status of polar bear conservation." *Aquatic Mammals* 20: 113–24.

Resilience Alliance 2004. *http://www.resalliance.org/programdescription/* (January 2004).

Reiersen, L., S. Wilson, and V. Kimstach. 2003. "Circumpolar perspectives on persistent organic pollutants: the Arctic Monitoring and Assessment Programme." In *Northern Lights Against POPs: Combating Toxic Threats in the Arctic*, edited by D. Downie and T. Fenge, 60–86. Montreal: McGill-Queen's University Press.

Rodon, T. 1998. "Co-management and self-determination in Nunavut." *Polar Geography* 22(2): 119–35.

Rothwell, D. 1996. *The Polar Regions and the Development of International Law*. Cambridge: Cambridge University Press.

Ross, K. 2000. *Environmental Conflict in Alaska*. Boulder: University of Colorado Press.

Selin, H. 2003. "Regional POPs policy: the UNECE CLRTAP POPs Protocol." In *Northern Lights Against POPs: Combating Toxic Threats in the Arctic*, edited by D. Downie and T. Fenge, 111–32. Montreal: McGill-Queen's University Press.

Shearer, R., and S. Han. 2003. "Canadian research and POPs: the Northern Contaminants Program." In *Northern Lights Against POPs: Combating Toxic Threats in the Arctic*, edited by D. Downie and T. Fenge, 41–59. Montreal: McGill-Queen's University Press.

Singleton, S. 1998. *Constructing Cooperation: the Evolution of Institutions of Co-Management*. Ann Arbor: University of Michigan Press.

Watt-Cloutier, S. 2003. "The Inuit journey towards a POPs-free world." In *Northern Lights Against POPs: Combating Toxic Threats in the Arctic*, edited by D. Downie and T. Fenge, 256–67. Montreal: McGill-Queen's University Press.

White, G. 2001. "And now for something completely northern: Institutions of governance in the Territorial North." *Journal of Canadian Studies* 35(4): 80–99.

Wiig, Ø., E.W. Born, and G.W. Garner, eds. 1995. Polar Bears. Occasional Paper of the IUCN Species Survival Commission (SSC) No. 10, IUCN, Gland, Switzerland.

Young, O. 2002. *The Institutional Dimensions of Environmental Change: Fit, Interplay and Scale*. Cambridge, MA: MIT Press.

CHAPTER 12

ADAPTIVE CO-MANAGEMENT OF ARCTIC CHAR IN NUNAVUT TERRITORY

Allan H. Kristofferson (Department of Fisheries and Oceans)
Fikret Berkes (University of Manitoba)

INTRODUCTION

The Inuit of Canada's new Territory of Nunavut are descendents of the people who made a living by following the resources of the land and the sea from one season to the next. Their continued existence is proof that their traditional management methods were indeed successful. The twentieth century has led to an end to this nomadic lifestyle. The contemporary Inuit are primarily located in settlements and live in a mixed economy that consists of wage employment, transfer payments, and subsistence resource harvests (Berkes and Fast 1996). Resource harvesting remains an important activity for economic, cultural, and nutritional reasons (Myers *et al.,* this volume).

Harvesting of fish, in particular Arctic char, near coastal communities, is still actively pursued. In addition to subsistence fisheries, small-scale commercial and recreational fisheries exist in various parts of Nunavut where resources permit. In some communities, such as Cambridge Bay on the south shore of Victoria Island (Figure 12.1), there exists an abundance of char beyond that needed to satisfy subsistence needs. This resulted in the development of a small commercial fishery in the early 1960s. With the development of this fishery came government regulation required by the laws of the land. Harvest levels were assigned on the basis of limited scientific data, but proved to be less than effective, as evidenced by declines. Recent research has led to a better understanding of sea run Arctic char populations in the area and the complexity of their stock structures. What is needed now is a more effective approach to the management of this important resource.

The Inuit of Cambridge Bay have developed a working relationship with Department of Fisheries and Oceans (DFO) managers, a relationship that has evolved over a number of years and led to the establishment of an informal

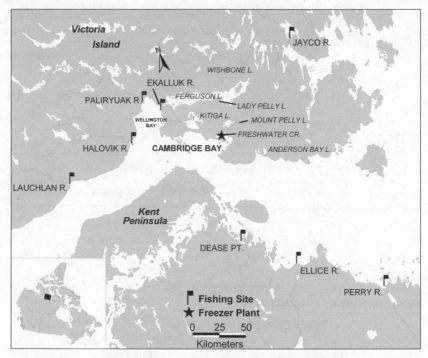

Figure 12.1 Fishing locations for sea run Arctic char.

co-management approach. Since then, under the Nunavut Land Claims Agreement of 1993, legislated co-management has been instituted in Nunavut Territory. This provides an opportunity to build upon past co-operative management experience.

In this chapter, we make the argument that adaptive co-management is the next step in the evolution of resource management with Arctic char and perhaps other species. Adaptive co-management systems are flexible community-based systems of resource management, tailored to specific places and situations and supported by, and working with, various organizations at different levels. Folke and others (2002, 20) define adaptive co-management as a process by which institutional arrangements and ecological knowledge are tested and revised in a dynamic, ongoing, self-organized process of learning-by-doing. Adaptive co-management is typically carried out by networks of actors sharing management power and responsibility (Olsson *et al.* 2004). It combines the *dynamic learning* characteristic of adaptive management (*e.g.,* Holling 1978; Walters 1986) with the *linkage* characteristic of co-operative management (*e.g.,* Pinkerton 1989).

We explore adaptive co-management as a practical method of dealing with the biological complexities of Arctic char stocks in a collaborative, learning-by-doing management approach that incorporates both traditional knowledge and scientific information. The objective of this chapter is to illustrate how such an adaptive co-management system can work. Further, this approach is not unique to the case study area but can probably be applied to the management of other Arctic char stocks in Nunavut.

The Arctic Char Resource

The Arctic char has a circumpolar distribution. In Canada it is found in Newfoundland and Labrador, north along the Ungava Peninsula to Hudson Bay, throughout the islands of the Arctic Archipelago and west to the Mackenzie River (McPhail and Lindsey 1970; Scott and Crossman 1973). It occurs as both a migratory (sea run) form and a nonmigratory form resident in lakes throughout the species range (Johnson 1980), and is widespread in Nunavut Territory.

The Arctic char spawns in fresh water in fall. The eggs incubate over winter and hatch in spring. The young then spend the first stage of their life entirely in fresh water. When they reach a size of about 150–200 mm, those that become sea run make their first migration to the sea, returning in fall to escape freezing. Generally, this pattern is repeated each year until they reach sexual maturity (Johnson 1980). Spawners appear to home to natal spawning grounds with a high degree of fidelity, resulting in the establishment of discrete stocks both between and within river systems (Kristofferson 2002). Spawning does not appear to take place in consecutive years, and there is evidence that Arctic char may not return to their home stream during nonspawning years.

The Arctic char is highly valued by the Inuit and is an excellent food source. In many places, there are competing demands for the species. On the one hand, populations are increasing and so is the general demand for more Arctic char to meet subsistence needs, even though demand for fish as dog food has declined since the 1970s (Usher 2002). On the other hand, economic opportunities are limited in Nunavut, and Arctic char fisheries, both commercial and sport, offer the promise of economic gain if they can be developed. These factors are contributing to an increased demand on Arctic char. However, the Arctic char cannot sustain heavy exploitation because of its relatively slow growth, low fecundity, and infrequent spawning and must be managed very carefully (Scott and Crossman 1973; Johnson 1980).

Management approaches that have been applied to Arctic char fisheries in Nunavut include the traditional one employed by the Inuit prior to government management and the conventional fishery management approach employed by government for northern commercial fisheries. Neither of these approaches deals successfully with the challenges of contemporary Nunavut, given the increase in the Inuit population, the need to develop a cash economy, and the biological complexity of the Arctic char resource itself. Thus, there is a need to develop and implement a new approach that can accommodate the new circumstances and provide sustainability over the long term. A short discussion of the previous management approaches is useful to provide a framework for the development of such a new management approach. The focus is on the Arctic char fishery in the Cambridge Bay area.

Traditional Management

The Arctic ecosystem is characterized by long cold winters, short cool summers, low annual biological production, and a general paucity of food resources. Human inhabitants had to utilize adaptive processes and survival strategies to ensure their existence over the long term (Balikci 1968). For example, the Inuit of Pelly Bay, formerly called the Netsilik Eskimos, followed an annual migration cycle. In winter they relied on seals out on the sea ice. In summer they moved inland, harvesting seals along shore and occasionally hunting caribou. In early autumn they fished for Arctic char using the stone weir or *saputit*. In late autumn the Netsilik fished for char through the thin river ice. In winter, they moved again onto the sea ice to pursue the seal (Balikci 1968). The Arctic char was a very important food source, and most harvesting took place during the autumn upstream migrations. In areas where Arctic char were abundant, starvation was rare (Balikci 1980).

The Inuit of the Cambridge Bay area, formerly called Copper Eskimos, had a seasonal economic cycle similar to that of the Netsilik (Damas 1968), using a mix of fish, marine mammals, and terrestrial mammals. Survival required that critical decisions be made to relocate if food sources went into decline in any particular area. The Inuit had accumulated a great deal of ecological and environmental expertise on a local level that provided them with a basis for this decision making (Riedlinger and Berkes 2001).

There are few studies on the traditional fishery management techniques of the Inuit of Nunavut Territory. Perhaps the most detailed study of subsistence fisheries in northern Canada comes from the James Bay area. Berkes (1999) summarizes a traditional fishery management approach employed by the Chisasibi Cree fishers of James Bay, for lake whitefish (*Coregonus clupeaformis*) and cisco (*C. artedi*), studied over a period of some fifteen years and reported through a series of research papers. The management strategy had three essential components. The first was to concentrate fishing effort on aggregations of fish, the second was to pulse fish intensively for a burst of time and then move on, and the third was to use methods that resulted in the harvest of a wide range of fish sizes. All three strategies were driven by the fishers' ability to detect declines in catch per unit of effort, and using this as an indicator of when to move on.

These strategies allowed Cree fishers to maximize their catch per unit of effort, making the best possible use of their time, selecting from among a range of possible resources. But at the same time, by moving on to other areas, to other fish stocks, and perhaps to other resources, they were able to conserve the existing stocks. Indeed, the distribution of the harvesting effort in space and time is not only used by the Cree and the Inuit but is one of the most commonly used traditional management practices throughout the world. Often, this practice takes the form of rotational use of fishing areas, harvesting one area intensively for a time but then lifting the fishing pressure completely, allowing the resource to replenish itself (Berkes 1999). There is strong evidence from the work of Johnson (1976) that the Inuit of the Central Arctic rotated their Arctic char fishing areas.

The third fishing strategy employed by the Cree (they use methods that result in the harvest of a range of fish sizes) produces another important ecological effect. Catching a range of sizes of fish instead of concentrating on the large (reproductive) ones would allow escapement of some of the spawners, thus ensuring the perpetuation of the stock. Modelling studies showed that the thinning of populations by the use of a mix of gill net mesh sizes (as the Cree fishers use) conserves population resilience, as compared to the wholesale removal of the older age groups by single large mesh size. The use of a mix of mesh sizes appears to be more compatible with the natural population structure than the use of a single large mesh size alone. Using a traditional Cree fishing strategy, models showed many reproductive year-classes remaining in the population. At the same time, the reduction of the overall population density likely increases productivity by stimulating growth rates and earlier maturation in the remaining fish, and helps the population renew itself (Berkes 1999, 125).

The Inuit of the Central Arctic seem to have practised management methods similar to those of the Cree fishers discussed above. By fishing for Arctic char at the *saputit* during the autumn upstream migration, they maximized their return for effort because Arctic char were present in great abundance in the upstream migration and were very vulnerable to capture in the shallow Arctic rivers. The char were also in prime condition after a summer of feeding in the sea and thus presented an ideal energy-rich food source. Arctic char of a variety of sizes were captured (Balikci 1980), thus allowing escapement of some of the potential spawners. By detecting declines in resource abundance, they would relocate to other systems, allowing the area to recover so that it could be fished again. This management approach of rotating fishing areas ensured the survival of both the Inuit and the fish.

Traditional management systems, such as the James Bay Cree fishery (Berkes 1999), tend to be adapted to the local area, and resource users themselves are the "managers." Allocation decisions are not made individually, and compliance is by social sanctions. These systems tend to have a large moral and ethical context, and there is no separation between nature and culture. Knowledge is primarily qualitative and data are diachronic in nature, that is, a long time-series of local information.

The Inuit of the Cambridge Bay area lived the traditional way of life until about 1946–47, following food sources through the seasons. The construction of the LORAN navigation beacon station at Cambridge Bay at that time served to create a wage economy which led to a concentration of Inuit in the settlement and a significant change from the traditional way of life. This event coincided with a decline in fox fur market, and the relatives of those employed drifted into Cambridge Bay for extended visits. This led to a further concentration of people in the community, ending their traditional lifestyle of living off the land by moving with the seasons (Abrahamson 1964).

Table 12.1

DISTINGUISHING CHARACTERISTICS OF INUIT TRADITIONAL MANAGEMENT PRACTICE FOR
ARCTIC CHAR VS. CONVENTIONAL SCIENTIFIC MANAGEMENT PRACTICE

The two sets of characteristics may be read as opposites, or they may be read as potential complementarities.

Inuit traditional management practice	Conventional management practice
Local knowledge of fish biology, *e.g.*, spawning areas, migration times	Universal knowledge of char biology applied locally
Diachronic information	Synchronic data
Qualitative observations related to management decision-making, *e.g.*, monitoring of catch per unit of effort, relative strength of spawning runs, fat content of fish	Quantitative data on population size by use of counting weir, age-specific growth rates, spawning sizes and frequencies, tagging to determine migrations
Indirect management by rotating fishing areas and spreading out fishing effort in space and time	Management by annual harvest quotas on assumed discrete stocks
Social enforcement of accepted, proper Inuit practice	Tools: quotas, gillnet mesh sizes, closed seasons
Sharing by social agreement and convention	Allocation decisions made by distant authorities
Enforcement by social mechanisms and, under the 1993 Nunavut Land Claims Agreement, through co-management mechanisms	Enforcement by the laws of the land, Federal Government fishery-related acts and regulations

Conventional Management

Conventional management, in contrast to traditional management, is based almost exclusively on scientific information and methods, using primarily quantitative data. These data are often synchronic in nature. That is, they are simultaneously observed, with little time depth. Conventional management takes a reductionist approach to the resource, and ecological complexities and uncertainties are often ignored. Resource managers are not the users themselves, and allocation decisions are made at a distance from the community. Such an approach leads managers in the direction of tighter government controls over fisheries. Such top-down management, over time, may become unworkable (Holling and Meffe 1996).

The conventional management approach was applied in the development of the commercial fishery for Arctic char in the Cambridge Bay area. Government established fishing areas, a harvest limit or quota, fishing seasons and, ultimately, a minimum mesh size limit (139 mm) for gillnets used in the commercial fishery (Barlishen and Webber 1973; Kristofferson and Carder 1980; Kristofferson *et al.* 1984). There were no regulations limiting the subsistence harvest. Table 12.1 summarizes some of the major features of the two kinds of approach to fishery management.

The Arctic char resource in nearby Freshwater Creek (Figure 12.1) was once abundant, which was why Inuit gathered seasonally at this "fair fishing place." However, the concentration of Inuit in the settlement of Cambridge Bay also meant increased concentration of fishing pressure in the area near the community, rather than rotational use over a wider geographic area. In 1961, the fishing co-operative was formed to begin the commercial exploitation of Arctic char in the area, and the first commercial fishery took place at nearby Freshwater Creek. However, the Freshwater Creek Arctic char fishery was showing signs of depletion (Barlishen and Webber 1973) because it was already supporting a large subsistence fishery as well as a non-native recreational fishery. Angling for Arctic char provided community residents with a much-needed pastime after a long cold winter. Therefore, the commercial fishery was relocated to the Ekalluk River in 1962 (Abrahamson 1964).

At the outset, an annual quota of 18,000 kg was allocated to the Ekalluk River commercial fishery (Barlishen and Webber 1973). This river-specific quota remained in effect until 1967, when area fishers petitioned the federal government for an area quota for Wellington Bay. The intention was to allow commercial fishing to take place at the Lauchlan, Halovik, and Paliryuak rivers that flow into Wellington Bay, as well at the Ekalluk River. An area quota of 46,000 kg was subsequently allocated to Wellington Bay. In 1967, the fishery reported sales of $28,904 and a net operating loss of $3,112. Fishers received $9,324 for their fish, while labour for processing earned $1,321 (Barlishen and Webber 1973).

The economic constraints of developing a commercial fishery in this area were severe. Float-equipped aircraft were used to transport the catch from the fishing sites to Cambridge Bay, and the frozen product was flown to markets in the South. In order to make a profit, fishers had to maximize the harvest and minimize the overhead. This led to a concentration of fishing effort at the Ekalluk River from 1967 to 1969. The result was a serious decline in the average size of Arctic char in the catch at Ekalluk River by 1969. The average weight of Arctic char taken in the Ekalluk River commercial fishery in 1963 was 3.9 kg. This had dropped to 1.4 kg by 1969. Consequently, the commercial fishery at the Ekalluk River was closed in 1970.

Following the closure of the Ekalluk River commercial fishery, river-specific quotas were put in effect and remain so to the present. This was based on the assumption that each river supported a discrete stock of Arctic char (Kristofferson et al. 1984). Gillnets, with a minimum mesh size of 139 mm, still predominate in the fishery, but a weir, now adopted for commercial harvesting purposes, has been used periodically at three sites (Jayco, Ekalluk, and Halovik rivers). This is a traditional-style weir but made of modern material (conduit pipe). There is no minimum size limit for Arctic char taken in the weir, although experience has shown that the larger char are selected. These river fisheries appear to have been sustained over the years. In 2003, Kitikmeot Foods Ltd., which runs the fishery, reported a harvest of 42,000 kg of Arctic char with sales of $450,000.

A total of twenty-four fishers received $140,000 for their efforts. The processing plant employed another sixteen people, who earned a total of $130,000 that year (C. Schindel, pers. comm.). However, in light of subsequent studies, the Arctic char resource, on a stock-by-stock basis, might well have been utilized in a less-than-effective manner.

Managing Arctic Char with Complex Stocks

Conventional fishery management, as used in the Canadian North and elsewhere, assumes that harvest quotas can be assigned on discrete stocks, whether for Atlantic cod or for Arctic char. There are a number of shortcomings of this approach, as criticized over the years (Charles 2001). It does not take into account ecosystem interactions such as predator-prey and competition relations; it does not take into account the year-to-year environmental variability; and it does not take into account complexities in stocks. Does the "discrete stock" assumption of conventional management hold in the Cambridge Bay fishery?

The study by Kristofferson (2002), which has revealed that multiple stocks of Arctic char spawn and overwinter within individual river systems, has complicated the management challenge. Significant differences in morphology (Figure 12.2) and trace elements (strontium) in otoliths (Figure 12.3) were found among aggregations of spawners both within and between river systems. This information supports the current river-specific harvest limits. However, the trace element data (Figure 12.3) indicate that fall upstream migrations are comprised of an admixture of Arctic char from the different resident stocks, as well as itinerant Arctic char from other river systems that migrate in only for overwintering purposes.

Such complexities pose a challenge to the conventional management approach, which is based on the assumption that the fishery is targeting a homogeneous stock at each fall fishing site. Random samples are taken each year from the commercial harvest that is carried out on these upstream migrations. Length and age data gathered over successive years are examined annually to determine the response of the stock to certain harvest levels. These random samples likely have no biological meaning because the harvest is comprised of Arctic char from more than one stock and the proportional contribution of each stock to the fishery is unknown. Such data would not be sensitive to a decline of smaller, more vulnerable stocks, and larger stocks could be harvested at less than optimal levels. Thus, utilizing these data for monitoring and modelling purposes is likely to give spurious results. Clearly, there is a need to manage these Arctic char fisheries as mixed-stock fisheries, and to develop appropriate techniques to do so.

A number of techniques have been used to estimate stock composition in the conventional management of mixed-stock fisheries. Examples of these techniques utilize stock differences based on morphology (Messinger and Bilton 1974; Cook 1982; Fournier et al. 1984; Friedland and Reddin 1994), enzyme electrophoresis (Utter and Ryman 1993), mitochondrial DNA (Bermingham et al. 1991) and nuclear DNA (Galvin et al. 1995). These and other techniques should be investigated in

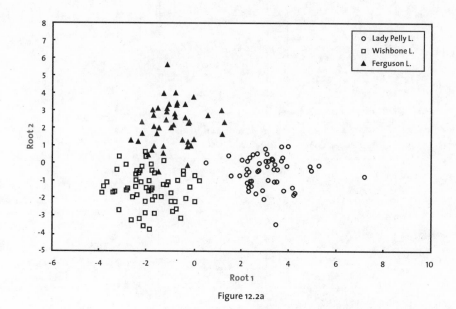

Figure 12.2a

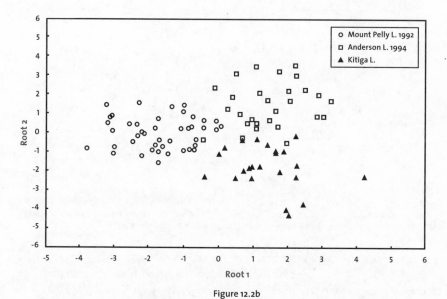

Figure 12.2b

Delineation of three discrete stocks of Arctic char based on morphology (Discriminant Function Analysis) (a) within the Ekalluk River system and (b) among three different river systems (from Kristofferson 2002).

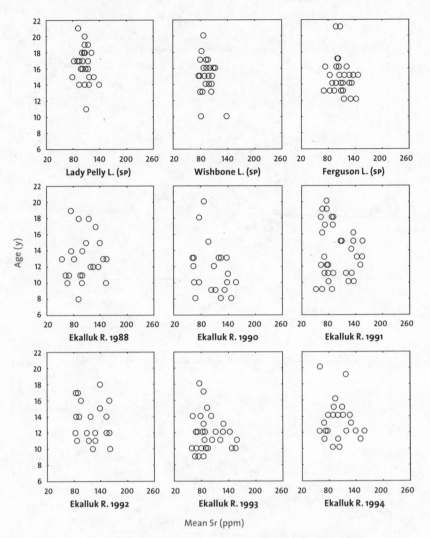

Figure 12.3 Mean strontium concentration in otoliths of three groups of spawners (upper row) compared with those from six upstream runs (below). The greater variation in samples from upstream runs is indicative of mixed stocks. (from Kristofferson 2002).

terms of their usefulness for managing mixed-stock Arctic char fisheries in the study area. However, even when this is done, the current conventional management strategy alone will likely not deal adequately with stock complexity and environmental uncertainty.

Evolution of Co-management in the Cambridge Bay Fishery

In many cases, fisheries need to be managed on a small ecological scale, taking into account local ecological factors such as habitat and local populations that

The community of Cambridge Bay, Nunavut, located on the south coast of Victoria Island.
Photo by D.K. McGowan, Department of Fisheries and Oceans.

are central to the health of the whole ecosystem. Fisheries management needs to be designed to fit this smaller scale by allowing resource users to take more responsibility for management and by utilizing their local knowledge of the resource (Berkes *et al.* 2001a). This can be accomplished through co-management, defined as a sharing of power and responsibility between the state and resource users in the management of natural resources (Pinkerton 1989).

Co-management as a process is flexible and participatory, and provides a forum for rule making, conflict management, power sharing, leadership, dialogue, decision making, negotiation, knowledge generation and sharing, learning, and development among resource users, stakeholders, and government (Berkes *et al.* 2001a). Co-management allows passing of responsibilities to resource users who then become accountable for their decisions. It can use fishers' own local knowledge, so that they become active participants in the development of management plans.

In Canada, almost all of the Arctic char resource is found in areas under land claims agreements: the Nunavut Territory, the Inuvialuit Settlement Region of the Northwest Territories, the Ungava region (which is under the *James Bay and Northern Quebec Agreement*), and the Labrador coast (where a land claims agreement is nearly finalized). The settlement of land claims in these areas has formalized resource co-management (Berkes *et al.* 2001b), and almost all Arctic char stocks are under joint jurisdiction. The details of sharing of jurisdiction for fisheries management can be found in specific sections of the *Nunavut Land Claims Agreement* (1993), the *Inuvialuit Final Agreement* (1984), and the other agreements.

The counting weir located upstream on Freshwater Creek.
Photo by D.K. McGowan, Department of Fisheries and Oceans.

Well before the *Nunavut Land Claims Agreement* came into effect in 1993, a form of co-management had developed between the Government of Canada, Department of Fisheries and Oceans (DFO), and the residents of Cambridge Bay. Initially, the local participation in management was limited to the employment of Inuit as technicians in management work.

During the late 1970s and early 1980s, DFO staff used a weir to enumerate the upstream migration of Arctic char at various commercial fishing sites in the Cambridge Bay area. Local Inuit were hired to assist in these projects and became familiar with this counting technique. Presentations of the results of these projects to community members contributed to a better understanding of what could be accomplished with this technique. Community members were well aware of the dwindling Arctic char resource in the nearby Freshwater Creek. Through the local *Ekaluktutiak Hunters and Trappers Association* (now called *Hunters and Trappers Organization*), they approached DFO with a request to enumerate the upstream migration of Arctic char in Freshwater Creek as a first step toward rehabilitation of the stock. DFO complied with this request and the upstream migration was enumerated by weir in 1982. This included a tagging program to determine the level of exploitation, harvest by fishery (recreational, subsistence), and the seasonal distribution of the Arctic char of Freshwater Creek.

The 1982 weir project counted 9,961 Arctic char in the upstream migration (McGowan and Low 1992), and 1983 returns on the 808 Arctic char tagged in 1982 revealed an exploitation rate in excess of 12 per cent. A study at nearby Nauyuk Lake (Johnson 1980) indicated that this was excessive for Arctic char stocks in the area. The estimated total harvest in 1983 was just under 2,000 Arctic char. A creel census taken in 1983 (Carder 1991), combined with tag returns from the various fisheries, revealed that about 46 per cent of the harvest was taken by the recreational fishery, 50 per cent by the subsistence fishery, and 4 per cent by the commercial fishery. In the following years, 86 per cent of all tag recoveries (N = 163) were made in Freshwater Creek, the sea near Cambridge Bay, or nearby Greiner Lake. The small number of Arctic char counted in the 1982 assessment convinced the residents of Cambridge Bay to develop a recovery plan for the Freshwater Creek Arctic char stock. Although it took time to implement, it appears to have been somewhat successful.

The evolution of the informal co-management in this fishery did not occur in a planned way. It occurred through the mutual recognition of a problem. The results of the tagging study revealed that both the recreational and subsistence fisheries were targeting the Freshwater Creek Arctic char stock. This provided information to the community and government that there was a need to reduce the harvest of both fisheries. The government responded by reducing the catch and possession limit for non-native sport fishers, and the community implemented a ban on its own subsistence gillnet fishing.

The locations where tagged char were captured provided the information needed to delineate the area where fishing pressure had to be reduced. The periodic counts of the upstream migration and the increase in migrant char observed in these counts provided information to community residents that their recovery program appeared to be successful. The periodic monitoring of the harvest provided evidence of compliance with the fishing restrictions. In essence, government and the community discovered through experience that each had critical information necessary to address the problem.

They found that by working together, they were able to accomplish what has been interpreted as a recovery of the Freshwater Creek Arctic char stock. In terms of power sharing, government had the authority to reduce daily catch and possession limits for non-native fishers, and they did so, from four Arctic char per day and seven in possession, to one per day. Government has no regulation on the subsistence fishery, but the community put a moratorium on subsistence gillnetting into effect, and ensured compliance through community sanctions. The informal co-management at Cambridge Bay is summarized in Table 12.2. As a footnote, although the number of Arctic char counted in the 1994 upstream migration was less still than that counted in 1991, the average size of char in the run had increased by 1994, as had the proportion of char of reproductive size.

Legislated Co-management under the Nunavut Agreement

Clearly, the sharing of management power is not a new concept to the residents of Cambridge Bay. But in any case, joint management has been instituted across the North through land claims agreements (Berkes and Fast, this volume). The legislated co-management put in effect in 1993 by the *Nunavut Land Claims*

Table 12.2

A SUMMARY OF INFORMATION-SHARING AND DECISION-MAKING BETWEEN DFO AND
THE COMMUNITY OF CAMBRIDGE BAY THAT REPRESENTS AN INFORMAL
CO-MANAGEMENT APPROACH TO THE ARCTIC CHAR FISHERY AT FRESHWATER CREEK

Government	Community
Weir count 1982 (9 961)	Community participation in creel census
Tagging program 1982 (N=808)	Moratorium on subsistence gillnets (1988)
Creel census 1983	Community monitoring to comply with moratorium on gillnets
Weir count 1988 (36 933)	Door-to-door harvest survey (1992, 93,94)
Reduce recreational limit (1 char daily) 1991	Continue dialogue with DFO
Weir count 1991 (39 559)	Continue dialogue with DFO
Weir count 1994 (26 150)	Concern by community on lower count

Agreement (NLCA) has provided the Inuit with the legal arrangements necessary to establish rights over natural resources, including fisheries. International experience suggests that a legal arrangement is necessary if co-management is to be durable and successful (Pomeroy and Berkes 1997; Berkes *et al.* 2001a).

Under Nunavut agreement, *Article 5, Wildlife,* the Principles (5.1.2 NLCA) recognize, among other things, that "there is a need for an effective system of wildlife management that complements Inuit harvesting rights and priorities, and recognizes Inuit systems of wildlife management that contribute to the conservation of wildlife and protection of wildlife habitat." The Principles also recognize the "need for an effective role for Inuit in all aspects of wildlife, including research," and that "Government retains the ultimate responsibility for wildlife management." The *Objectives* (5.1.3 NLCA) recognize, among other things, the creation of a wildlife management system that "fully acknowledges and reflects the primary role of Inuit in wildlife harvesting," and "invites public participation and promotes public confidence, particularly amongst Inuit."

The Nunavut Wildlife Management Board (NWMB) was established as the main instrument of wildlife management in the Nunavut Settlement Area (5.2.1 NLCA). It is a nine-member board with four Inuit representatives, four government representatives, and a chairperson nominated by the NWMB. While it recognizes that Government has the ultimate responsibility for wildlife management (5.2.33 NLCA), the NWMB functions in a variety of different ways. It participates in research, conducts the Nunavut Wildlife Harvest Study, rebuts presumptions as to need, establishes, modifies, or removes levels of total allowable harvest, ascertains and adjusts basic needs level, allocates resources to other residents and existing operations. It also deals with priority applications, makes recommendations as to allocation of remaining surplus, establishes, modifies or removes non-quota limitations, sets trophy fees, and any other function required by the agreement.

Recognizing the ability and the right of the Government of Canada to carry out its research function, the NWMB also has a role to play in research as outlined in 5.2.37 of the NLCA. This includes identifying research requirements and deficiencies pertinent to wildlife management, identifying relevant persons and agencies to undertake wildlife research, and promoting the employment of Inuit and Inuit organizations in research.

The Department of Fisheries and Oceans is charged with the development of Integrated Fisheries Management Plans (IFMP) with resource users across Canada. The mandate for the development of IFMPs also holds in areas where settled land claims exist. These IFMPs are based on the principles of co-management. Inuit traditional knowledge or *Inuit Qaujimajatuqangit* (IQ) has traditionally been a part of Inuit systems of fisheries management. The DFO recognizes the role of IQ in all aspects of fisheries management, including research, and the need to incorporate it into IFMPs wherever possible.

DISCUSSION AND CONCLUSIONS

Adaptive management is a relatively new management approach that has developed out of concern with uncertainty in fishery and wildlife management (Holling 1978). It advocates learning from management successes and failures, and relies on systematic feedback learning. It utilizes common-sense logic that emphasizes learning-by-doing and it eliminates the barrier between research and management. Adaptive management can be viewed as a rediscovery of traditional systems of knowledge and management. Although there are differences between the two, adaptive management is, in a sense, the scientific analog of traditional ecological knowledge because it integrates uncertainty into management strategies and it emphasizes practices that confer resilience (Berkes *et al.* 2000).

Adaptive and conventional resource management differ primarily in approach and scientific methodology (McDonald 1988). Conventional resource managers attempt to simplify complex relationships in harvesting systems, accumulating large quantities of data that form the basis of conservative harvesting policies until a better biological understanding can be achieved. Adaptive managers, on the other hand, acknowledge uncertainty and attempt to identify key relationships in an ecosystem that can provide a measure of how the resource responds to various management practices. While both management approaches recognize the need for management, they differ in their perception of the role of biological uncertainty (McDonald 1988). The conventional approach assumes that biological uncertainties can be resolved through research and modelling. The adaptive management approach recognizes the inherent uncertainty of ecological systems and emphasizes the need to learn from experience and experimentation to deal effectively with uncertainty (Walters 1986).

Co-management is adaptive because it is based on learning through information sharing among stakeholders, leading to problem solving in a stepwise matter and to iterative improvements in management (Berkes *et al.* 2001a). This is indeed what has happened through experience at Freshwater Creek in Cambridge Bay. The case study also indicates another key element of co-management.

An essential ingredient for successful co-management seems to be the establishment of a level of trust among all involved. Government fisheries personnel spent a number of years in Cambridge Bay working on river systems other than Freshwater Creek and involved residents in the field studies, and residents began to see the value of this work. Community meetings were held explaining the results of the studies as they became available. People got to know one another as individuals and knowledge was shared freely. Because co-management is adaptive, allowing participants to adjust their activities based on results obtained and lessons learned, modifications to the recovery plan took place as new data became available.

Community members understand their own situations better than outsiders do, and can devise and administer regulatory mechanisms that are often more appropriate than those imposed by external regulations. While government could, and did, restrict the harvest of char by non-native anglers in Freshwater Creek through changes in regulations, no such mechanism existed to limit the subsistence harvest. This was accomplished through community sanctions. Community involvement may give fishers a sense of ownership that often translates into greater compliance with management measures over the long term. This apparently happened in the Cambridge Bay experience.

Legislative change under the *Nunavut Land Claims Agreement* now sets the stage for formalizing this informal arrangement and building on it. Can the collaborative management that evolved in Cambridge Bay be characterized as adaptive co-management in the sense used by Folke *et al.* (2002) and Olsson *et al.* (2004)? The informal arrangement certainly has some of the elements of adaptive co-management. But the arrangement can be improved through more effective collaboration (as now legally required) and more systematic adaptive management that incorporates active learning into the management design (Walters 1986). An outline for an adaptive co-management process for the Arctic char resource in the Cambridge Bay area is presented in Table 12.3. It follows the three cyclical phases in the adaptive management process presented by Walters (1986), which includes identifying a range of management alternatives, developing key management indicators, and the design and implementation of effective monitoring systems. While there is no single "correct" method of implementing adaptive management (Holling 1978), it is necessary to develop a plan that is specific to the system, and then to implement it. Success may be achieved over time, essentially by learning-by-doing.

The study by Kristofferson (2002) provided information that has revealed a level of complexity in stock structuring of sea run Arctic char that was previously suspected but not proven. Currently, there is no way to manage Arctic char in the Cambridge Bay area on a stock-by-stock basis. Even if stock admixtures can be taken into account, the cost of the data needed would be prohibitive.

Table 12.3

STEPS TO IMPLEMENT AN ADAPTIVE CO-MANAGEMENT PLAN FOR ARCTIC CHAR IN THE CAMBRIDGE BAY AREA

Headings follow McDonald (1988).

Phase	Implementation Action
Dialogue	Conduct a community presentation to outline the study results and the problem. Include a discussion of conventional management, identification of management goals and the need for an alternative approach. Develop a shared understanding of the management problem to be solved. Document Inuit understanding of the problem and how they perceive the resource. Discuss commonalities and differences
Field Study and Analysis	Identify the need to collect and analyze additional information to provide a better understanding of the biological relationships within the ecological system that relate to key questions posed by management goals. Incorporate traditional ecological knowledge such as identification of additional spawning grounds within river systems. Develop methods to determine the relative contributions of different stocks to a mixed fishery.
Design of Alternative Management Actions	Explore alternative management options jointly such as pulse fishing, use of different gillnet mesh sizes, weirs, timing of fishery, that can be tested within the range of predictive outcomes.
Monitoring and Assessment of Management Actions	Analyze management actions in relation to outcomes predicted by ecological theory. Identify key indicators in the system (index netting of spawning aggregations) to ensure the quality of the monitoring system. Maintain continuous dialogue with fishers to assess their "gut feelings" of responses of stocks to each management action.
Evaluation	Determine likely impacts of alternative management options (modelling) in view of the different approaches taken. Jointly identify key questions posed by the management options which initiates subsequent rounds of the adaptive management process.

The conventional management approach, as it has been applied to this fishery, has proven to be less than effective in light of this additional complexity. An adaptive approach can utilize different methods such as rotational pulse fishing, removing a range of sizes of Arctic char with variable mesh size gillnets or weirs, and fishing over the duration of upstream runs to spread effort over as many stocks as possible, if there is temporal segregation of returning stocks. As many spawning sites as possible need to be identified in each river system fished, and this can often be accomplished using traditional ecological knowledge (Inuit IQ). Periodic monitoring of these spawning assemblages can be used to assess the effects of a particular management method on individual stocks. Changes in the management plan can then be implemented if data indicate a decline in any particular stock.

An adaptive co-management approach, implemented under current legislation in effect throughout most of the distribution area of Arctic char in Canada, offers a potentially effective way to manage the Arctic char resource, while simultaneously providing optimum socio-economic benefits to resource users. It may provide an opportunity to combine traditional ecological knowledge with the scientific research that will ultimately lead to a better understanding of biological complexities and ways of dealing with ecological uncertainties. It will also provide users with the incentive to utilize the resource in the best manner possible because of their partnership in management, ultimately contributing to more effective resource management in the Cambridge Bay area and elsewhere throughout the Territory of Nunavut.

REFERENCES

Abrahamson, G. 1964. "The Copper Eskimos, an area economic survey." Department of Indian and Northern Affairs. 194 pp.

Balikci, A. 1968. "The Netsilik Eskimos: adaptive processes." In *Man the hunter*, edited by R.B. Lee and I. DeVore, 72–82. Chicago: Aldine.

———. 1980. "Charr fishing among the Arviligjuarmiut." In *Charrs: salmonid fishes of the genus Salvelinus.*,edited by E. K. Balon, 7–9. The Hague: Dr. W. Junk.

Barlishen, W.J., and T.N. Webber. 1973. A history of the development of commercial fishing in the Cambridge Bay area of the Northwest Territories. Unpublished Report for the Federal-Territorial Task Force report on Fisheries Development in the Northwest Territories. 37 pp.

Berkes, F. 1999. *Sacred ecology: Traditional ecological knowledge and resource management.* Philadelphia and London: Taylor and Francis.

———, and H. Fast. 1996. "Aboriginal peoples: The basis for policy-making towards sustainable development." In *Achieving Sustainable Development*, edited by A. Dale and J.B. Robinson, 204–64. Vancouver: UBC Press.

———, J. Colding, and C. Folke. 2000. "Rediscovery of traditional ecological knowledge as adaptive management." *Ecological applications* 10(5): 1251–62.

———, R. Mahon, P. McConney, R. Pollnac, and R. Pomeroy. 2001a. Managing small scale fisheries. Ottawa: International Development Research Centre.

———, J. Mathias, M. Kislalioglu, and H. Fast. 2001b. "The Canadian Arctic and the Oceans Act: the development of participatory environmental research and management." *Ocean and Coastal Management* 44: 451–69.

Bermingham, E., S.H. Forbes, K. Friedland, and C. Pla. 1991. "Discrimination between Atlantic salmon (*Salmo salar*) of North America and European origin using restriction analysis of Mitochondrial DNA." *Can. J. Fish. Aquat. Sci.* 48: 884–93.

Carder, G. 1991. "Creel census and biological data taken from the sport fishery for Arctic charr, *Salvelinus alpinus* (L.), at Freshwater Creek, Northwest Territories, 1981–1983." *Can. Data Rep. Fish. Aquat. Sci.* 851: iv + 13 pp.

Charles, A. 2001. Sustainable fishery systems. *Fish and Aquatic Resources Series 5.* Oxford: Blackwell Science.

Cook, R.C. 1982. "Stock identification of sockeye salmon (*Oncorhynchus nerka*) with scale pattern recognition." *Can. J. Fish. Aquat. Sci.* 39: 611–617.

Damas, D. 1968. The diversity of Eskimo societies. *Man the hunter*, edited by R.B. Lee and I. DeVore, 111–17. Chicago: Aldine.

Folke, C., S. Carpenter, T. Elmqvist *et al.* 2002. *Resilience for sustainable development: Building adaptive capacity in a world of transformations.* Paris: ICSU, Rainbow Series No. 3. *http://www.sou.gov.se/mvb/pdf/resiliens.pdf*

Fournier, D.A., T.D. Beacham, B.E. Riddell, and C.A. Busack. 1984. "Estimating stock composition in mixed stock fisheries using morphometric, meristic and electrophoretic characteristics." *Can. J. Fish. Aquat. Sci.* 41: 400–408.

Friedland, K.D., and D.G. Reddin. 1994. "Use of otolith morphology in stock discriminations of Atlantic salmon (*Salmo salar*)." *Can. J. Fish. Aquat. Sci.* 51: 91–98.

Galvin, P., S. McKinnell, J.B. Taggart, A. Ferguson, M. O'Farrell, and T.F. Cross. 1995. "Genetic stock identification of Atlantic salmon using single locus minisatellite DNA profiles." *J. Fish. Biol.* 47(supp. A): 186–199.

Holling, C.S. 1978. Adaptive environmental assessment and management. Wiley International Series on Applied Systems Analysis, vol. 3. Chichester, UK: Wiley.

—— and G.K. Meffe. 1996. "Command and control and the pathology of natural resource management." *Conservation Biology* 10: 328–37.

Inuvialuit Final Agreement. 1984. Indian and Northern Affairs, Canada. 113 pp.

Johnson, L. 1976. "Ecology of arctic populations of lake trout, *Salvelinus namaycush*, lake whitefish, *Coregonus clupeaformis*, arctic char, *S. alpinus*, and associated species in unexploited lakes of the Canadian Northwest Territories." *Journal of the Fisheries Research Board of Canada* 33: 2459–88.

——. 1980. "The Arctic charr, *Salvelinus alpinus*." In *Charrs: Salmonid fishes of the genus Salvelinus*, edited by E.K. Balon, 15–98. The Hague: Dr. W. Junk.

Kristofferson, A.H. 2002. "Identification of Arctic char stocks in the Cambridge Bay Area, Nunavut Territory, and evidence of stock mixing during overwintering." Ph.D. dissertation, University of Manitoba. 255 pp.

——, and G. Carder. 1980. "Data from the commercial fishery for Arctic charr, *Salvelinus alpinus* (Linnaeus), in the Cambridge Bay area of the Northwest Territories, 1971–1978." *Can. Data Rep. Fish. Aquat. Sci.* 184: v + 25 pp.

——, D.K. McGowan, and G.W. Carder. 1984. "Management of the commercial fishery for anadromous Arctic charr in the Cambridge Bay area, Northwest Territories, Canada." In *Biology of the Arctic charr*, edited by L. Johnson and B.L. Burns, 447–61. Proceedings of the International Symposium on Arctic charr, Winnipeg, May 1981. Winnipeg: University of Manitoba Press.

McDonald, M. 1988. "An overview of adaptive management of renewable resources." In Traditional knowledge and renewable resource management in northern regions, edited by M.M.R. Freeman and L.N. Carbyn, 65–71. Occasional Publication No. 23. IUCN Commission on Ecology and the Boreal Institute for Northern Studies.

McPhail, J.D., and C.C. Lindsey. 1970. "Freshwater fishes of north-western Canada and Alaska." *Fish. Res. Board Can. Bull.* 173: x + 381 pp.

McGowan, D.K., and G. Low. 1992. "Enumeration and biological data on Arctic charr from Freshwater Creek, Cambridge Bay area, Northwest Territories, 1982, 1988, and 1991." *Can. Data Rep. Fish. Aquat. Sci.* 878: iv + 23 pp.

Messinger, H.B., and H.T. Bilton. 1974. "Factor analysis in discriminating the racial origin of sockeye salmon (*Oncorhynchus nerka*)." *J. Fish. Res. Board Can.* 31: 1–10.

Nunavut Land Claims Agreement. 1993. Tungavik and the Department of Indian Affairs and Northern Development, Ottawa. 279 p.

Olsson, P., C. Folke and F. Berkes 2004. "Adaptive co-management for building resilience in social-ecological systems." *Environmental Management* 34: 75–90.

Pinkerton, E., ed. 1989. *Cooperative management of local fisheries: new directions for improved management and community development.* Vancouver: UBC Press.

Pomeroy, R.S., and F. Berkes. 1997. "Two to tango: the role of government in fisheries co-management." *Marine Policy* 21: 465–80.

Riedlinger, D., and F. Berkes. 2001. "Contributions of traditional knowledge to understanding climate change in the Canadian Arctic." *Polar Record* 37: 315–28.

Scott, W.B., and E.J. Crossman. 1973. "Freshwater fishes of Canada." *Fish. Res. Board Can. Bull.* 184: xiii + 955 pp.

Usher, P.J. 2002. "Inuvialuit use of the Beaufort Sea and its resources, 1960–2000." *Arctic* 55 (supp. 1): 18–28.

Utter, F., and N. Ryman. 1993. "Genetic markers and mixed stock fisheries." *Fisheries* 3(8): 11–21.

Walters, C. J. 1986. *Adaptive management of renewable resources.* New York: McGraw-Hill.

CHAPTER 13

UNPACKING SOCIAL LEARNING IN SOCIAL-ECOLOGICAL SYSTEMS:

CASE STUDIES OF POLAR BEAR AND NARWHAL MANAGEMENT IN NORTHERN CANADA

Alan Diduck (University of Winnipeg)
Nigel Bankes (University of Calgary)
Douglas Clark (Wilfrid Laurier University)
Derek Armitage (Wilfrid Laurier University)

As northerners search for appropriate responses to rapid environmental change and look for tools to guide nature-society interactions along sustainable trajectories, social learning (*i.e.,* learning at collective levels) provides a promising frame of reference. First, it can help organizations, governments, and communities adapt in basic ways to changing social and environmental conditions. Additionally, and perhaps more importantly, it is a way for such groups to generate positive change through collective expressions of human agency. Moreover, in the face of uncertain futures, it can illuminate legitimate endpoints and help craft means for reaching those endpoints.

This chapter provides a social learning analysis of the Nunavut Wildlife Management Board's (NWMB) responses in two recent management crises. The NWMB is a co-management board created under the terms of the Nunavut Land claims agreement. The chapter examines the board's introduction of community-based management (CBM) of narwhal in five communities, and a subsequent problem that arose when the fishery was closed in one of those communities. It also reviews the NWMB's polar bear mandate and considers the board's response in a crisis engendered by the recognition that one population of bears, the M'Clintock Channel (MC), was 50 per cent smaller than estimated. (For ease of reference, a list of acronyms is provided at the front of the book.)

The first part of the chapter contextualizes our work in a general way in the growing literature on social learning in resource and environmental management. Part two summarizes our theoretical framework. The third part presents the narwhal case study, and the fourth presents the polar bear case. The cases were chosen because, while subject to different legislative regimes, they involved the same co-management authority (the NWMB), comparable resource types (large marine mammals), and similar management crises (relating to population estimates and harvesting levels).

Part five of the chapter is a comparative analysis of the cases. The analysis describes evidence of social learning processes, using constructs from the theory. It also identifies systemic and institutional features that enabled or inhibited such learning, relying on a grounded or inductive approach. Part five provides a largely theoretical exploration and is based principally on secondary documentation from selected sources (mostly NWMB and government records). Nevertheless, it provides a good starting point for more detailed intensive case studies of social learning dynamics in co-management boards created by other northern land claims agreements. The chapter concludes with a review of the implications for management, and considers policy and practice reforms that could encourage social learning for sustainability in the North.

SOCIAL LEARNING PERSPECTIVES IN RESOURCE MANAGEMENT

Discussions of social learning often start with reference to the politics and epistemology of John Dewey, who argued that public policy decisions should be viewed as a series of experiments. He argued that, guided by the principles of scientific inquiry and bounded by democratic debate, experimental politics would yield progressive social improvement (Dewey and Sidorsky 1977). In the planning literature, Freidmann's (1987) transactive model built on Dewey's notion of learning by doing, and elaborated the dynamics of interpersonal relations and the institutional arrangements conducive to learning at collective levels. In the organizational development literature, Argyris and Schön (1978), Senge (1990), and others examined social learning ideas in the context of corporate governance. Their work advanced systems thinking in social contexts, presented innovative and coherent theoretical constructs, and furnished rich empirical evidence of key ideas. As discussed below, our analysis is founded on concepts from this literature. Further important contributions to the social learning literature have been made in sociology, psychology, politics, and other disciplines. Excellent reviews of the literature can be found in Argyris and Schön (1978), Friedmann (1987), Levitt and March (1988), and Parson and Clark (1995).

In resource and environmental management, seminal publications by Holling (1978), Walters (1986), and Lee (1993) applied notions of learning by doing to large-scale management interventions. Nelson and Serafin's (1996) civics approach modelled resource management functions (*e.g.*, planning, assessment, monitoring) as mutual learning processes. Webler *et al.* (1995) helped focus attention on the learning outcomes of deliberative public involvement programs. Recently, social learning studies have identified properties that enable or hinder learning by international institutions in the management of global environmental risks (Haas 2000), described participatory agent-based modelling as a forum for social learning and for fostering changes toward sustainability in the water sector (Pahl-Wostl 2002), and illuminated connections between social capital and social learning (Fien and Skoien 2002). Yet other research has applied social learning frameworks in analyzing policy changes (Brown 2000; Fiorino 2001), multilateral negotiations (Eckley 2002), sustainability indicators (Shields *et al.* 2002), international development initiatives (Dyck *et al.* 2000), and non-formal education (Krasny and Lee 2002).

In addition, social learning is an important part of the emerging framework of social-ecological resilience (Gunderson and Holling 2002; Walker *et al.* 2002; Folke *et al.* 2002; Berkes *et al.* 2003). In resilience thinking, society and nature are viewed as interconnected, complex, adaptive systems. The core concept of resilience explicitly includes as a defining characteristic the capability for learning (along with capacity for adaptation, ability to absorb change, and capacity for self-organization).

THEORETICAL FRAMEWORK

As noted earlier, our analysis is founded on concepts from the organizational development literature, specifically the theory of action framework of social learning (Argyris 1977; Argyris and Schön 1978; Argyris 1993). We adopted this framework because it accommodates human agency, links individual and social learning, and describes a process for generating innovative change. Without providing a comprehensive summary of the theory, the ensuing discussion introduces constructs that form our analytic framework.

Learning is viewed as a process of detecting and correcting error, and occurs under two conditions. The first is when intentions match outcomes of action, and the second is when intentions and outcomes do not match. Single-loop learning occurs when matches happen, or when mismatches are corrected by changing one's strategy or behaviour while preserving basic values and norms. Double-loop learning occurs by correcting mismatches by first changing or supplementing existing values and norms, and then changing strategies or behaviour (Figure 13.1). Learning occurs at both individual and social levels, but individuals are the agents for social collectives. Therefore, social learning does not occur until individuals encode what they have learned in social memory. The media of social memory include public maps (*e.g.*, legislation, regulations, licences, bylaws, informal rules) and private images (*i.e.*, mental models of self in relation to others and in relation to the social collective). The key processes of the double-loop social learning dynamic are:

- detecting the mismatch between intention and outcome;
- investigating the source of the mismatch;
- developing alternatives for avoiding future mismatches;
- identifying conflict over competing visions or goals;
- resolving that conflict;
- implementing the preferred alternative;
- evaluating the results;

- modifying practice and theory accordingly, including fundamental goals, norms and assumptions; and,
- embedding the modified practice and theory in the images and maps of social memory.

Following a summary of both case studies, the theory of action framework of social learning will be utilized to explore and highlight the type, form, and direction of learning in narwhal and polar bear management regimes in Nunavut. Preliminary observations on policy implications and questions for further research will also be offered.

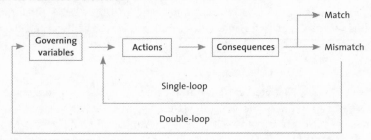

Figure 13.1 Double- and single-loop learning. Governing variables are underlying values of the system that can be inferred from observing the actions of organization agents. (Argyris 1993)

COMMUNITY-BASED MANAGEMENT OF NARWHAL
Background and Context

Narwhal have long been important to eastern Arctic Inuit communities both as a food source (especially the muqtuk) and for the prized tusks of the male narwhal. The balance of the animal would historically have been used for dog food, but these days is used less intensively or is discarded. There is no suggestion in the literature that Inuit harvesting had a significant effect on the sustainability of the resource, and the narwhal was not a major target of European whaling between the 1850s and the First World War, which focused instead on the bowhead whale.

Nevertheless, the federal government introduced community quotas for narwhal harvesting in 1977 (SOR/76-471, Narwhal Protection Regulations), apparently out of a concern that Inuit harvesting levels were increasing primarily to secure the tusk as a source of ivory. The regulations replaced an earlier approach that provided an individual quota for each Inuk. The consensus, however, is that there was no real scientific basis for the community quotas, which were set more or less arbitrarily. The ability of communities to harvest up to the level of the quota was highly variable, dependent on such factors as ice conditions and migration routes. In some years, for example, some communities had no opportunity to harvest any narwhal, whereas in other years the narwhal were close to the community but harvesting ended when the quota was reached, even though the community had not satisfied its needs. Moreover, quotas could not be carried over from one year to the next, and neither were they transferable to other communities. Inuit dissatisfaction was compounded by an appreciation that the narwhal populations of the eastern Arctic were shared with Greenland and yet there were no (or much laxer) restrictions on narwhal harvesting by hunters in that jurisdiction (Bankes 2003).

In 1993, the NWMB was established under the terms of the Nunavut Final Agreement (NFA), a comprehensive land claims agreement. The NFA not only treated the NWMB as central in the management of wildlife, it had as one of its informing imperatives the transfer of responsibilities and control to local communities. Soon after the NWMB was established, it was faced with several requests from communities to vary narwhal quotas. The NWMB found it difficult to evaluate these requests on an *ad hoc* basis and resolved instead to find a more principled way to deal with narwhal management issues. The NWMB also wanted to use this as an opportunity to return control of harvesting to the communities. A three-year community-based management system was therefore introduced in 1999 in the communities of Repulse Bay, Arctic Bay, Qikiqtarjuaq (formerly Broughton Island), and Pond Inlet. Kugaaruk (formerly Pelly Bay) was later added to the project. The NWMB (2002) described community-based management (CBM) as:

> ... a system of wildlife management characterized to date by a removal of formal annual quotas and a transfer of initial management responsibility away from the NWMB and Government, directly to a community.

The CBM project was explicitly framed as an experimental process, and included an internal review at the end of its initial three-year period. The project was spearheaded by the NWMB, but it had the support of the organizations that would be key to implementation: the Hunters and Trappers Organizations (HTOs) for the individual communities, the regional wildlife management organizations (RWOS), the Nunavut Tunngavik Inc. (NTI), and the federal Department of Fisheries and Oceans (DFO).

NWMB played an important role of gatekeeper and standard setter by establishing and applying the criteria for communities to participate in CBM. These requirements were for: (i) communities to establish a reporting system for all narwhal struck, landed, and lost; (ii) hunters to obtain and complete a narwhal tag for all narwhal landed; and (iii) HTOs to make bylaws or rules to regulate hunting by members. The objectives of the rules had to be to ensure effective management and conservation, ensure education and proper training of harvesters, minimize loss and wastage, ensure humane and effective hunting practices, and maximize the safety of hunters (NWMB 1999a).

To assist communities in securing eligibility for CBM, the NWMB developed a series of briefing notes as well as draft narwhal hunting rules that could be adapted and adopted by HTOs as they saw fit, subject to NWMB approval of the final product. HTOs, therefore, were expected (s. 5.7.3 of the NFA) to regulate harvesting practices and techniques among members and to allocate and enforce the community's entitlement to quota stocks. By the same token, RWOs were expected to assume similar responsibilities for sharing regional entitlements among communities, especially for shared wildlife stocks (s. 5.7.6). To aid in their enforcement responsibilities, HTOs and RWOs were expected to develop bylaws to discipline members (s. 5.7.12). The NWMB was expected to provide adequate funding for the operation of HTOs and RWOs (s. 5.7.13).

As to the remaining actors, NTI was the key entity on the Inuit side of the claim responsible for representing Inuit interests. DFO was responsible for administration of the *Fisheries Act* and the Marine Mammal regulations. While much of the implementation of the CBM model was, by its very nature, left to the communities, one of the things that the Fisheries Minister did was relax the quota requirements of the regulations. In addition, DFO had to make narwhal tags available on demand. Finally, while each of the NWMB, HTOs and RWOs, and DFO had an individual role to play, they also worked collectively, along with NTI, to prepare communities for the adoption of CBM by visiting communities and conducting workshops to discuss CBM requirements.

The Management Crisis

Within a year of its trial implementation, the outcome of the CBM process became a matter of significant concern to NWMB, DFO, some community members, and at least one environmental non-governmental organization (NWMB 1999b; 2000a; 2000b; World Wildlife Fund 2001). Estimated total annual mortality from 1999 to 2001 exceeded historic quotas for all communities for which data were available (with the exception of Pond Inlet in 2001) (Tables 13.1 to 13.3).

Important concerns related to waste of the resource (suggested by the high struck/lost levels), low levels of utilization of the meat, increased commercialization of the narwhal hunt (for the tusk), harvesting methods (shooting by rifle before harpooning), and the overall sustainability of these harvest levels (evidenced largely by comparison with the more or less arbitrary former quota levels). As well, some DFO officials were concerned that an increased harvest was inconsistent with Canada's position in ongoing discussions with Greenland as to management of the shared population that neither party should alter its management approach pending the outcome of these discussions. Finally, DFO representatives noted that struck/loss reporting was not always satisfactory (thereby undermining one of the key premises for DFO's support for lifting quotas) and that hunters were not always providing sampling information to assist in determining stock affiliation. In October 2000, DFO decided to close the Qikiqtarjuaq narwhal fishery, relying on the minister's power to issue emergency orders (s.5.3.24).

The closure created significant conflict among the organizations involved in the CBM project, but it did not result in a return to the centrally controlled, rigid quota system previously enforced by the DFO, nor the discontinuation of the community-based experiment itself. Rather, the 'crisis' provoked by the situation in Qikiqtarjuaq provided motivation for the NWMB, DFO, and Nativak

Table 13.1
1999 HARVESTING DATA

Community	Historic Community Quota	Struck & Landed	Struck & Escaped	Struck & Sunk	Estimated Total Hunting Mortality
Pond Inlet	100	130	14	16	146–160
Qikiqtarjuaq	50	81	30	25	106–136
Repulse Bay	25	156	68	30	54–63
Arctic Bay	100	101	?	?	?
Kugaaruk	10	0	?	?	?

Table 13.2
2000 HARVESTING DATA

Community	Historic Community Quota	Struck & Landed	Struck & Escaped	Struck & Sunk	Estimated Total Hunting Mortality
Pond Inlet	100	166	21	10	176–197
Qikiqtarjuaq	50	137	79	40	177–256
Repulse Bay	25	49	9	5	54–63
Arctic Bay	100	101	?	?	?
Kugaaruk	10	30	?	?	?

Table 13.3
2001 HARVESTING DATA

Community	Historic Community Quota	Struck & Landed	Struck & Escaped	Struck & Sunk	Estimated Total Hunting Mortality
Pond Inlet	100	63	5	27	90–95
Qikiqtarjuaq	50	89	8	9	98–106
Repulse Bay	25	100	38	21	121–159
Arctic Bay	100	134	20	4	138–158
Kugaaruk	10	41	18	8	49–67

HTO (Qikiqtarjuaq) to identify their different goals and agendas, the sources of management conflict, and the mechanisms required to resolve those conflicts.

A subsequent review of the closure order (required by s.5.3.24 of the NFA), along with the formal evaluation of the CBM project, identified a need for an enhanced and diversified knowledge base concerning narwhal. As stated above, the actual harvest and struck/loss levels led to conservation concerns, yet the foundation for these concerns was not completely clear. It was difficult, for example, to allege (at anything other than an intuitive level) that the harvest was unsustainable, simply because harvest levels under CBM were so much higher than those under the old quota rules. Those quotas were set more or less arbitrarily and not on the basis of population estimates and recruitment rates. In addition, it soon became apparent that there were major gaps in the knowledge base for narwhal. The DFO acknowledged, for example, that its estimates of narwhal populations were based on aerial surveys conducted in areas of known whale concentrations. Some of these surveys were dated, and in any event, they did not account for narwhal that were submerged and beyond view at the time of the survey, narwhal that were outside the survey area, or narwhal that were missed by observers because of ice conditions or because of poor visibility. Moreover, hunters generally believed that the stocks were larger than estimated by DFO and also believed that narwhal reproduced more frequently. Similarly, it emerged that while DFO science was based on the idea that all narwhals in the Baffin Region belong to a single stock, more recent research suggested that there were a number of different stocks or aggregations that made up the Baffin Bay narwhal population. All of this led the NWMB to conclude that it was essential to improve the state of knowledge of narwhal and that this endeavour must include the development of traditional knowledge (*Inuit Qaujimajatuqangit* [IQ]) studies of Baffin narwhal.

The closure order review and the evaluation also identified a need for better communication between DFO and local communities. There had, of course, been extensive consultation between DFO and the NWMB and the communities before the introduction of CBM. DFO had also communicated its concerns about the high struck/loss rates for narwhal. Nevertheless, the actual issue of the closure order came as a shock to the community, which felt that it had responded to DFO concerns by requiring that all narwhal be harpooned before they were shot.

Subsequent to the review and evaluation, in spring 2002, the NWMB decided to continue with CBM for another year, at which time a final review and determination of the future of the CBM process would be made. The changes in CBM that occurred, or are currently being considered, include community adoption of informal or pseudo quotas, and possible adoption of five-year rolling quotas. These are thought to be more suited to the needs of long-lived populations for which the harvest may be highly variable due to natural conditions, including migration patterns and ice conditions (NWMB 2001).

The final results of the narwhal CBM system have not yet been determined. Yet it is valuable from both theoretical and practical perspectives to explore the learning processes and enabling conditions evident in the initial iterations of the system described above. However, before doing that, we turn our attention to the NWMB's polar bear mandate, and to its response during a management crisis concerning the M'Clintock Channel (MC) bear population.

MANAGEMENT OF POLAR BEARS
Background and Context

Polar bear management in Canada and internationally is coordinated by a network of government agencies and co-management bodies, and is closely linked with academic and government research programs. The two important interagency organizations involved in polar bear management in Canada are the Federal-Provincial Polar Bear Technical Committee (PBTC) and the higher-level Federal-Provincial Polar Bear Administrative Committee (PBAC). The shape of the polar bear management network has been strongly influenced by Canada's obligations under the International Agreement for the Conservation of Polar Bears and their Habitat. This 1973 agreement between Canada, the United States, the former Soviet Union, Norway, and Denmark was spurred by international concern about rapidly increasing harvests of polar bears. The agreement is widely recognized as a success and is considered to have been instrumental in the establishment of effective polar bear conservation regimes and research programs throughout the Arctic (Fikkan et al. 1993; Prestrud and Stirling 1994). Complementing the PBTC and the PBAC at the international level is the IUCN's Polar Bear Specialist Group (PBSG). The PBSG was originally formed in 1965 and continues to be an active forum for international coordination of conservation efforts.

At the domestic level, the legal regime for harvesting polar bears varies from jurisdiction to jurisdiction. In Nunavut, the regime is based upon the territorial Wildlife Act (Government of Northwest Territories 1988) and regulations (Government of Northwest Territories 1990; 1992) as modified by the NFA. In other words, it is a territorial responsibility rather than a federal responsibility, although the federal Canadian Wildlife Service maintains an active research program on at least one population harvested in Nunavut and, through the PBTC and PBAC, a strong role in management decisions nationwide. The current Wildlife Act of Nunavut was inherited from the Northwest Territories, but a Bill to replace that Act was introduced in the Nunavut legislature in 2003 (Government of Nunavut 2003). The new Bill contains a series of provisions designed to enhance the role of traditional knowledge. Within the Government of Nunavut, responsibility for the Wildlife Act falls to the Department of Sustainable Development (DSD). In the post-NFA environment, there is an allocation of responsibilities between the NWMB and DSD that is similar to the allocation of responsibilities that we have already noted in the context of narwhal. The minister responsible for DSD has the authority to exercise the exceptional powers of disallowance and emergency decisions already described in the context of the narwhal fishery.

Under the current system in Nunavut, a person may not hunt without a licence and a tag, and tags are issued for a particular polar bear management zone designated under the regulations (Figure 13.2). The number of tags issued for a particular area constitutes an overall quota, but that quota is sub-allocated to individual communities. This sub-allocation is based upon historical practice and negotiations, and recorded in polar bear management agreements between DSD and the relevant HTOS. Communities may "lend" or "trade" their quota to other communities with quota in the same management zone. Quota calculations have long been designed to achieve a target of maximum sustained yield and encourage preferential harvesting of males. A single community may have quota entitlement with respect to a number of different populations, but tags are issued and may only be used with respect to a specific population. Quotas and variations to quotas are determined annually on the basis of actual harvesting numbers and are confirmed by the NWMB. Tags are issued to HTOS, and the HTO allocates tags within its community. An HTO may issue a tag to a non-resident for a sport hunt. Further, it decides what proportion of its community allocation sports hunters may harvest.

Sport hunting of polar bears (largely by Americans and Europeans, guided by experienced Inuit hunters) provides important economic opportunities for Nunavut communities, but sport hunter interest in that hunt is in large measure dependent upon the ability of the prospective hunter to be able to import the trophy to his or her home jurisdiction if the hunt is successful. While polar bears are listed in the Convention on International Trade on Endangered Species Appendix II, and trade therefore requires the issuance of import and export permits, the United States (U.S.) took additional measures in 1972 through its *Marine Mammal Protection Act* (MMPA) to further restrict the import of trophies to the U.S. Congress amended the MMPA in 1994 to allow for the issuance of permits to authorize the import of sport-hunted trophies, but only where the Secretary of the Interior is able to certify that certain conditions can be met, including that: (i) the exporting jurisdiction (Canada) has a monitored and enforced sport hunting program consistent with the purposes of the international agreement, and that (ii) Canada has a sport hunting program based on scientifically sound quotas ensuring the maintenance of the affected population stock at a sustainable level. The U.S. Fish and Wildlife Service (US FWS) has interpreted these conditions as requiring, among other things, the existence of a management agreement signed by all user groups where the harvesting of a particular population is shared by more than one community or jurisdiction and prescribing "scientifically sound quotas" (US FWS 1997).

In 1997, following extensive review of the Canadian regulatory scheme, the US FWS agreed to list (*i.e.*, approve) five of the twelve (now thirteen) identified Canadian polar bear populations: Southern Beaufort Sea, Northern Beaufort Sea, Viscount Melville Sound (subject to a harvesting moratorium), Western Hudson Bay, and M'Clintock Channel (MC). Two additional populations were subsequently defined and added in 1999. With the exception of Gulf of Boothia,

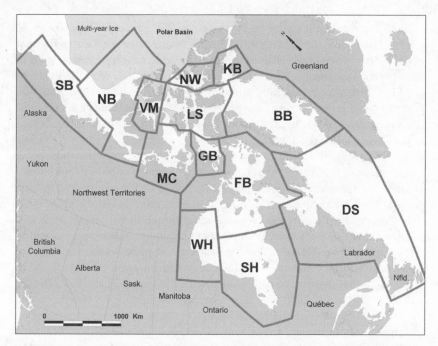

Figure 13.2 The thirteen polar bear management zones in Canada, which are based upon the boundaries of discrete populations of bears.

(Legend: BB = Baffin Bay, DS = Davis Strait, FB = Foxe Basin, GB = Gulf of Boothia, KB = Kane Basin, LS = Lancaster Sound, MC = M'Clintock Channel, NB = Northern Beaufort Sea, NW = Norwegian Bay, SB = Southern Beaufort Sea, SH = South Hudson Bay, VM = Viscount Melville Sound, WH = Western Hudson Bay)

(Source: Dean Cluff, NWT Resources and Economic Development.)

the only populations that remained unlisted were those that Nunavut shares with Greenland or another Canadian province, and for which there did not exist a joint management agreement. The authorization for the MC population was subsequently revisited and withdrawn following the events described below (US FWS 2001a, 2001b).

The Management Crisis

The communities of Talyoak, Gjoa Haven, and Cambridge Bay hunt the MC population, which was originally surveyed between 1972 and 1978 as part of a more geographically extensive population study (Furnell and Schweinsburg 1984). At that time, this study area was thought to represent a single "Central Arctic Islands" bear population, but subsequent research has demonstrated otherwise. That study area and the population estimate were subsequently subdivided into the MC population and portions of three other populations: Gulf of Boothia, Lancaster Sound, and Viscount Melville Sound (Taylor and Lee 1995; Taylor *et al.* 2001). Accounts of the partitioning of the population estimate and the setting of a quota for the MC population differ.

The formal account in the US FWS's rule-making process suggests the following. First, while the existing estimate for the combined MC/Gulf of Boothia population had been 1,081 bears, this was adjusted to nine hundred for each of the discrete populations to take account of the known bias of non-representative sampling. Second, consultations with Inuit hunters resulted in a reduction of the original estimate from nine hundred to seven hundred (US FWS 2001b) (This is consistent with the account of the IUCN PBSG (Lunn *et al.* 2002; 26).) In later discussions (2000) within the NWMB, Government of Nunavut officials acknowledged logistical difficulties with the survey and suggested that the original raw estimate of the MC population was revised upward based on community consultations, perhaps setting "the stage for the possible problem we are seeing today" (NWMB 2000a, per Atkinson). In any event, all accounts agree that the estimate information was problematic (although in fairness, Furnell and Schweinsburg (1984) were very candid about the limitations of their sampling and estimation procedures), and that all subsequent quota determinations were based on the lower figure of seven hundred. And it was on that basis, and in recognition that there was a community agreement in place, and that hunting had been at a two-male-to-one-female harvesting ratio for several years, that US FWS accepted the population for listing.

A new study of the MC population was initiated in 1998, and early results (NWMB 1999b) were not encouraging, as researchers were having difficulty finding enough bears for the survey. By 2000, Government of Nunavut biologists were presenting revised estimates of between 238 and 399, with 288 as the best preliminary estimate. The Government of Nunavut subsequently revised its current best estimate to 367 (US FWS 2001a). Whatever the number, it was clear that there had been a dramatic decline in this population if indeed the original survey was accurate. It was also clear that the current quota of thirty-two bears was not sustainable. These results were communicated to the NWMB and also to the US FWS.

The US FWS responded by initiating an emergency rule-making procedure to revoke the listing of the MC population on the grounds that the population was no longer being managed at a sustainable level. US FWS clearly regarded a quota cut (which the NWMB had decided upon), rather than a moratorium on harvesting, as an inadequate response to the problem, especially in light of the premise that the historical baseline for the population was nine hundred bears, from which there had been a precipitous decline. The discussions within the NWMB took a somewhat different course.

First, the NWMB considered a range of possible outcomes, based on the data presented by DSD. One option was to reduce the quota, thereby allowing some limited hunting to continue but delaying rebuilding of the population. Another option was the imposition of a moratorium on harvesting. Following discussion, the board adopted a two-year proposal to reduce the quota to twelve bears for the 2000/2001 harvest year followed by a complete moratorium for 2001/2002 – effectively a quota of six bears per year for each of the two years. In reaching this conclusion, the board rejected the advice of its director of wildlife that a moratorium "was probably the best option" (NWMB 2000d), and seemed to be persuaded by the importance of proceeding incrementally and with the concurrence of the communities, recognizing that an immediate moratorium might produce hardship and resentment.

Second, there was much discussion within the NWMB as to how this crisis had occurred (NWMB 2001, 39, per Koonoo; NWMB 2000d, 5). Was it the result of last-minute changes that had been made to the results of the survey in the mid-1970s? If so, what had occasioned those changes? Was it possible that original estimates were completely wrong, in which case a target of returning to 700–900 animals was not realistic? Was the decline due to the fecundity of the MC population being lower than that of the adjacent Gulf of Boothia population? Was it due to over-hunting? What were the implications of climate change on ice conditions, and therefore on bear-seal interactions? Had there been an out-migration of bears from this area? Would the problem have been identified earlier had greater attention been paid to traditional knowledge and the observations of hunters?

Third, whether coincidentally or not, the Government of Nunavut at about this time proposed to change the manner in which it determined quotas for given populations by adopting what it described as a risk-based model rather than a maximum sustained yield model. In brief, the new approach contemplates that communities would be more directly involved in selecting appropriate harvest targets in light of possible scenarios. Such scenarios would be generated by computer models such as RISKMAN (developed by a Government of Nunavut polar bear biologist), which allow managers and stakeholders to evaluate the outcomes of different harvest strategies given available quantitative population data and known or estimated uncertainties within that data (*e.g.*, McLoughlin *et al.* 2003).

Fourth, members of the NWMB were concerned to explore the consequences of the US FWS decision for affected communities. The community most seriously affected was Gjoa Haven, since it had no alternative population from which to harvest. This led to suggestions that other communities might be prepared to loan Gjoa Haven quota rights to the Gulf of Boothia population (NWMB 2001, Meeting 30). In the end, the healthy state of the Gulf of Boothia population allowed DSD and the NWMB to increase the quota for that population and to give Gjoa Haven a quota of three (Government of Nunavut 2002).

Fifth, members of the NWMB expressed a variety of other concerns. Some of these concerns related to the costs of ongoing polar bear research needed to support harvesting based on maximum sustained yield models, especially in relation to other priorities, while other concerns related to the perceived complexity of both the current flexible quota system of the management agreements as well as the Government of Nunavut's new proposals. Throughout, board members emphasized the need to take account of traditional knowledge in making further assessments of this population.

Finally, it is clear that all of the discussion within the NWMB occurred within the shadow of the MMPA and concerns as to the U.S. reaction. This was a consistent theme in the NWMB minutes: there are constant references to the MMPA and US FWS decision making, and also concerns that a misstep by the board in this one case might trigger distrust of its role in managing other populations.

DISCUSSION AND COMPARATIVE ANALYSIS

As both case studies illustrate, efforts to manage narwhal and polar bear reveal a complex array of institutional, organizational, socio-cultural, and ecological challenges. Framing the complex management milieu associated with both species, for example, are numerous value conflicts associated with resource use and protection, quotas and the mechanisms with which they are established, and the roles of Western science and traditional knowledge. In the exploration of these issues, therefore, a number of questions will be addressed: To what extent has the NWMB's management of narwhal and polar bear exhibited evidence of learning? What type of learning has taken place (*i.e.,* single-loop and/or double-loop learning)? What mechanisms, and systemic and institutional features, enabled or inhibited that process of learning? And finally, what are the implications for planning, management, and the policy reforms required to encourage social learning for sustainability in the Canadian North?

The CBM project was a management innovation, and provides evidence of the double-loop learning dynamic. Key double-loop processes (identifying and resolving normative conflict, and modifying precepts of theories-in-use) were evident in implementation of the project, including the NWMB's response to the Qikiqtarjuaq crisis. These processes were not easy, nor are they complete. At their heart, however, is an effort to bridge different management paradigms and knowledge systems that have come into conflict, and which require communities, regional wildlife organizations, the NWMB, and the DFO to test their management norms and modify key assumptions. In addition, the NWMB's review of the CBM project reflects the evaluation phase of the double-loop learning dynamic, and indicates an internal commitment to and monitoring of its past decisions. As well, the review was done in a participatory or community-based manner, which was conducive to incorporating fresh perspectives and diverse values into the existing management system.

In contrast, the NWMB's response in the MC polar bear crisis, while involving some attention to conflict identification, was essentially a short-term single-loop adaptation. Like other organizations in the larger polar bear management network, the NWMB has struggled with key double-loop processes (identifying and resolving normative conflict, and modifying the precepts of its theories-in-use), and has relied on the single-loop dynamic (in seeking to increase effectiveness). This is apparent in how the board responded to the MC population crisis. The response (which arguably was quite reasonable) was a case in which value and normative conflict were identified (at least by the NWMB) but not resolved. As well, it is a case in which the NWMB relied on a single-loop response to derive a short-term solution to a pressing problem. The board engaged in fundamental learning processes, *e.g.*, detecting key errors linked to the 1973 estimates, investigating likely sources of the errors, developing response options (*i.e.*, reducing quotas or instituting a moratorium), and discussing the consequences of the various options (with a focus on the effects of the US FWS delisting). It also began discussions suggesting it was identifying sources of normative conflict (*e.g.*, its questioning of maximum sustained yield models and its critique of the quota systems), but in the end, it did not engage fully in the double-loop dynamic.

Differences between the cases can be found in the political, legal, and institutional conditions that shaped the learning processes described above. The CBM project and the Qikiqtarjuaq crisis fit squarely within the mandate of the NWMB and within the context of the NFA, both of which resulted from the resolution of profound historical and value conflict. That is, CBM was forged in a context that favoured sweeping political and social change, particularly in a direction toward community-based governance models. Further, the NWMB created a non-threatening environment in which it was possible to work through the new rules and to discuss how change should be managed. Additionally, the NWMB was committed to seeing CBM succeed – not only because of its application to the management of other cetaceans, but also because of its potential application to other quota species, such as walrus and polar bear. And perhaps most importantly, the introduction of CBM was explicitly couched as an experiment. The key players at all levels recognized from the outset that there would need to be adjustments as CBM was put into practice. As well, the time-limited nature of CBM's introduction anticipated the need for a review. The NWMB was also quite explicit in treating the process as an opportunity for learning, and in a sense it set explicit learning goals by requiring communities that wished to participate to develop a set of bylaws or hunting rules that met certain minimum standards. The reporting system required as part of the CBM process, and which offered the potential to obtain records not only of those animals taken but also a record of those animals that were struck and either escaped or sank, implies a commitment to monitoring and experiential learning. Therefore, all the key players recognized that this was a collaborative experiment from which all parties had to see some gains. But this also meant that there was a strong, shared commitment to making CBM work.

In contrast, the NWMB's polar bear mandate and its response in the MC crisis were grounded firmly in an established legislative and policy regime, few aspects of which were meant to encourage major social and political change. Moreover, the MC crisis was influenced strongly by decisions taken under foreign legislation, namely the U.S. *Marine Mammal Protection Act*. Consequently, the planning and decision-making environment during the MC crisis was not conducive to evaluating long-standing goals, identifying alternatives, and discussing points of conflict (particularly those that spanned value systems). From a resource management perspective, this highlights both the potential and the limitations of enabling legislation and policy. In the narwhal case, the legislative and policy regime created a safe and legitimate climate for management innovation, experimentation (and failure), and double-loop learning. In the polar bear case, the regime (although it shared important components with narwhal governance) encouraged incremental adjustments in existing technologies and end points, and single-loop learning.

Another essential difference in the cases was the extent to which the NWMB's polar bear and narwhal mandates were based in traditional or scientific knowledge systems. Relatively speaking, scientific research plays a larger role in polar bear management than it does in narwhal management. This is due to the strong scientific capacity found at most levels of the polar bear management network, including the NWMB. For most of the past four decades, sustainability of harvest has been the major conservation concern, and a comprehensive system of assessing populations and assigning (and monitoring) sustainable quotas has been operating for over twenty years now. Considered from a strictly experimental perspective, such efforts would reflect a relatively high degree of adaptiveness because quotas are adjusted regularly to compensate for changes in the harvest or apparent changes in the population detected during subsequent surveys. These attributes have facilitated important and adaptive single-loop processes in Canada's polar bear management system, but the very efficiency of the system could now be preventing organizations in the network (such as the NWMB) from engaging in the double-loop dynamic. Polar bear management has an enviable track record in the wildlife management field, and given that the network of practitioners is grounded in orthodox wildlife science, acceptance of diverse values and goals has been difficult. Hence, the network's capacity for profound double-loop learning has been constrained.

Yet other important differences between the narwhal and polar bear cases relate to the nature and degree of risk the NWMB faced in engaging in double-loop processes. Overall, higher levels of risk affecting polar bear management likely reduced the board's willingness to engage in double-loop processes. For example, polar bear scientists and managers have deeply embedded values and very strong personal and professional affiliations with their science and with the overall management network (and often for very good reasons since the standard of their work is extremely high). Due to the 1973 international agreement and individual biologists' successful leverage of that to start and maintain research and management programs, polar bear scientists have significant vested interests in the institutional network they created. They form a powerful epistemic community with no parallel in the narwhal management network. For polar bear scientists and managers, therefore, there was likely a higher level of political risk than for their narwhal counterparts in adopting fundamental change in research strategies or management approaches. Similarly, there were various socio-cultural risks in the polar bear case that were not present in the narwhal CBM project. In fact, there were likely socio-cultural risks associated with *not* experimenting with new narwhal management approaches, given the past management problems and concerns expressed by Inuit over their rights under the land claim.

CONCLUSIONS

As illustrated in both the narwhal and polar bear cases, the process of learning may take many forms – depending on the organizations, the individuals, and the management context. For managers, practitioners, and researchers concerned with issues of learning and sustainability, therefore, it is important to elucidate those mechanisms and/or conditions that serve to impede or foster the learning process, and in particular double-loop learning processes. What can we learn from the narwhal and polar bear cases to help guide resource management policy and practice in northern Canada? The review of the two cases has revealed three broad mechanisms and/or conditions that have played a fundamental role in shaping the learning process:

1   *The emergence of an enabling political and institutional framework*: Foremost, the two cases reveal the importance of an institutional framework that permits – through legislative means and new management mechanisms – greater opportunity to directly address conflicts over competing visions or goals, and an arena in which to resolve that conflict. In the case of narwhal CBM, the NFA has resulted in the distribution of authority among a greater number of interests, and led to the creation of new management entities (*e.g.,* NWMB) that provide a more favourable forum for conflict identification and resolution than under the previous management regime. As a result, there are strong connections between the NFA and the double-loop learning process evident in CBM. Although the same basic institutional conditions apply in the polar bear case, there are several complicating factors, including extra-territorial pressures exerted through the US MMPA, and as illustrated, the status accorded to the scientific community. However, the existence of the NFA institutional framework, and the process of on-going socio-political change, is likely to exert further influence on the current polar bear management approach.

2   *A willingness to experiment and receptivity to risk*: As is evident in the comparison of the polar bear and narwhal cases, opportunities for double-loop learning appear to be in many respects a function of perceptions about risk. Although connected in part to the emergence of an institutional and political framework in which conflict identification and resolution is more likely to be addressed, issues of ecological, social, political, and scientific risk permeate the two cases in different ways. In the narwhal case, there are

arguably fewer risks involved, and therefore, greater willingness to experiment and challenge management norms. In the case of polar bear management, the political and scientific risks are perceived to be more significant, and serve to dampen a willingness to experiment with new management models and opportunities for double-loop learning. At the level of both individuals and organizations, the degree of 'entrenchment' of interests further mediates perceptions of risk. As illustrated, there is a greater disconnect between the individuals involved in the narwhal CBM process and the outcomes of the management strategy. The same cannot be said of polar bear management where individuals, their careers and the management outcomes create a much tighter loop.

3 *A shift in the dominant management worldview or model and the corresponding integration of different knowledge sources and frameworks*: In the case of narwhal management, an institutional framework that encouraged conflict identification and resolution, along with a greater tolerance for risk, has contributed to modifications of the dominant management worldview. This modification of the management worldview, moreover, included requirements for the integration of different knowledge sources and frameworks. The creation of opportunities in which different perspectives can be expressed, and in ways that are closely connected to the management process, is fundamental to the modification of theories-in-use, and serves to encourage a 'check' of basic management goals, norms, and assumptions. This has begun to occur in the narwhal case and has contributed to the double-loop learning process that is characteristic of the CBM process. In contrast, the predominant discourse of polar bear management remains positivist and science-oriented.

The circumstances that have shaped opportunities for double-loop learning in the narwhal and polar bear management cases as outlined above, while illuminating, represent preliminary insights into the connections between management conflict and the learning process. In an effort to further identify preconditions for double-loop learning in complex management circumstances, and continue the development and elaboration of a set of principles managers and practitioners can utilize as a reference, much research remains to be done. For example, the preliminary conclusions highlighted above deserve additional research. Rephrased as propositions, these conclusions provide a useful starting point for further, and comparative, analyses of polar bear, narwhal, and other similar cases. The outcomes of this type of research should result in the development of a detailed and empirically tested set of principles to encourage social learning in complex management contexts.

There are many other outstanding questions, some of which are specific to the case studies, while others have more general implications. Where single-loop learning has been dominant to date – as in the case of polar bear management – when and how does a need for double-loop learning become apparent? Does a crisis in respect of one population (the MC population) really constitute adequate grounds to propose an overhaul of the system? It may be possible to achieve key environmental or conservation objectives in the context of a single-loop learning process. In the long term, nevertheless, a failure to meet the critical social objectives that shape sustainability (*e.g.,* equity, empowerment, participation) will likely serve to undermine management efforts. Therefore, if change is desired, how could the adaptive strengths of existing management networks be preserved during a period of double-loop learning, and not be abandoned simply for the sake of change? What kinds of change would actually be beneficial?

A key theme in many of these questions is the issue of learning across scales (see also Berkes *et al.,* this volume). For example, at what level is social learning most prevalent? Are the procedures established by the NFA (and other land claims agreements) more important than the particular powers accorded to co-management institutions when it comes to social learning? With respect to narwhal CBM, is the learning that has occurred confined to Nunavut-based managers? Does it extend to the DFO Winnipeg office? Does it extend to the head office? Will the double-loop learning process evident in the case of narwhal extend to the management of other resources in the region, or to other regions? Important questions at the scale of the individual also require further analysis, for example, to what extent is personal chemistry important in social learning? The contributions of individuals in the polar bear case suggest that it can be significant, but under what specific circumstances? How important is staff continuity, or board member continuity, when institutions such as claims-mandated co-management boards have only recently been established? And finally, how can the lessons that emerge from these different learning processes be constructively transferred?

As these two cases illustrate, there are many questions that require further analysis. Understanding the processes and structures associated with learning in complex management contexts remains a significant challenge – as does the identification of transferable lessons from one context to another. In the long term, however, the concern with social learning will likely prove central to the development of the management capacity necessary to guide nature-society interactions along sustainable trajectories.

REFERENCES

Argyris, C. 1977. "Double loop learning in organizations." *Harvard Business Review* 55 (September–October): 115–25.

———. 1993. *On organizational learning.* Cambridge, M A: Blackwell Business.

Argyris, C., and D.A. Schön. 1978. *Organizational learning: a theory of action perspective.* Reading, M A: Addison-Wesley.

Bankes, N. 2003. "Implementing the fisheries provisions of the Nunavut Claim: recapturing the Resource?" *Journal of Environmental Law and Practice* 12: 141–204.

Berkes, F., J. Colding and C. Folke, eds. 2003. *Navigating social-ecological systems: Building resilience for complexity and change.* Cambridge: Cambridge University Press.

Brown, L.M. 2000. "Scientific uncertainty and learning in European Union environmental policy making." *Policy Studies Journal* 28(3): 576–97.

Dewey, J., and D. Sidorsky. 1977. *John Dewey: the essential writings.* New York: Harper & Row.

Dyck, B., J. Buckland, H. Harder, and D. Wiens. 2000. "Community development as organizational learning: the importance of agent-participant reciprocity." *Canadian Journal of Development Studies* 21 (special issue): 605–20.

Eckley, N. 2002. "Dependable dynamism: lessons for designing scientific assessment processes in consensus negotiations." *Global Environmental Change Part A* 12(1): 15–23.

Fien, J., and P. Skoien. 2002. "'I'm learning … how you go about stirring things up – in a consultative manner': social capital and action competence in two community catchment groups." *Local Environment* 7(3): 269–82.

Fikkan, A., G. Osherenko, and A. Arikainen.1993. "Polar bears: the importance of simplicity." In *Polar politics: creating international environmental regimes*, edited by O. Young and G. Osherenko, 96–151. Utica, N Y: Cornell University Press.

Fiorino, D.J. 2001. "Environmental policy as learning: a new view of an old landscape." *Public Administration Review* 61(3): 322–34.

Folke, C., S. Carpenter, T. Elmqvist *et al.* 2002. *Resilience for sustainable development: Building adaptive capacity in a world of transformations.* Paris: I C S U, Rainbow Series No. 3. *http://www.sou.gov.se/mvb/pdf/resiliens.pdf*

Friedmann, J. 1987. *Planning in the public domain: from knowledge to action.* Princeton, NJ: Princeton University Press.

Furnell, D.J., and R.E. Schweinsburg. 1984. "Population dynamics of central Canadian Arctic polar bears." *Journal of Wildlife Management* 48: 722–28.

Government of Nunavut. 2002. Press release. Interim Polar Bear Hunting Quota announced for Gjoa Haven. May 1, 2002.

———. 2003. Bill 35. Wildlife Act.

Government of the Northwest Territories. 1988. Wildlife Act. Revised Statutes of the Northwest Territories, chapter W-4.

———. 1990. Wildlife Management Polar Bear Area Regulations. Revised Regulations of the Northwest Territories, chapter W-13.

———. 1992. Big Game Hunting Regulations. Revised Regulations of the Northwest Territories, chapter R-019-92.

Gunderson, L., and C.S. Holling, eds. 2002. *Panarchy: understanding transformations in human and natural systems.* Washington, D C: Island Press.

Haas, P.M. 2000. "International institutions and social learning in the management of global environmental risks." *Policy Studies Journal* 28(3): 558–75.

Holling, C.S. 1978. *Adaptive environmental assessment and management.* New York: John Wiley and Sons.

Krasny, M.E., and S.-K. Lee. 2002. "Social learning as an approach to environmental education: lessons from a program focusing on non-indigenous, invasive species." *Environmental Education Research* 8(2): 101–19.

Lee, K.N. 1993. *Compass and gyroscope: integrating science and politics for the environment.* Washington, DC: Island Press.

Levitt, B., and J.G. March. 1988. "Organizational learning." *Annual Review of Sociology* 14: 319–40.

Lunn, N.J., S. Schliebe, and E. Born, eds. 2002. *Polar bears: proceedings of the 13th working meeting of the IUCN/SSC Polar Bear Specialist Group, Nuuk, Greenland.* Occasional Paper of the IUCN Species Survival Commission No. 26, IUCN, Gland, Switzerland.

March, J.G., and J.P. Olsen. 1976. "Organizational learning and the ambiguity of the past." In *Ambiguity and choice in organizations*, edited by J.G. March and J.P. Olsen, 54–68. Oslo: Universitetsforlaget.

McLoughlin, P.D., M.K. Taylor, H.D. Cluff, R.J. Gau, R. Mulders, R.L. Case, and F. Messier. 2003. "Population viability of barren-ground grizzly bears in Nunavut and the Northwest Territories." *Arctic* 56(2): 185–90.

Nelson, J.G., and R. Serafin. 1996. "Environmental and resource planning and decision making in Canada: a human ecological and a civics approach." In *Canada in transition: results of environmental and human geographical research*, edited by R. Vogelsang, 1–25. Bochum: Universitatsverlag Dr. N. Brockmeyer.

NWMB (Nunavut Wildlife Management Board). 1999a. Meeting Minutes, Meeting 22, May 1999.

———. 1999b. Meeting Minutes, Meeting 24, November 1999.

———. 2000a. Meeting Minutes, Meeting 25, March 2000.

———. 2000b. Meeting Minutes, Meeting 26, September 2000.

———. 2000c. Meeting Minutes, Meeting 27, November 2000.

———. 2000d. Meeting Minutes, Conference Call 54, December 2000.

———. 2001. Meeting Minutes, Meeting 30, September 2001.

———. 2002. Letter from the NWMB to the Minister of Fisheries and Oceans, April 9, 2002.

Pahl-Wostl, C. 2002. "Towards sustainability in the water sector – the importance of human actors and processes of social learning." *Aquatic Sciences* 64(4): 394–411.

Parson, E.A., and W.C. Clark. 1995. "Sustainable development as social learning: theoretical perspectives and practical challenges for the design of a research program." In *Barriers and bridges to the renewal of ecosystems and institutions*, edited by L.H. Gunderson, C.S. Holling, and S.S. Light, 428–60. New York: Columbia University Press.

Prestrud, P., and I. Stirling. 1994. "The International Polar Bear Agreement and the current status of polar bear conservation." *Aquatic Mammals* 20(3): 113–24.

Senge, P.M. 1990. *The fifth discipline: the art and practice of the learning organization.* New York: Doubleday.

Shields, D.J., S.V. Solar, and W.E. Martin. 2002. "The role of values and objectives in communicating indicators of sustainability." *Ecological Indicators* 2(1–2):149–60.

Taylor, M. and J. Lee. 1995. "Distribution and abundance of Canadian polar bear populations: a management perspective." *Arctic* 48(2): 147–54.

Taylor, M.K., S. Akeeagok, D. Andriashek, W. Barbour, E.W. Born, W. Calvert, H.D. Cluff, S. Ferguson, J. Laake, A. Rosing-Asvid, I. Stirling, and F. Messier. 2001. "Delineating Canadian and Greenland polar bear (Ursus maritimus) populations by cluster analysis of movements." *Canadian Journal of Zoology* 79: 690–709.

US FWS (United States Fish and Wildlife Service). 1997. Importation of Polar Bear Trophies From Canada Under the 1994 Amendments to the Marine Mammal Protection Act; Final Rule. Federal Register 62(32): 7301–7331.

———. 2001a. Import of Polar Bear Trophies From Canada: Change in the Finding for the M'Clintock Channel Population and Revision of Regulations in 50 CFR 18.30. Emergency Interim Rule. Federal Register 66(7):1901–1907.

———. 2001b. Import of Polar Bear Trophies From Canada: Change in the Finding for the M'Clintock Channel Population. Affirmation of emergency interim rule as final rule. Federal Register 66(194):50843–50850.

Walker, B., S. Carpenter, J. Anderies, N. Abel, G. Cumming, M. Janssen, L. Lebel, J. Norberg, G.D. Peterson, and R. Pritchard. 2002. "Resilience management in social-ecological systems: a working hypothesis for a participatory approach." *Conservation Ecology* 6(1): 14-. *http://www.consecol.org/vol6/iss1/art14* (15 August 2002).

Walters, C.J. 1986. *Adaptive management of renewable resources.* New York: McGraw Hill.

Webler, T., H. Kastenholz, and O. Renn. 1995. "Public participation in impact assessment: a social learning perspective." *Environmental Impact Assessment Review* 15(5): 443–63.

World Wildlife Fund. 2001. "Nunavut narwhal and beluga reduced by over harvesting." *Arctic Bulletin* 2.01: 21.

CHAPTER 14

EXPLORING THE ROLES OF LAW &
HIERARCHY IN IDEAS OF RESILIENCE:
REGULATING RESOURCE HARVESTING IN NUNAVUT

Nigel Bankes (University of Calgary)

*... the key to regional ecosystem management resides in a nested set of dy-
namic policy settings, in which the natural variability and diversity of the
ecosystem itself should guide policy targets over management domains that
are periodically adjusted ... the hierarchy of policy should be reversed with
regional agents given free rein to monitor and adapt to the pulse of the region,
but only within the nested hierarchy of global system values. Now to find a
bureaucratic organization ... that permits such flexibility would really add
to the arsenal of progressive environmental policy, both north and south.*

— Sanderson 1995, 390

INTRODUCTION

Social systems and natural systems interact in complex, dynamic, and adaptive ways. Ideas of resiliency offer us a way of thinking about
that interaction in a manner that focuses on maintaining the social-ecological
system within a desired domain of attraction. Resiliency thinking suggests that
we direct our attention to the capacity of the system to cope with and adapt
to change rather than on controlling change, or on increasing the productive
capacity of the ecological system (Folke *et al.* 2002). Resiliency is an important
concept in the search for sustainability, not because resiliency is itself a desirable
social goal (consider the tremendous resiliency exhibited by social caste systems,
a hierocracy [Holling *et al.* 2002b, 96]), but because it may help us maintain
desired ecosystem states and equitable social arrangements. Our commitment
to resilient systems is therefore conditional and consequentialist.

Much of the writing on ideas of resiliency is explicitly multidisciplinary, but
the discipline and role of law is frequently absent or understated (Gunderson
and Holling 2002; Folke *et al.* 2002; Gunderson *et al.* 1995). Rose (2002) makes
a similar point about related problems of commons research. To the extent that

291

law is a key means of social control, and to the extent that any efforts at managing ecosystems must, of necessity, be concerned with managing the people who interact with the ecosystem (Lee 1999), this omission seems surprising.

The failure to consider the role of law is not universal, and some writers on resilience draw attention to the role of law. For example, Alcorn and Toledo (1998, 216) note the importance of state protection, through law, of community-based tenurial systems as part of the claim that the "sustainability of larger-scale systems depends on renewal cycles in the dynamic, local systems that maintain the fine-scale variability arising from adaptations to local environmental and social conditions." Others draw attention to the fact that changes in the law create opportunities, and, in particular, opportunities for experimentation (Trosper 2003; Westley 2002). In emphasizing the role of law in creating change, or instigating opportunities for change, these authors are reflecting on the role that law can play in the back-loop portion of Holling's adaptive cycle (Berkes *et al.* 2003).

This chapter seeks to build on this work by offering some preliminary contributions, from the discipline of law, to the resilience project. If the goal is to understand ideas of change and resilience within complex systems, including the interaction between natural systems and social systems, then, at a minimum, we need to understand the role that law may play in this process. Law is, after all, a central, and in many cases a defining, feature of all complex social systems; it is itself a complex social system. Laws will ordinarily embody the values of the different societies that they serve and the values thus embodied will likely be deeply held and enduring rather than ephemeral and trivial.

In some cases law can be transformative and an instrument of social change, *e.g.,* domestic and international human rights laws. In other cases, law serves to entrench the *status quo* and serves as an obstacle to change, *e.g.,* property law, which generally protects existing entitlements – that which is proper (but even property law has transformative potential [Macpherson 1978] and may impose obligations on owners [Freyfogle 1996]). Law is often at the forefront of interactions between social systems and natural systems. Consider that it is law that sets the terms and conditions for the protection of species at risk, that sets the parameters of environmental assessments, that determines the degree of protection afforded to protected areas, and that fixes open and closed seasons for harvesting and the terms of quota arrangements.

The general question that we might like to answer could be "What is the role of law in fostering resilience?" or "Does law foster resilience or increase vulnerability?" But it is evident that these are overly simplistic and general questions for a number of reasons. In the first place, it is difficult to separate law from the society which gives rise to it – laws and society co-evolve (Levin 1999, 34). But it is also misleading to talk in the singular about law and legal system; for the reality is that there are multiple laws and legal systems and that these systems are increasingly coupled and nested, the one within the other. We can see this in the increasing interaction between domestic norms and international norms, but we are also familiar with interacting systems in the context of customary norms and positive state norms, the interactions of public law and private law, ideas of private ordering (the law of contract) and public ordering (the law of torts), and the interrelationships, at least within common law systems, between statute law and judge-made law (case law). Laws are scaled at both vertical and horizontal levels, and there is frequently considerable duplication, overlap, and redundancy (*sensu*, Landau 1969).

So it seems that we must limit the ambition of our inquiry and focus instead on some limited aspects of resilience thinking and indeed on some limited aspects of legal systems. A central idea of ecological resilience (as opposed to engineering resilience [Holling 1996]) is the capacity to learn and adapt. A linked social and ecological system is more likely to be resilient if it can build and increase its capacity for learning and adaptation. This chapter explores the capacity to adapt in the context of the institutional systems that have been created in northern Canada over the last decade or so for the management of wildlife or, more accurately, human-wildlife interactions. These institutions were created as part of the settlement of Aboriginal land claims, and the relevant commitments are recorded in Aboriginal land claims agreements. A land claims agreement or modern treaty is a complex document with a complex juridical status. It represents a contract between the parties (the Crown and the Aboriginal people concerned), but it may also have statutory force and is protected by the Constitution.

My original hunch in looking at the institutional systems created by these agreements was that they would prove to be rigid. I suspected that they would likely lack resilience because they would lack the capacity to adapt to changing conditions, to changing understandings of ecosystem function, and indeed to changing conceptualizations of the relationships between humans and the environment. I thought that there was a risk that the constitutional status of these agreements, combined with the level of detail at which they were drafted, while offering much sought-after protection for indigenous peoples, might turn out to be a millstone. In sum, I thought that there was a risk that these agreements had frozen learning at a particular point in time and had created rigid hierarchies that were not well adapted for long-term survival.

While it is far too early to pass judgment on the long-term adaptability of these institutions, I now think that my original hunch was misleading. There are reasons for thinking that these institutions, notwithstanding their constitutional status, offer considerable potential for adaptability. More generally, they may go some way to developing the form of institutions identified by Sanderson in the epigraph to this chapter and identified by others as equally desirable (Wilson 2002), that is to say, "institutional structures that match ecological and social processes operating at different spatial and temporal scales" (Folke *et al.* 2002). The balance of the chapter seeks to explore these claims by focusing on the structure of the institutions created by land claims agreements. It does so by focusing on ideas of hierarchy and the cross-scale interactions (*sensu* Young 2002) between different elements of these institutions. It proceeds as follows. The next part of the chapter explains some key terms, describes the importance of hierarchical organization in natural systems and social systems, and then offers some observations on law as a hierarchical form of organization. The third part of the chapter examines the regulatory scheme that has been developed for the regulation of the fishery in Nunavut. The chapter considers the regulatory scheme that was in place prior to the settlement of the Nunavut land claim, but also looks at the scheme developed in the Nunavut Land claims agreement (NLCA 1993). The NLCA flipped or profoundly transformed the existing hierarchical organization of wildlife management and thereby created new cross-scale interactions. The fourth part of the chapter draws comparisons between the pre- and post-NLCA regimes, and the final part of the chapter offers some brief conclusions and suggestions for further research.

IDEAS OF RESILIENCE AND HIERARCHY

Complex systems are generally organized hierarchically and partitioned into nearly decomposable (or independent) subsystems (Wilson 2002). Within each subsystem, rates of interaction are high, but lower as between subsystems. There are advantages associated with hierarchical organization. Herbert Simon's parable of the two watchmakers demonstrates the considerable efficiency implications of hierarchical organization (Simon 1973), and Levin reminds us that the modular organization of hierarchies offers buffering against cascades of disaster (Levin 1999). The choice of, or identification of, a system is subjective and scale-dependent. As we change scales or perspective we observe that individual systems are nested within other systems to which they are interconnected horizontally and vertically.

For ecologists, hierarchies result from the interactions between fast and slow processes across spatial scales (King 1993; Holling *et al.* 2002b). Climate change offers an example of a slow process; budding, photosynthesis, and then decomposition of leaves an example of a fast process. Slow processes act as outside constraints. For example, a forest stand moderates the climate within the stand to narrow the range of temperature variation. Different processes act at different spatial levels. Fire, for example, operates at a meso-scale and affects stands in a forest, or an entire forest, rather than just, say, needles.

This description of hierarchies suggests, at first glance, that they must be static structures, since the slow-moving, large-scale processes will always constrain the behaviour of faster and smaller-scale processes. But those concerned with the adaptive cycle of systems wish to guard against this view (Holling *et al.* 2002b). In particular, they suggest that during the phases of creative destruction (omega) and renewal (alpha) of the Holling cycle of renewal, each level of a system's structure and processes can be reorganized and the reorganization may transform the hierarchy (revolt) (*e.g.*, a small fire at the tree level becomes a major conflagration affecting a whole stand). The re-shuffling of the hierarchy may create the possibility of new system configurations, but those new configurations will still be influenced by the remembrance of the old. In the case of natural systems, the memory will be provided by such things as the natural seed bank, surviving species, and remaining physical structures.

The dynamic potential of the renewal cycles "nested one within the other and across space and time scales" led Holling and colleagues to reject the term "hierarchy" and to coin a new phrase, "panarchy." By using the term "panarchy," the authors hoped to capture "an image of unpredictable change and [to draw upon] notions of hierarchies across scales to represent structures that sustain experiments, test results and allow adaptive evolution"(Holling *et al.* 2002a, 5). The key feature of the panarchy concept, therefore, is its emphasis on the inter-action between different systems and different processes (nested cycles), which creates possibilities for innovation. This leads to a more fluid idea of hierarchy and rejection (or at least qualification) of "the rigid, top-down" use of the term that seems to be part of ordinary usage and indeed accepted within discussions of social systems. Do the insights that led to the idea of panarchy for natural systems also illuminate our thinking about social systems? Are there features of existing structures that we think of as hierarchies that encourage interaction, innovation, adaptation, and learning? Or are social hierarchies simply versions of rigidity traps? Are there varieties of hierarchical systems and can existing hierarchies be modified? What role does law play in fostering the development of hierarchies, or indeed more horizontal systems, so that they encourage in-teraction, innovation, adaptation, and learning?

The imagery of panarchy encourages us to look more deeply at the cross-scale interactions within nested social systems. Here we adopt Young's (2002, 264) terminology of vertical and horizontal interactions. For Young there is a horizontal interaction where there is interaction at the same level of organiza-tion (*e.g.,* between trade regimes and environmental regimes operating at the international level) and a vertical interaction where the interaction occurs across levels of social organization (*e.g.,* interactions between local systems of tenure and national regulatory systems). Ideas of hierarchy emphasize the vertical rather than the horizontal interactions, but the nested nature of social systems requires us to consider the horizontal interactions as well. The structure and routing of these interactions is one factor in the ability of management systems to respond to environmental feedbacks. The term environmental feedback is used here to refer to information about the results of human interactions with the environment. Feedbacks provide opportunities to change behaviour and may be positive (and therefore reinforce that behaviour) or negative (and therefore suggest the need for behavioural changes or rule changes). The loca-tion of the capacity to respond to feedbacks with rule changes is a function of institutional structure, but here we face a conundrum. For, as Folke *et al.* (1998) observe, while the record shows that local level institutions learn and develop the capacity to respond to environmental feedbacks faster than do centralized agencies, if institutions are too decentralized then the feedback between the user groups of different resources or between adjacent areas may be lost. Wilson

(2002) posits that the answer is to be found in designing multiscale institutions whose organization and activities parallel the organization and activities of the natural system so as to increase the likelihood of acquiring the information necessary for learning.

It is conventional wisdom that legal systems are organized hierarchies and rigid hierarchies at that. While the notion of hierarchy within ecosystems is explainable in terms of interactions between fast and slow processes across spatial scales, the notion of hierarchy in law has a more formal and normative meaning more closely related to the ordinary dictionary meaning of the term. In law, the idea of hierarchy is concerned with the interrelationship between conflicting norms and the associated trumping rules for resolving those conflicts.

A conventional account would emphasize that the Constitution is the supreme law and that all laws inconsistent with the Constitution are of no force or effect. Similarly, the decisions of senior courts bind lower courts. That account might also emphasize that statutes determine the validity of regulations and ministerial orders and that municipal bylaws will be of no force or effect to the extent that they conflict with provincial statutes. The account recognizes that there are different subsystems of rules based on different centres of power: municipalities, provinces/states, federal governments (Hogg 2001). But these systems are not isolated and there are vertical and horizontal interactions between these different systems.

A legal system is a nested hierarchy. Changes in the Constitution may trigger cascading changes in statutes, regulations, and particular decisions. The enactment of a constitutionalized bill of rights such as the Canadian Charter of Rights and Freedoms, or the constitutional protection of Aboriginal and treaty rights, will bring about such a cascading change.

In addition to the formal account it is also possible to discuss hierarchies in law by drawing upon ideas of fast and slow processes and ideas of spatial scales rather than ideas of conflict and trumping. The process of constitutional (as opposed to revolutionary) amendment is universally a slow process, hedged about with special rules which constrain the availability of future change. A Constitution may therefore capture a particular understanding at a particular point in time. While constitutional scholars would no doubt resist characterizing such constitution-building events as representing "frozen accidents" (Holling and Gunderson, 2002) they would have little difficulty characterizing them as "rare" or accepting the idea that their influence "can shape the future for long periods." Much faster processes would include the passage of statutes (some surviving decades, others, like property statutes, centuries and still others, like the *Income Tax Act*, subject to more or less continual renewal), regulations, and faster yet would be the issuance of ministerial orders (for example, variation orders opening and closing particular fisheries and locations) and the exercise of statutory discretions to make particular decisions. As an example of the latter consider the *National Parks Act*. The Act provides a very general framework for the use and management of Canada's national parks. One, and perhaps the key, guiding idea is that of ecological integrity. But such an idea is clearly not self-implementing and must be interpreted in the crucible of concrete decisions. The extent to which the slow process of legislative reform actually controls the fast process of ministerial decision making will be the result of the complex interaction between the specificity of the statutory direction and accompanying definitions and the extent of the judicial deference accorded to the decision maker's *interpretation* of the statutory direction. Another way to think about these processes in terms of more traditional legal categories is to sort them into legislative functions (general rule-making functions) and administrative functions (the application of general rules to particular cases).

Figure 14.1 shows this range of fast and slow processes (the vertical scale) along with a sense of geographical range (the area of application of the law or rule, the horizontal scale), using fisheries rule making as an example. This simplified diagram endeavours to include international levels of regulation as well as domestic regulation. Later portions of the chapter explore the interactions between these levels in some more detail and in the context of the Nunavut fishery.

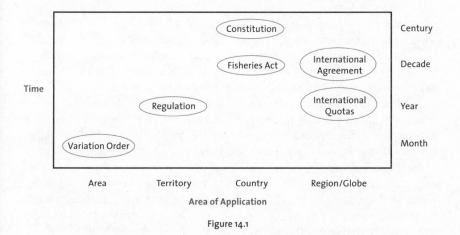

Figure 14.1

If it is conventional wisdom that legal systems are organized hierarchically, there is an equally conventional radical critique of the role of hierarchy (and therefore law) as dominance. According to this view, hierarchies disempower and foster inequality and the law encodes that disempowerment and dominance. Within this discourse, the term "hierarchy" is often used in a pejorative way, suggesting that existing hierarchies are necessarily undesirable and synonymous with the abuse of power. We can see this, for example, in various strands of the radical critique, including feminist literature (Smith 1993, 2), ecofeminist literature (Hessing 1993; Hughes 1995), and critical race theory literature (Harris

1993). While some versions of the critique seek to reverse the hierarchy (belying the claim that hierarchies are necessarily undesirable), others seek to abolish hierarchy and attain true equality through emphasizing the importance of relationships rather than trumping (Nedelsky 1993).

More mainstream critiques are equally likely to take a negative view of hierarchical social organizations. For example, there is often a direct association between the terms "hierarchy" and "command and control" as ways of organizing systems. Such systems are often contrasted with market-based systems in which the individual consumer is sovereign and with highly distributed systems such as the Internet. By implication, hierarchies are inefficient, rigid, maladaptive, highly connected, and vulnerable (Rifkin 2002).

Both the conventional account and the critique often fail to recognize that the term "hierarchical" is a relative term. We recognize this in ordinary discourse because we commonly speak of some hierarchies (*e.g.,* the military or the church) as being "rigid" forms of hierarchy or "very" hierarchical, whereas other forms of organization may be described as non-hierarchical (*e.g.,* universities) even though closer examination reveals some well-developed sense of hierarchy even within these flatter or more horizontal organizations. But there is also evidence of the same relative usage within the legal system. Thus the school of legal pluralism (the "situation in which two or more legal systems coexist in the same social field") (Merry 1988, 870), while often concerned with the relationship of inequality between indigenous systems of social ordering and those of settler/colonial societies, also recognizes that that dominance was not total. There were pockets of resistance and survival of the old methods of ordering (Harris 2001).

Much of the literature on pluralism is concerned not only to posit the existence of multiple systems of rules within a given society but to explore the relationships between different normative orders and the nature of the interactions. Those interactions include the ways in which state law may shape other normative orders (*e.g.,* custom may become part of state law or be administered by state actors), but also the ways in which customary normative orders may shape state law. State law may be facilitative and may make possible areas of private ordering. While the power relationship between these systems may be unequal and the state system dominant, it may be subverted at the margins.

This diversity of views allows us to conceive of law as something other than hierarchy or at least allows us to suggest that a rigid hierarchical view of law and legal systems is much too simplistic. A hierarchy will be very rigid if there are firm trumping rules and if the apex of the decision-making pyramid is isolated from the base. A hierarchy will be less rigid if it provides for communication and feedback between the different levels of the hierarchy; if it provides for reasoned, discursive resolution of conflict between different levels of the hierarchy; if it provides that in some cases different levels of the hierarchy will have the final say according to some principle of subsidiarity; and if trumping is simply one possible outcome when all efforts at collaborative solutions have failed.

There is all sorts of evidence that legal systems form less rigid hierarchies than might be supposed based on the conventional account. Consider the following:

- Conventional practices modify hierarchies. For example, while s. 90 of the Canadian Constitution still provides that provincial laws are subject to disallowance by the federal Governor General, it is inconceivable that such a power would be exercised now, and indeed we would characterize any attempt to do so as unconstitutional. The formal hierarchy has been modified by actual practice (Hogg 2001.)

- Not all legal systems exhibit a strong sense of hierarchy. For example, it is generally acknowledged that there is no hierarchy as between the two main sources of international law, treaty and custom (Kontou 1994). The horizontal nature of international society is often a source of criticism since the absence of a central law maker (legislature) within the international system leads, it is argued, to inefficiency and perhaps even rigidity (it may be hard to change multilateral treaty norms based upon consent).

- Ideas of hierarchy within legal systems break down when we view issues from the different perspectives of different legal orders. For example, from the perspective of some domestic legal systems (*e.g.,* that of Canada), a conflict between a domestic norm and an international norm should be resolved in favour of the domestic norm (Hogg 2001). Viewed from the perspective of international law the opposite conclusion would be reached.

- Federal systems are explicitly designed to divide sovereignty (and along with that, any notion of hierarchy) between different orders of government according to subject matter and relative perceptions of local and national importance. One of the traditional justifications of federalism is that it encourages and facilitates experimentation; experimentation by one unit of a federation provides resources and ideas for other units. Sub-units of federal states are linked horizontally but not vertically.

- Hierarchies only matter when there is a conflict. Most of the time there is no conflict and thus multiple levels of norms (international, federal, provincial and municipal, private law, and public law) may apply in a truly nested sense to any particular activity (Hogg 2001).

- Legal systems are typically composed of a complex of rules and more general principles. Although rules operate in an all-or-nothing way, principles do not and may have profound transformative effect

- slow variables that serve to modify the application of particular rules (Dworkin 1977; Reference re Secession of Quebec 1998).

• Whatever the sense of hierarchy exhibited by a legal system, the particular details of the hierarchy are always contingent. Hierarchies may be transformed or flipped. Consider, for example, the effect of the constitutional protection afforded to existing Aboriginal and treaty rights in 1982. Prior to 1982 one could think of Aboriginal rights as being either at the bottom of the hierarchy (*Sikyea* 1964) or as being located outside the hierarchy altogether. The result of constitutional protection was to reposition these rights much closer to the apex of the hierarchy. Ideas of transforming hierarchies strongly influenced the so-called Canada round of constitutional negotiations during the early 1990s that led to the text of the Charlottetown Accord. Although the Canadian people declined to approve that accord, one of its key ideas was the recognition of the Aboriginal right of self-government as an *inherent* right. Central to the idea of an inherent right is that it exists outside the ordinary hierarchy and does not therefore depend on a grant of authority from the apex.

• Hierarchies may also be modified. Nedelsky (1993) provides the example of the s. 33 of the Canadian Charter of Rights and Freedoms, the famous "notwithstanding clause," which allows Parliament and provincial legislatures to declare that a law shall apply "notwithstanding" specific provisions of the Charter. Section 1 of the Charter provides another example. This section confirms that no right is absolute and that constitutional rights have no necessary trumping effect notwithstanding their position in the hierarchy. The so-called justification test developed by the Supreme Court in *Sparrow* (1990) serves a similar office for Aboriginal rights protected by s. 35 of the Constitution and confirms as well that the transformation of rights of which we spoke in the last paragraph is less than complete. A key aspect of the justification test is the requirement of consultation, which draws attention to procedure and to the relationships between different elements of the polity rather than to trumping rules.

• The relationship between the executive (and its regulatory authorities and boards) and the courts, while often described in terms of hierarchy (the idea of the "superior" courts demanding and enforcing adherence to the "rule of law"), is actually characterized by a high degree of deference, which restrains the courts from interfering in the substance of decisions in which they have no expertise (Mullan *et al.* 2003).

Just as our interest in resilient systems is contingent and instrumental and not intrinsic (we do not value resiliency for its own sake), so also might be our interest in the hierarchical (or indeed horizontal) organization of social systems. As with resilience, our question should be, "Does this hierarchical arrangement help us maintain desired ecosystem states and equitable social arrangements? And if not, how might we modify or flip that hierarchy to accommodate these needs, and what role will law play?"

The account to this point, while abstract, does suggest that it will be useful to focus further inquiry on ideas of nested systems and the interrelationships between those systems – in other words the cross-scale connections, both vertical and horizontal, but especially the former. And in looking at those connections what do we want to know from a resiliency perspective? We might put the question this way: are the social institutions of these nested systems *and the connections between them* scaled in such a way as to tighten feedback loops and to maximize opportunities for social learning and adaptation? How does law help structure these connections and relationships?

While this is a very general question (perhaps excessively general), the question may be more tractable if we think of it in relative terms and in a more concrete context. I propose to do this in the context of the nested systems that have developed to regulate the Nunavut fishery, looking in particular at the transformation that has occurred since the settlement of the Nunavut (Inuit) land claim in 1993.

REGULATION OF THE NUNAVUT FISHERY

It is possible to identify four main historical periods in the exploitation of the fisheries resource in Nunavut (Bankes 2003). The first period is represented by the pre-contact period (regulation of an Inuit fishery by Inuit customary rules), and the second by the whaling period (an absence of regulation, the transformation of the fishery to an open access commons). The third era is characterized here by the continued exploitation of the fisheries resource, but increasingly under the terms of settler laws (*i.e.,* the laws of the colonist) rather than traditional, customary rules. The fourth period is represented by the period commencing in 1993 with the settlement of the Inuit Aboriginal land claim and the creation of a new co-management/co-jurisdiction authority the Nunavut Wildlife Management Board (NWMB). My focus here will be on the contrast between the third and fourth periods; the account of the first two periods is offered primarily for background.

The pre-contact period

In the pre-contact period Inuit, organized in miut groups with particular geographical affinities, harvested marine resources in accordance with customary rules and practices. The most important species for the Nunavut Inuit harvest were ringed seal, bearded seal (harp seal to a lesser extent), walrus, beluga, narwhal, and bowhead. Walrus and narwhal were valued for their tusks (ivory) as well as for food purposes. Marine fish (*e.g.,* cod and sculpin) were generally not important. Within the freshwater system, important species were char (land locked and sea-run) and lake trout (Freeman *et al.* 1976; Royal Commission 1986).

Methods of harvest varied with the species (Freeman *et al.* 1976; Nelson 1969). Char and trout were taken by spearing, by weirs, with hooks through the ice, and with nets. Marine mammals were harvested by harpooning, or later with rifles (both from umiaks or kayaks and from the floe edge), or by herding animals into shallow water.

The whaling period

The international whaling fleet was active in the eastern Arctic from as early as the beginning of the eighteenth century to about 1915. (Ross 1975; 1982; Eber 1989). Although the bowhead was the primary target species, Ross (1982, 9) notes that after the tremendous harvests of the 1820–1840 period and "the overhunting of bowheads brought declining yields," the whalemen turned to smaller cetaceans, including beluga, narwhal and other sea mammals, walrus, and various species of seal.

Inuit were employed in the fishery and indeed played essential roles and were recognized for those contributions (Mitchell 1996), but they were not recognized as owners of the resource whose permission might be required to engage in the harvest (NWMB 2000). In some cases Inuit were retained as contract whalers for American and Scottish whalers and played an important role in provisioning overwintering whaling camps with meat. This had an impact on the harvesting of other resources, including muskox and caribou (Ross 1975).

As the resource was exploited in a particular area, the whalers moved on, and by the late 1870s the industry was in decline. As the whaling industry declined due to shortage of whales and the introduction of substitutes (Ross 1975), whalers sought to extend the trading relationship with Inuit and trade for ivory from walrus, and skins and fur (*e.g.,* white fox) became an increasingly important part of the business (Eber 1989). Ultimately, however, the whalers simply never returned. According to Mitchell (1996, 83), "in all cases the retreat of the whalers left the Inuit, who had to some extent come dependent upon them, stranded."

The trading relationship founded upon whaling was important to Inuit. In particular, it provided access to a wide range of technology including the firearm and the needle. It also introduced the whaling boat, which replaced both the umiak and the kayak as the preferred vessel from which to engage in whale harvesting. While the effect of the whaling era on the transformation of Inuit society is still debated (Mitchell 1996), the effect of the harvest was to push bowhead from being relatively abundant to the brink of extinction.

What rules and institutions governed this intensive but relatively short-lived exploitation of the whaling resource? The whaling literature has very little discussion of law and institutions. When it does discuss law, it is the law pertaining to the relationship between whalers rather than the state regulation of whaling (Eber 1989; Ross 1975). The absence of *positive* state laws governing the regulation of whaling confirms the status of the resource as an open access common property resource. The perceived absence of law is misleading, for the "absence" of positive rules does not mean that there is no law: indeed the very absence of specific prohibitions and limits merely serves to affirm a set of liberties in the sense that one who takes whales, for example, does no wrong at least under the laws of the settler society (Singer 1982; Williams 1956).

The introduction of settler laws

The introduction of harvesting laws by the settler society in the third period only occurred as government authority moved north – a movement that coincides with the introduction of police, missionaries, and traders. The initial focus of such laws was not the regulation of the marine mammal fishery but the regulation of the muskox harvest (Ross 1975, 110). The precise point at which the fisheries resource came to be regulated is difficult to pinpoint both as a matter of positive law and in terms of its actual application on the ground. It seems to have happened at different times for different resources, occurring first for walrus (Walrus Protection Regulations, 1928) and much later for (the few remaining) bowhead, the smaller whales (the narwhal harvest was regulated 1971, beluga were regulated 1949, initially only for the tidal tributaries of Hudson Bay, Hudson Strait, James Bay, and Ungava Bay), seals, and other fishery resources. One assumes (and further research along these lines would be useful), that there would have been some geographical variation in on-the-ground application of these rules as Inuit moved off the land and into communities. But these details are less important for present purposes than some broad observations on the types of institutions and rules that were put in place as the harvest of the resource came to be regulated by settler society and which were in place in the years immediately preceding the Nunavut Land claims agreement (NLCA). Cumulatively, they served to dis-embed responsibility for resource management from the communities they supported (McCay and Jentoft 1998).

The following describes key elements of the rule system in the years preceding the NLCA. There are further details in Bankes (2003).

- The framework for the rules is provided by the interrelationship between the background rules of the common law and statute, the *Fisheries Act*. In the decades prior to the NLCA, regulations passed under the *Fisheries Act* comprehensively regulated access to the fishery.

- The harvesting of marine mammals (whales, walrus, seals) came to be regulated by the Marine Mammal Regulations (MMR), superseding earlier discrete regulations for specific categories of marine mammals (seals, narwhal, beluga, walrus, etc). These regulations prohibited harvest of bowhead by eastern Arctic Inuit and allocated a community level quota for harvesting narwhal, beluga, and walrus. The community level quotas replaced earlier methods of regulation based on individual quotas. The regulations prohibited harvesting of cetaceans by non-Inuit. Seal harvesting for other than commercial purposes was essentially unregulated. There was some relevant international regulation through the International Whaling Convention (1946), although Canada is no longer a party to that agreement. Trade in some marine mammals and the products thereof (walrus, narwhal) was regulated by the Convention on International Trade in Endangered Species (CITES) (1973) and national implementing legislation. International public opinion came to be seen as a significant factor in the harvesting of seals and cetaceans and as an important consideration in the marketing of products.

- The harvesting of other fish was regulated principally by the Northwest Territories Fishery Regulations (NWTFR). These regional regulations, also passed pursuant to the *Fisheries Act*, authorized Inuit to harvest fish without a licence for non-commercial purposes but provided that no person may engage in a commercial fishery for char or other species without a licence and then only within a designated body of water. The regulations prescribed quota limits for particular species (*e.g.*, char or lake trout) as well as gear restrictions for these individual bodies of water. Commercial licences were issued for particular bodies of water on an annual basis.

- The NWTFR and the MMR did not provide the whole picture. The Nunavut offshore fishery in the Davis Strait for turbot etc. was and is managed under the terms of the Atlantic Fishery Regulations, also passed pursuant to the federal *Fisheries Act*. There is a measure of international regulation of the offshore fishery insofar as the area in which the fishery is prosecuted falls within North Atlantic Fisheries Organization (NAFO). In addition, the Fisheries (General) Regulations (FGR) (and their predecessor provisions) provide DFO with significant incremental flexibility in the actual season-by-season and day-to-day management of the fishery by allowing the minister and delegates to issue variation orders to vary the application of the regulations (*e.g.*, to open or close bodies of water) and also to issue experimental licences.

But what happened to the customary rules and norms of the Inuit harvesters with the introduction of settler law? The question is a complex one and requires further empirical research. As a practical matter many community norms must have survived and continued in parallel with settler norms; pluralism prevailed as is the case with so many commons regimes (McCay 1978, 407; Ostrom 1990). These norms might continue to regulate the use of fish and marine mammals harvested, the manner of harvest, the use of harvesting locations, the sharing of the harvest, etc. The introduction of settler law did not somehow erase all other forms of community ordering. While the *Fisheries Act* and regulations conferred no *express* recognition on local, community-based HTOs, there was implied recognition in some cases through the institution of community level quotas in the MMR. The MMR did not prescribe how this community quota was to be divided within the community or among individual harvesters – they assumed that there were internal community norms for making these decisions. In other cases where both express and even implied recognition of customary institutions was lacking (the NWTFR and the FGR), the actual practice of fisheries officers on the ground showed considerable deference to communities in terms of decisions to issue commercial or experimental licences for particular bodies of water (Bankes 2001).

But what if there were a conflict? What if under community norms harvesting was lawful but under the MMRs unlawful? In such a case, the hierarchy between these orders was clear: the positive law of the settler society would be applied by a Canadian court to trump the customary norm. This conflict between federal fishery and wildlife rules and Aboriginal and treaty rights was played out across the country and always, ultimately, with the same result (Bankes 2003; Harris 2001; *Sikyea* 1964).

During this period Inuit became increasingly connected with, and dependent on, not only the Canadian national economy but the global economy. This connectivity carried with it a vulnerability to exogenous change, dramatically illustrated by the loss of the European seal market (Royal Commission 1986). Toward the end of this period, southern harvesters began to show increasing interest in the potential of northern fishery resources (shrimp, turbot) as their adjacent cod fishery encountered difficulty (Bankes 2003).

The post-NLCA period

The final period begins in 1993 with the settlement of the Inuit land claim with the federal government through the terms of the Nunavut Land claims agreement (NLCA). I have described in detail elsewhere (Bankes 2003) the effect of the NLCA on the pre-1993 regulatory regime for fisheries. For present purposes we can summarize the main changes under two headings: (1) enhanced protection for Inuit harvesting rights, and (2) enhanced role for Inuit in management of the fishery through the NWMB.

The NLCA makes a basic distinction between those stocks for which there are conservation concerns and for which the NWMB therefore needs to set a total allowable harvest (TAH), and those stocks for which (non-TAH) there is no conservation concern. If the stock is a non-TAH stock, Inuit have the right to harvest that stock up to their full level of social and economic need. Non-Inuit continue to require licences as before. Where the stock is a TAH stock, no person may harvest over the TAH; the Inuit right is a priority right to harvest up to the full basic needs level. Non-Inuit may only engage in the harvest of a TAH stock to the extent that there is a surplus. Those harvesting restrictions (quota and non-quota limitations, *e.g.,* net size) that were in place on the date the NLCA entered into force (to the extent that they were themselves lawful) were grandparented, but all new quota and non-quota restrictions require the approval of the NWMB.

The NLCA created the NWMB as a co-management and co-jurisdiction (rule-making) body. It has an equal number of Inuit and government (federal and territorial) members and a chair nominated by the members. As indicated in the previous paragraph, the NWMB has extensive powers to initiate and put in place important management rules – quota and non-quota limitations for the fishery. But the NWMB is not the only player in the post-NLCA world. The Department of Fisheries and Oceans continues to play an important role in a number of respects. First, NWMB decisions on fisheries matters are subject to disallowance by the minister. Second, the minister may still take emergency decisions to protect the fishery. Third, the actual issuance of licences, where necessary, continues to be the responsibility of the department. Fourth, the minister continues to have primary responsibility for those marine areas that are adjacent to Nunavut but which fall outside the actual settlement area. While these represent important continuing powers, it is important to emphasize that the NLCA imposes significant procedural limitations on the exercise of these powers. For example, any exercise of the disallowance power must be accompanied by reasons, along with an opportunity to the NWMB to respond. Any exercise of the emergency power requires the NWMB to examine and report on the exercise of the power. In addition to the continuing but limited authority of DFO, the Nunavut regime continues to be nested within relevant international regimes, including NAFO, CITES, and, in the case of polar bears, the Polar Bear Convention. Finally, the NLCA envisages that community-based hunters and trappers organizations (HTO) and regional wildlife organizations (RWO) will play important roles in wildlife and fisheries management and work in conjunction with the NWMB. In particular, the NLCA envisages that these organizations will sub-allocate quota entitlements within regions and communities and will develop regional and community norms to govern harvesting practices.

SOME COMPARISONS BETWEEN THE PRE- AND POST-NLCA REGIMES

Figure 14.2 offers schematic outlines of the pre- and post-NLCA regulatory schemes.

The two diagrams are organized to show the vertical lines of authority. The pre-NLCA diagram depicts a traditional hierarchical arrangement with the authority to fish derivable, at least in formal terms, from the *Fisheries Act* and its regulations and ultimately the existing constitutional order. The post-NLCA diagram recognizes that the hierarchy has flipped. The NLCA has constitutional status and harvesting rights owe their authority as much and more to the NLCA and the Constitution as they do to the terms of the *Fisheries Act* and regulations. In this diagram the NWMB emerges as a completely new player, but the diagram

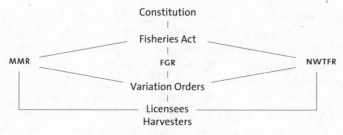

Figure 14.2a Pre-NLCA regulatory scheme.

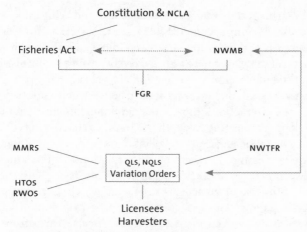

Figure 14.2b Post-NLCA regulatory scheme.

also explicitly recognizes the roles of community and regional organizations (the HTO and RWO).

How have cross-scale interactions changed as a result of the NLCA? The vertical cross-scale interactions are particularly useful as a means of exploring the scope of the modifications to the hierarchy. As we have seen, the formal pre-NLCA regime failed to accord explicit recognition to the role of communities and customary norms. The regulations therefore failed to prescribe vertical cross-scale interactions between the fisheries authorities and the communities; they concerned themselves exclusively with the state/subject relationship (McCay and Jentoft 1998, 26). Such linkages as developed were more informal (at the level of fisheries officer/community interaction) or the very formalized and *ad hoc* notification and consultation that might accompany new regulation-making exercises.

The formal pre-NLCA regime was equally silent on the question of horizontal linkages such as might occur between communities that might share the same resource, whether that resource was a char stream or a shared stock of beluga, narwhal or walrus. Similarly, there was no formal interaction between fisheries regulators and territorial wildlife regulators such as those responsible for large terrestrial mammals (caribou, muskox or even polar bear). Where populations were shared with other domestic jurisdictions (*e.g.,* walrus with Quebec and groundfish (turbot) and shrimp/scallop resources with northern Quebec and the Atlantic provinces) there is little evidence that there was much in the way of horizontal or vertical cross-scale interactions. The internationally shared offshore groundfish resources fall within the jurisdiction of NAFO, but the history recited in the turbot cases (NTI 1997; Bankes 2003) confirms the absence of vertical linkages, thereby depriving communities adjacent to the resource of opportunities for input into quota determination and allocation decisions.

Other marine resources (*e.g.,* narwhal stocks and beluga stocks) are shared with Greenland/Denmark, and while there are currently efforts underway under the auspices of the Canada/Greenland Joint Commission on Narwhal and Beluga (NWMB 1994) to achieve a greater level of co-operation, historically there has been little interaction at the horizontal level either between governments (Canada/Greenland) or, for the most part, between communities. The absence of cross-scale interactions impairs the development of negative and positive feedback loops connecting harvesting activities to harvesting entitlements.

The NLCA regime is far more explicit in its recognition of the importance of nested regimes and the need to facilitate both horizontal and vertical linkages between regimes and key actors. This is consistent with the experience elsewhere with co-management arrangements, which suggests the importance of land claims agreements as facilitating "a process of mutual social learning in which each side [government and resource users] learns from and adjusts to the other over a period of time." (Berkes 2002, 304) Whether this was an outcome that was actively contemplated when the NLCA was negotiated or whether (as seems more likely) this has simply emerged as an aspect of self-organization as the NWMB has evolved and grown into its responsibilities, the NLCA has helped to create "situations with the possibility of truly open and constructive deliberation" which are so important for resolving commons problems (McCay 2002, 383).

As the name implies, the jurisdiction of the NWMB is broad, covering not only all aspects of the fishery (marine mammals and fish) but also all terrestrial wildlife as well as plants. It therefore serves as a forum for bringing together a range of disciplines and management interests including traditional fisheries biologists from the federal Department of Fisheries and Oceans, along with bird biologists (from the federal Canadian Wildlife Service) as well as polar bear and large mammal biologists (from the Nunavut Department of Sustainable Development (DSD) (NWMB 1993–2003). In addition, these managers and disciplines must interact with holders of traditional knowledge (*Inuit Qaujimajatuqangit* – IQ) in efforts to implement the mandate of the NWMB. The interactions thus engendered have several implications. First, these horizontal interactions create the opportunity to build the climate of trust that is important for experimental or adaptive management. Second, knowledge gained from one experiment may prove to be directly or indirectly applicable in other management contexts because of the broad scope of board jurisdiction. For example, as Diduck *et al.* show in this volume, the NWMB is currently experimenting with "community-based management" for narwhal and beluga. The NWMB and the other actors involved very much understand that the experience so acquired may be useful in the management of terrestrial mammals as well as other marine mammals. Similarly, experience acquired through the negotiation of intercommunity polar bear management agreements and the assignment of a part of the community quota to guided sports hunts may be transferable to other contexts, including walrus. In sum, while the management efforts of the NWMB and the relevant government departments continue to be species-focused, the NLCA creates a forum for a broader learning environment across communities, government departments, species, and disciplines.

But the effect of the NLCA in flipping the hierarchy also has important implications for vertical cross-scale interactions. The dominant player in instituting new quota and non-quota limitations in the fishery is now the NWMB rather than DFO, but DFO has important continuing responsibilities, creating considerable uncertainty as to the respective roles of the NWMB and DFO. For the most part this uncertainty, while doubtless leading to inefficiency, has also proven positive insofar as it has encouraged learning and a willingness to experiment and to develop collaborative approaches to solving common problems. (Diduck et al, this volume) The deliberations within the NWMB serve to open up and make more transparent the ongoing management decisions that affect the fishery. Even the more conflict-charged special powers retained by the minister (the power of disallowance and the emergency rule-making power), which together suggest hierarchy modification rather than flipping, can only be exercised in a discursive context that emphasizes public reasoning and ongoing learning. Not only do these powers provide the opportunity to address the "nested hierarchy of global system values," but they also provide the opportunity to address issues that transcend the scale of NWMB jurisdiction thereby paralleling (*sensu* Wilson 2002) the natural system.

It would be a mistake to see the NWMB and the NLCA as driving all the cross-scale interactions that we can now observe. Some such interactions were already well in place before the NLCA entered into force. These interactions would include informal intercommunity connections as well as the more formal domestic and international arrangements for polar bear management described by Diduck *et al.* in this volume. It would equally be a mistake to think that the NLCA has created an environment in which all the appropriate cross-scale linkages to help tighten feedback loops has emerged. This is clearly not the case. There continue to be serious cases of over-harvesting of char stocks, whether those stocks are specific to a particular community or shared with other communities. (Welch 1995). In the case of the turbot resource, the NWMB's jurisdiction is much more attenuated, since the stocks lie for the most part outside the Nunavut Settlement Area. While the NLCA endeavoured to build some cross-scale linkages for such stocks through the creation of a formal consultative role for the NWMB, this did little to effect change in the allocation of turbot quotas until the Nunavut Tunngavik resorted to litigation (Bankes 2003).

Much also remains to be done to build appropriate cross-scale linkages where populations are shared by Nunavut and another jurisdiction. While the NLCA contains important and specific provisions (Article 40) to deal with wildlife stocks shared by Inuit and other Aboriginal resource users (notably Inuit from northern Quebec [Nunavik]), as well as more general provisions calling for Inuit involvement in international negotiations that may affect them (Part 9 of Article 5 of the NLCA), appropriate cross-scale arrangements with Greenland over shared resources are almost completely lacking. This observation applies across the whole spectrum of shared resources, from such high profile species as polar bears, narwhal, and beluga to bird species such as murres, auks, and eider and harlequin duck. In sum, while a hierarchy-flipping event like the NLCA (or the famous *Boldt* decision in the Unites States; Lee 1995, 222) may create important opportunities for learning and for building both vertical and horizontal cross-scale interconnections at a domestic level, it is less likely to have that effect at an international level.

This suggests that while it may be possible to generate crisis-like events through changes in the law that may serve as important triggers for renewal within a domestic system, it is harder to imagine legal changes at the international level serving a similar function (other than perhaps binding decisions of an arbitral body or the International Court of Justice), principally due to the more horizontal nature of the international legal system. Learning opportunities at the international level may require "true" crises such as a population crash or a major power outage such as that which occurred in northeastern United States and Canada in the summer of 2003 (Wolfe and Homer-Dixon 2003).

CONCLUSIONS

> An institution's success in minimizing the cost and difficulty of observation and analysis depends principally on its ability to capture the feedback in the system that it governs. To do this well, the organization of institutions must take on a hierarchical structure that reflects the patchy, multiscale hierarchical structure of the natural system. At each level in the hierarchy, institutions must be "positioned" so that their boundaries correspond as much as possible in terms of scale and location to the boundaries of strong interactions in the biological system. There must be connections (information flows) between locations at the same scale and between higher and lower scales as in the ecosystem.
> — Wilson 2002, 352

This chapter has endeavoured to explore the role of law and hierarchies in building resilience. A commitment to both resilience and hierarchy is contingent and consequentialist. In each case we want to know if our current institutional structures facilitate maintenance of desired ecosystem states and socially and ecologically sustainable development. We want to know if they are structured so as to build and increase the capacity for learning and adaptation. There are no absolute answers to these forms of questions.

The NLCA was designed to foster socially and ecologically sustainable development. The preamble to the NLCA recites that the agreement was designed, *inter alia*, to foster the "cultural and social well-being of Inuit" and the wildlife article of the agreement includes in its objectives the creation of a wildlife management system that both implements principles of conservation and serves and promotes the long-term economic, social, and cultural interests of Inuit harvesters (NLCA 1993). The legal technique chosen to achieve these objectives, a constitutionalized land claims agreement, transformed the old formal hierarchy that developed as settler laws were introduced. The NLCA protects Inuit harvesting rights and established their priority. It also entrenches an Inuit role in making new rules for the fishery (co-jurisdiction) as well as in ongoing management decisions (co-management). The NLCA did not completely flip the hierarchy (DFO and the Minister retain roles that allow them to protect global system values), but it did represent a significant transformation of the hierarchy.

Canada's Oceans Strategy stresses the importance of developing institutional governance mechanisms that are coordinated and collaborative and which provide a "framework for integrated, horizontal ocean governance" (Fisheries and Oceans Canada 2002, 17). This chapter suggests that nested hierarchical structures are not necessarily problematic and that horizontal models are not necessarily desirable. The key is to find an appropriate balance that facilitates the development of appropriate cross-scale interconnections – both horizontal and vertical.

ACKNOWLEDGMENTS

Thanks to my fellow members of the resilience group of the node, and especially Fikret Berkes, Doug Clark, and Derek Armitage, for their encouragement and support.

REFERENCES

Alcorn, J., and V. Toledo.1998. "Resilient Resource Management in Mexico's Forest Ecosystems: the contribution of property rights." In *Linking Social and Ecological Systems: Management Practices and Social Mechanisms for Building Resilience,* edited by F. Berkes and C. Folke, 216–49. Cambridge: Cambridge University Press.

Bankes, N. 2001. Discussion Paper of Fisheries Regulations for Nunavut. A Report prepared for the Nunavut Regulatory Review Committee. Unpublished. On file with the author and the Nunavut Wildlife Management Board.

———. 2003. "Implementing the Fisheries Provisions of the Nunavut Claim: Re-Capturing the Resource." *Journal of Environmental Law and Practice* 12: 141–204.

Berkes, F. 2002. "Cross Scale Institutional Linkages: Perspectives from the Bottom Up." In *The Drama of the Commons,* edited by E. Ostrom *et al.,* 293–321. Washington, DC: National Academy Press.

———, J. Colding, and C. Folke. 2003. *Navigating Social-Ecological Systems: Building Resilience for Complexity and Change.* Cambridge: Cambridge University Press.

Dworkin, R. 1977. "Is Law a System of Rules?" In *The Philosophy of Law,* edited by R. Dworkin, 38–65. Oxford: Oxford University Press.

Eber, Dorothy. 1989. *When the whalers were up north: Inuit memories from the eastern Arctic.* Kingston: McGill-Queen's University Press.

Fisheries and Oceans Canada. 2002. *Canada's Oceans Strategy: Our Oceans Our Future.* Ottawa: Fisheries and Oceans Canada.

Folke, C., F. Berkes, and J. Colding. 1998. "Ecological practices and social mechanisms for building resilience and sustainability." In *Linking Social and Ecological Systems: Management Practices and Social Mechanisms for Building Resilience,* edited by Fikret Berkes and Carl Folke, 414–36. Cambridge: Cambridge University Press.

———, Steve Carpenter, Thomas Elmqvist, Lance Gunderson, CS Holling, Brian Walker, Jan Bengtsson, Fikret Berkes, Johan Colding, Kjell Danell, Malin Falkenmark, Line Gordon, Roger Kasperson, Nils Kautsky, Ann Kinzig, Simon Levin, Karl-Göran Mäler, Fredrik Moberg, Leif Ohlsson, Per Olsson, Elinor Ostrom, Walter Reid, Johan Rockström, Hubert Savenije and Uno Svedin. 2002. Resilience and Sustainable Development: Building Adaptive Capacity in a World of Transformations. Scientific Background Paper on Resilience for the process of The World Summit on Sustainable Development on behalf of The Environmental Advisory Council to the Swedish Government. *http://www.sou.gov.se/mvb/pdf/resliens.pdf*

Freeman, Milton, Alan C. Cooke, and Fred H. Schwartz, eds. 1976. *Inuit Land Use and Occupancy Project.* Ottawa: Ministry of Supply and Services.

Freyfogle, Eric. 1996. "The Construction of Ownership." *University of Illinois Law Review* 173–87.

Gunderson, Lance H., and C.S. Holling, eds. 2002. *Panarchy, Understanding Transformations in Human and Natural Systems.* Washington: Island Press.

Gunderson, Lance H., C.S. Holling, and S. Light, eds. 1995. *Barriers and Bridges to Renewal of Ecosystems and Institutions.* New York: Columbia University Press.

Harris, Cheryl. 1993. "Whiteness as Property." *Harvard Law Review* 106: 1709–91.

Harris, Douglas. 2001. *Fish, Law and Colonialism: The Legal Capture of Salmon in British Columbia.* Toronto: University of Toronto Press.

Hessing, Melody. 1993. "Women and Sustainability: Ecofeminist Perspectives." *Alternatives* 19(4): 14–21.

Holling, C.S. 1996. "Engineering Resilience versus Ecological Resilience." In *Engineering within Ecological Constraints*, edited by P. Schulze, 31–43. Washington, DC: National Academy Press.

—— and L. Gunderson. 2002. "Resilience and Adaptive Cycles". In *Panarchy, Understanding Transformations in Human and Natural Systems*, edited by L. Gunderson and C.S. Holling, 25–62. Washington: Island Press.

——, L. Gunderson, and D. Ludwig. 2002a. "In Quest of a Theory of Adaptive Change." In *Panarchy, Understanding Transformations in Human and Natural Systems*, edited by L. Gunderson and C.S. Holling, 3–22. Washington: Island Press.

——, L. Gunderson, and G. Peterson. 2002b. "Sustainability and Panarchies." In *Panarchy, Understanding Transformations in Human and Natural Systems*, edited by L. Gunderson and C.S. Holling, 63–102. Washington: Island Press.

Hogg, Peter. 2001. *Constitutional Law of Canada*, 4th ed. Scarborough: Carswell.

Hughes, Elaine. 1995. "Fishwives and Other Tails: Ecofeminism and Environmental Law." *Canadian Journal of Women and the Law* 8: 502–30.

Kennedy, Duncan.1982. "Legal Education and the Reproduction of Hierarchy." *Journal of Legal Education* 32: 591–615.

King, Anthony. 1993. "Considerations of Scale and Hierarchy." In *Ecological Integrity and the Management of Ecosystems*, edited by Stephen Woodley, James Kay, and George Francis, 19–45. St. Lucie Press.

Kontou, Nancy. 1994. *The termination and revision of treaties in the light of new customary international law*. Oxford: Oxford University Press.

Landau, Martin. 1969. "Redundancy, Rationality, and the Problem of Duplication and Overlap." *Public Administration Review* (July/August): 346–58.

Lee, Kai. 1995. "Deliberately seeking sustainability in the Columbia River Basin." In *Barriers and Bridges to Renewal of Ecosystems and Institutions*, edited by L. Gunderson *et al.*, 214–38. New York: Columbia University Press.

——. 1999. "Appraising adaptive management." *Conservation Ecology* 3(2). *http://www.consecol.org/vol3/iss2/art3*

Levin, S. 1999. *Fragile Dominion*. Cambridge: Perseus Books.

McCay, Bonnie J. 1978. "Systems Ecology, People Ecology and the Anthropology of Fishing Communities." *Human Ecology* 6(4): 397–422.

——. 2002. "Emergence of Institutions for the Commons: Contexts Situations and Events." In *The Drama of the Commons*, edited by E. Ostrom et al, 361–402. Washington DC: National Academy Press.

—— and S. Jentoft. 1998. "Market or Community Failure? Critical Perspectives on Common Property Research." *Human Organization*: 57(1): 21–29.

Macpherson, C.B. 1978. "Liberal Democracy and Property." In *Property, Mainstream and Critical Positions*, edited by C.B. Macpherson, 199–207. Toronto: University of Toronto Press.

Merry, Sally Engle. 1988. "Legal Pluralism." 22(5) *Law and Society Review* 22(5): 869–901.

Mitchell, Marybelle. 1996. *From Talking Chiefs to a Native Corporate Elite: the Birth of Class and Nationalism among Canadian Inuit*. Montreal: McGill-Queen's University Press.

Mullan, David, ed. 2003. *Administrative Law: Cases Text and Materials*, 5th ed. Toronto: Emond Montgomery.

Nedelsky, Jennifer. 1993. "Reconceiving Rights as Relationship." *Review of Constitutional Studies* 1: 1–26.

Nelson, Richard K. 1969. *Hunters of the Northern Ice*. Chicago: University of Chicago Press.

NLCA. 1993. *Nunavut Land claims agreement*. Ottawa: NWMB. 1993–2003. Minutes of NWMB Meetings. *http://www.nwmb.com/english/*

NTI 1997. Nunavut Tunngavik Inc v. Canada (Minister of Fisheries and Oceans) 149 DLR (4th) 519, varied 162 DLR (4th) 625 (Federal Court of Appeal).

NWMB. 1994. Minutes of NWMB Meeting #4, 9–11 August 1994, Lake Harbour, NT.

———. 1997. Minutes of NWMB Meeting #16, 16–19 November 1997, Salliq (Coral Harbour), NT.

———. 2000. *Final Report of the Inuit Bowhead Knowledge Study*. Iqaluit: Nunavut Wildlife Management Board.

Ostrom, Elinor. 1990. *Governing the Commons: The Evolution of Institutions for Collective Action*. Cambridge: Cambridge University Press.

Reference re Secession of Quebec, [1998] 2 *Supreme Court Reports* 217.

Rifkin, Jeremy. 2002. *The hydrogen economy: the creation of the worldwide energy web and the redistribution of power on earth*. New York: J.P. Tarcher/Putnam.

Rose, Carol. 2002. "Common Property, Regulatory Property, and Environmental Protection: Comparing Community Based Management to Tradeable Environmental Allowances." In *The Drama of the Commons*, edited by E. Ostrom *et al.*, 233–57. Washington, DC: National Academy Press.

Ross, Gillies. 1975. *Whaling and Eskimos, Hudson Bay, 1860–1915*. Ottawa: National Museums of Canada.

———. 1982. *Distribution of the kills of bowhead whales and other sea mammals by Davis Strait whalers, 1829–1910*. Calgary: Arctic Pilot Project.

Royal Commission. 1986. Report of the Royal Commission, *Seals and Sealing in Canada*. Ottawa: Ministry of Supply and Services.

Sanderson, Steven. 1995. "Ten Theses on the Promise and Problems of Creative Ecosystem Management in Developing Countries." In *Barriers and Bridges to Renewal of Ecosystems and Institutions*, edited by Lance H. Gunderson, C.S. Holling and S. Light, 375–90. New York: Columbia University Press.

Sikyea. 1964. *Supreme Court Reports* 642–46.

Simon, Herbert. 1973. "The Organization of Complex Systems." In *Hierarchy Theory: The Challenge of Complex Systems*, edited by Howard H. Pattee, 1–27. New York: George Braziller.

Singer, Joseph. 1982. "The Legal Rights Debate in Analytical Jurisprudence from Bentham to Hohfeld." *Wisconsin Law Review*: 975–1059.

Smith, Patricia. 1993. *Feminist Jurisprudence*, New York: Oxford University Press.

Sparrow. 1990. [1990] 1 *Supreme Court Reports* 1075.

Trosper, R.L. 2003. "Policy transformations in the US forest sector, 1997-2000: implications for sustainable use and resilience." In *Navigating Social-Ecological Systems: Building Resilience for Complexity and Change*, edited by Berkes, F., J. Colding, and C. Folke, 328–51. Cambridge: Cambridge University Press.

Welch, Buster. 1995. "Marine Conservation in the Canadian Arctic: A regional overview." Northern Perspectives. 23(1): 5–9.

Westley, Frances. 2002. "The Devil in the Dynamics: Adaptive Management on the Front Lines." In *Panarchy, Understanding Transformations in Human and Natural Systems*, edited by L. Gunderson and C.S. Holling, 333–60. Washington: Island Press.

Williams, Glanville. 1956. "The Concept of Legal Liberty." *Colorado Law Review* 56: 1129–50.

Wilson, James. 2002. "Scientific Uncertainty, Complex Systems and the Design of Common-Pool Institutions." In *The Drama of the Commons*, edited by E. Ostrom *et al.*, 327–53. Washington, DC: National Academy Press.

Wolfe, Sarah, and Thomas Homer-Dixon. 2003. "The Matrix of Our Troubles." *Globe and Mail*, 16 August.

Young, Oran. 2002. "Institutional Interplay, the Environmental Consequences of Cross-scale Interactions." In *The Drama of the Commons*, edited by E. Ostrom *et al.*, 263–91. Washington, DC: National Academy Press.

GOVERNANCE, POLICY, & FUTURE DIRECTIONS

CHAPTER 15

THE RETURN OF THE "VIKINGS":

THE CANADIAN-DANISH DISPUTE OVER HANS ISLAND
– NEW CHALLENGES FOR THE CONTROL OF THE
CANADIAN NORTH

Rob Huebert (University of Calgary)

The Vikings have returned to Canada and are trying to take over Canadian territory! They have not come in the traditional long boats but are arriving in modern, ice-strengthened frigates. In the summer of 2002, the Danish government sent the frigate *Vaedderen* to patrol the waters between the northwest corner of Greenland and Ellesmere Island. This was then repeated in the summer of 2003 when the Vaedderen's sistership the *Triton* arrived in these waters. While the Danes have been regularly sending warships to this area, the 2002 and 2003 voyages are notable for where the vessels specifically sailed and what they did. In a dispute few Canadians were aware of until very recently, the governments of Canada and Denmark contest the ownership of Hans Island, a small island located at the northernmost tip of western Greenland (see map on inside cover of book). It is a small and almost insignificant island that shows up on few maps. Canada maintains that the island is Canadian, while Denmark contends that it belongs to Greenland and therefore to Denmark. Because the warships sailed into the disputed waters surrounding the island and landed sailors on it, the Canadian government took the step of issuing a diplomatic protest to the Danish government. There is also some evidence to suggest that the Danes have been landing on the island for some time now. If this is the case, the Danish move would represent the first challenge to Canadian northern land territory since the late 1930s (Franckx 1993, 96).

When this chapter was first written, few Canadians had heard of the dispute. However, during the spring of 2004 the *National Post* examined the issue in detail. As a result, most Canadians are now well aware of the "return of the Vikings!" Furthermore, the publicity has also raised the political attention that Canadian decision makers had previously given the issue.

This chapter in some ways represents the "ugly duckling" of the work of the Integrated Node of the Ocean Management Research Network (OMRN). While the other chapters in this book demonstrate the manner in which the principles of integrated management are beginning to be utilized in the Canadian North, this chapter will demonstrate that in the international area, the Canadian government has not been successful. Rather, this chapter will show that the Canadian response to international challenges has been primarily of a reactive and minimalist nature. The issue of Hans Island is important by itself, but it is even more important in terms of illustrating an overall lack of strategy when it comes to the issue of Canadian control of Arctic regions in the face of international challenges.

Thus, the question that arises is why Hans Island is important enough to cause the Danish government to take the effort to challenge Canada, and for that matter, why has Canada laid claim to it? In and of itself, the island is remote and uninhabited. However, its location affects the manner by which the maritime boundary is determined between northern Greenland and Canada. In turn, this international boundary takes on significance for four reasons. First, these waters may contain important fish stocks, including turbot and shrimp. The boundary will affect the northern divisions of these resources. Secondly, it has been reported that Greenland Inuit have been crossing over to Ellesmere Island to engage in illegal Polar Bear hunts. The Canadian Rangers have been dispatched to Alexandra Fjord on Ellesmere Island but have not caught any of the alleged hunters.[1] If it proves to have been a long-term habit, it is conceivable that the Greenland home rule government could argue that the hunt is an established right. Thus, any other boundary dispute between Canada and Denmark could exacerbate that situation.

Third, the impact of climate change is expected to cause substantial warming of the polar region. Thus, while the region today is remote and inhospitable, this could change rapidly as the region warms. There is also the possibility that the island and/or waters immediately surrounding it may have oil reserves. In the late 1970s and early 1980s, Dome petroleum undertook research on the island. While the studies seemingly were not specifically to determine the presence of oil in the area, it is noteworthy that Dome selected Hans Island as a site for ice research. It is not known if oil was found (Rolston and McDorman 1988, 30). At the same time, the Danish government has begun to increase its seismic exploration of the waters off the west coast of the Greenland in 2002.[2]

Before Canadians see this as a return of the attacking, pillaging "Vikings," the issue needs to be put into perspective. The island is small and very inhospitable. If Canada was to lose its claim, the impact would be to produce a minor change in the boundary line between Canada and Greenland. Nevertheless, a negative outcome for Canada has broader ramifications than mere ownership of the island. First, Canadians need to be concerned about the precedent that will be established by an unfavourable resolution of this dispute. Most Canadians may believe that the territorial integrity of Canada is complete and without challenge. However, beyond the issue surrounding Hans Island there are now five other current or potential boundary disputes in the Canadian North (Gray 1994, 131–44). First, the United States still maintains that the Northwest Passage is an international strait and not internal waters as Canada claims.[3] Second, the northern maritime boundary between Alaska and the Yukon is disputed (Fogarassy 1991, 2–3). Third, the possibility exists that Canada and Russia may have overlapping claims for the continental shelf in the high Arctic. Canada has issued a diplomatic note to the effect that it does disagree with the Russian claim, though Canada has yet to determine its own claim (Macnab 2003, 12). With the Canadian ratification of the United Nations Law of the Sea Convention, Canada now has ten years to delimit all of its continental shelves, including those found in the North (Macnab 2003, 9–13). A fourth possible dispute could arise when Canada delimits the boundary of the continental shelf that it shares with Greenland. Finally, it is also possible that Canada will need to delimit the continental shelf it shares with the United States between Alaska and the Yukon. Since Canada and the United States already disagree about the boundary dividing their respective Exclusive Economic Zone (EEZ) and territorial seas, it can be expected that the division of the continental shelf will also be disputed.

Given the importance of precedence in international law, the unfavourable settlement to Canadian interests of any of these disputes will have ramifications on the others. If Denmark wins its case regarding Hans Island, or even if the Canadian response is seen as weak, other states may come to regard the Canadian ability to protect its northern interests as marginal. Thus the true significance of the issue lies in the manner in which Canada is able to manage its Arctic region in the face of international challenges.

The Canadian response to the Danish claim also provides the opportunity to analyze Canadian northern foreign policy in the context of integrated management. While the other chapters offer evidence of the ability of the Canadian government to utilize principles of integrated management concerning domestic issues, the Canadian response to the Danish government can be characterized as reactive and minimalist. The recent incident involving the Danish vessel(s) highlights the problem of Canada's ability to know and defend its interests in the North or to articulate an overall policy response (Canada's Coastlines 2003, 13–58). Canada has a limited ability to know what is happening in its northern regions and to respond. These limitations have forced Canadian officials to take a position of restrained engagement. Discussions with Danish officials have been directed toward temporizing the situation and not toward building support for the Canadian position.

HANS ISLAND AND THE VOYAGE OF THE VAEDDEREN AND TRITON

The dispute between Canada and Denmark can be traced to the period of the European exploration of the Arctic. One of the core issues of the dispute is the determination of who first claimed the island. According to Kenn Harper, it was the Americans who were the first non-Inuit to visit the island (Harper 2004, 5). Specifically, it was the expedition led by Charles Francis Hall that first discovered the island between 1871 and 1873. Harper has suggested that it was named after an Greenlandic Inuit guide who went by the name Hans Hendrik and whose Inuit name was Suerssaq (Harper 2004, 5). It should be noted, however, that there are other reports that suggest the Danes claim the island is named after King Hans, the ruler of Sweden, Norway, and Denmark from 1455 to 1513 (Wattie and Humphreys 2004, A6). Thus the source of the name is still not clear. After its discovery, it was primarily American explorers such as Robert Peary who frequented the area. The crux of the Danish claim can then be traced to the American decision in 1916 to surrender all claims to northern Greenland in exchange for the Danish West Indies (now the Virgin Islands) and $25 million (Harper 2004, 5). However, when the Americans made this transfer, there was no specific mention of Hans Island. The Canadian claim to the island is based on the fact that the discovery of the island was made by the British explorer George Nares on his voyage north in 1875–76 (Officer and Page 2001, 121–23).[4] Britain passed on its rights to the Arctic in 1888 to Canada, which presumably included Hans Island. However, as in the case of the American transfer in 1916, there is no specific mention of Hans Island. In 1953 the first modern determination of the location of the island was made by Eric Fry of the Topological Society of Canada (Gray 1994, 143).

From the 1960s to the beginning of the twenty-first century there is little evidence that either Canada or Denmark did anything to consolidate their respective claims. While evidence is sketchy, it appears that while both countries had made specific claims over the island by 1970, they agreed not to let it detract from their ongoing relationship as northern neighbours (Rolston and McDorman 1988, 30). During the early 1970s Canada and Denmark agreed to enter into negotiations to determine the maritime boundary between Greenland and Ellesmere Island. This was based on the anticipation that the expected negotiations for the third conference of the United Nations Law of the Sea Conference would probably extend coastal state control over their maritime regions. Specifically, it was becoming apparent that some form of control would be granted up to two hundred nautical miles from the coast (Churchill and Rowe 1988, 133–34). Thus the intention of the Canadian and Danish governments was to determine the boundary that would be created.

The negotiations went well and the two governments were able to provide an acceptable delimitation between Ellesmere Island and Greenland. However, the line was not extended the entire length of the northern EEZ. Instead both sides agreed not to engage in resource development in this region until international law was more firmly established and/or either side wished to undertake resource exploitation in the region (U.S. Office of the Geographer 1976, 8).

Nevertheless there was one other gap in the line. As the dividing line proceeds north, it stops just short of Hans Island and then resumes immediately north of the island (U.S. Office of the Geographer 1976, 8). While there is no mention of the island in the agreement, it is clear that the two sides could not agree on how to demarcate the island. As such they appear to have decided to simply ignore it for the purposes of the 1973 agreement (U.S. Office of the Geographer 1976, 8).[5]

The two counties have also been able to co-operate in exploring the region. In 1971 Canada and Denmark mounted a joint operation that included a Canadian icebreaker, the *Louis St. Laurent*, that travelled north through the Robeson Channel to the Lincoln Sea (Maginley and Collin 2001, 152). At the time, this was the most northern maritime expedition by a modern non-Soviet ship. Once again the co-operative nature of the exercise clearly indicates that neither country considers their disagreement to be significant enough to halt the voyage, even though it sailed very close to the island.

In the 1980s there were several incidents regarding Hans Island that resulted in diplomatic protests. In the late 1970s and 1980s, Dome Petroleum used the island as a test bed to examine the impact of ice-fixed structures. This resulted in a diplomatic protest by the Danish government (Rolston and McDorman 1988, 30). On 28 July 1984, the Danish minister responsible for Greenland visited the island and left a Danish flag (Rolston and McDorman 1988, 30). The Canadian government responded by issuing a diplomatic protest. In 1988, a Danish inspection crew reached the island by ship and planted another Danish flag. The Canadian government once again protested this by issuing a diplomatic note (DFAIT 28 December 2002). There are also media reports that the Danes also landed on the island in 1995 to again raise their flag (Wattie and Humphreys 2004, A6). However, the details of this voyage are not known, nor what the Canadian response was to that specific visit.

Despite these incidents both states have seemingly been able to accept the disagreement. It is somewhat surprising and disquieting that the Danish government has now decided to begin to support its claim with renewed vigour by visiting the island with a warship and possible landings. There is a problem in determining the exact nature of the voyage of the *Vaedderen, Triton*, or any other such voyages that may have recently taken place. It is equally unclear why the Danish government is now attempting to force the issue with this new intensity. The isolated and remote location of the voyages makes it difficult to know with a high degree of certainty as to what did happen.

What is known is that the Canadian government received an official note requesting diplomatic clearance for the *Vaedderen* on 16 July 2002, for entry into the waters surrounding Hans Island. The request was interesting in that it was one of three such requests coming from the Danish government at roughly the same time. Besides requesting clearance for the *Vaedderen*, the Danish government also requested clearance for the Faeroe fishery inspection vessel *Brimil* (DFAIT 30 July 2002) to enter St. John's and for the seismic research vessel MV *Professor Polshkov* to engage in research in the Davis Strait (DFAIT 30 July 2002).

The response of the Canadian government was to grant the clearance at least for the *Vaedderen*[6] entry into Canadian waters but also to issue a diplomatic note of protest regarding the Danish intent to land on the island without seeking permission. The wording of the note remains classified, but it appears that the

Danish flag on Hans Island.

Danes were only asking permission to be in Canadian waters but not on the island. In developing the diplomatic note, Canadian officials did comment that since the *Vaedderen* was a governmental vessel, it could be exempted by the Governor-in-Council from the regulations of the *Arctic Waters Pollution Prevention Act* because it is "owned or operated by a sovereign power ..." (Coast Guard 1 August 2002). This particular correspondence then went on to state that "[i]nformation should be *requested* from the responsible authorities that will confirm the vessel's compliance with the Arctic Shipping Pollution Prevention Regulations" (Coast Guard) (Italics added).This passage provides the Canadian government with a means by which to "allow" governmental ships to not be bound by the Canadian regulatory regime established under the *Arctic Waters Pollution Prevention Act*. In short, while it is unknown what is in the Canadian protest, it is clear that the Canadian government was going to allow the Danish voyage to occur and to exclude it from the regulation regime established by the *Arctic Waters Pollution Prevention Act*.

The Danish frigate the *Vaedderen* sailed from Thule, Greenland. From here it sailed north in the beginning of August 2002. It is not publicly known when it entered Canadian waters, but at one point the vessel proceeded up the Nares Strait, the waterway that divides northern Ellesmere Island and the northwest tip of Greenland. The vessel then entered the waters surrounding Hans Island on 13 August. The vessel landed its sailors on the island and they raised the Danish flag.[7] At least fifteen sailors were landed on the island. They raised a new flagpole and Danish flag and placed a plaque noting their arrival and that the island belonged to Denmark (Wattie and Humphreys 2004, A6). This plaque was added to similar ones left in 1988 and 1995 by preceding Danish expeditions.

A second voyage was made by the *Vaedderen* sistership the *Triton* in the summer of 2003, In and around 1 August, the *Triton* reached Hans Island. Once again there is no indication that the Danish government requested permission. However, declassified Canadian government documents make it clear that Canadian officials deemed the voyage as a violation of Canadian sovereignty over the island. The Department of National Defence was mandated to undertake a sovereignty flight over the vessel. The intent of the flight was to keep the *Triton* under observation when it sailed to Hans Island and to demonstrate a Canadian presence. Efforts were also taken to place a Canadian naval officer on board the *Triton* in the name of NATO military co-operation (DND E-mail 11 July 2003). Not surprisingly, this initiative was not accepted by the Danish Navy.

In the days leading up to the voyage of the *Triton* to Hans island, the Department of National Defence identified the vessel as the "arctic surveillance #1" mission (DND E-mail 31 July 2003). In order to monitor the activities of the *Triton*, an Aurora CF-140 long-range patrol aircraft was tasked to overfly the vessel. Its primary goal was to photograph the ship at Hans Island and any activity associated with the visit (DND E-mail 1 August 2003). This was to provide "a sovereignty counterpoint to the Danish assertion of their sovereignty on and around the contested island" (DND E-mail 21 July 2003).

The *Triton* arrived at Hans Island on 1 August. It remained there for two days and left the area on 3 August. The Canadian Aurora reached Hans Island early on 3 August, two hours before the *Triton* left (DND Communication "Purple 706 AMP 01" 3 August 2003). The aircraft conducted operations over the area for one hour and reported that a Danish flag was flying from the island.

MEDIA ATTENTION

One of the more recent developments regarding this dispute has been the publicity that it received in the spring of 2004. Following a series of front page articles that appeared in one of Canada's national newspapers, the *National Post*, senior Canadian and Danish officials were required to make public statements regarding their countries' respective positions.

Prior to these stories, there had been at least two other stories that had attracted some attention in the early 1980s. The first story on Hans Island first appeared in a local Greenland newspaper, *Hainang* (Harper 2004, 2) The story was picked up by the national media in Denmark and Canada. It is possible that this story was the catalyst for the 1984 visit to Hans Island by Tom Høyem, the Danish minister responsible for Greenland (Harper 2004, 4). In December 2002 another article on the situation appeared in the Canadian media. Written by the author, it was an op-ed that appeared in the *Globe and Mail* (Huebert 2002). The focus of the article was to raise awareness of the dispute. However, it provoked only limited discussions.

The most significant media attention on this dispute began in the spring of 2004. In March 2004, a series of articles appeared in the *National Post*. The catalyst of the stories was the announcement that Canadian forces personnel were to conduct a sovereignty land patrol in the Canadian high Arctic. In examining the story, Adrian Humphreys of the *National Post* interviewed the author as to what were the main sovereignty challenges facing Canada. In outlining the four issues that were highlighted earlier in this chapter, Humphreys then wrote a story that highlighted the nature of the dispute (25 March 2004, 1).In the story, Reynald Doiron, a spokesperson for the Department of Foreign Affairs, stated:

> Our position is simple: Hans Island is a part of Canadian national territory and no assertion by the Danish ambassador or any Danish official here or over there detracts from the absolute sovereignty Canada enjoys over the island. (Humphreys 25 March 2004)

In a second story that appeared the next day, the *National Post* reporters were able to contact the Danish ambassador to Canada, who stated that while Denmark is willing to begin negotiations to resolve the issue, the island was Danish (Humphreys and Wattie 26 March 2004, A1).

As a result of the stories, Canadian political leaders began to offer comments on the Canadian position. Stockwell Day, the Conservative foreign affairs critic, raised the issue in Parliament. He noted that the recall of Alfonso Gagliano due to the sponsorship scandal had left Canada without an ambassador to Denmark. He also suggested that Canada did not have the military capability to stop the Danes because of liberal budget cuts to the Department of National Defence (Humphreys 30 March 2004, A1). Aileen Carroll, Minister for International Cooperation, responded by restating the Canadian government's position that Hans Island was "part of the national territory of Canada" (Humphreys 30 March 2004, A1).

The next day the Canadian Minister of Foreign Affairs, Bill Graham, stated in Parliament that "this government will not surrender any sovereignty of any of Canada's lands in the Arctic or anywhere else in the world. We are telling Denmark, clearly, Hans Island is Canada's."(Wattie and Humphreys 31 March 2004, A6). At the same time a Danish spokesperson made it clear that while the issue was not a major crisis, Denmark was not going to renounce its claim. Peter Taksoe-Jensen, head of international public law with the Danish Ministry of Foreign Affairs, stated: "Since we think of it as part of Danish territory, we feel officials have that right [to visit the island]. It is only natural that Danish officials go there from time to time." (Humphreys 30 March 2004, A1). However, he added that Denmark would take issue if Canada were to land their service personnel on the island.

Within a few days officials from both countries were attempting to downplay the seriousness of the dispute. Alex Jensen, a Danish navy spokesperson, stated that "there is no crisis, there is no dispute." (CP Wire 1 April 2004, A10). This position was echoed by Erik Rosenstand, a spokesperson with the Canadian

Embassy in Copenhagen, who stated, "It's an old story. There is nothing dramatic about all this." Both spokespeople also denied earlier reports that a Canadian diplomat had been summoned to the Danish Foreign Ministry over the issue. (CP Wire 1 April 2004, A10).

While this was the last public statement of officials from either Canada or Denmark, there were still numerous Canadian editorials on the issue. There was even a student protest that involved the burning of Danish pastries in front of the Danish consulate in New Brunswick (Humphreys 16 April 2004, A6). While this protest was obviously not serious, it does serve to highlight some of the more interesting responses to the dispute. The general tone of most editorials suggested that the Canadian government needed to give greater attention to defending its interests in the North (Calgary Herald 29 March 2004, A12; Winnipeg Free Press 30 March 2004, A12; and Toronto Star 5 May 2004, A16). Others suggested that the issue was either insignificant or a plot by the Canadian government to divert attention away from other more pressing issues (Petrie 27 March 2004, B1; Worthington 2004, 14). While some of the editorials illustrated misunderstandings by some of the writers, they did indicate a widespread national interest in the issue.[8]

In summary, the attention provided by the Canadian media has been instrumental in raising the profile of the issue. In Canada, all three of Canada's principle foreign and defence relations ministers were required to make statements on the dispute. The Danish government also was forced to make public statements about its position regarding the island. Previous attempts by the author to obtain such a statement had been ignored.

DANISH AND CANADIAN CAPABILITIES COMPARED

Having examined the nature of the dispute, this chapter will now assess the capabilities of Canada and Denmark in protecting and promoting their respective positions. First and perhaps most importantly, the dispute demonstrated the ability of the Danish Navy to mount such an expedition if it decides to do so. The Danish capacity to send warships high into the Arctic region is an almost unique capability. Very few other navies have vessels that can sail this far north. The *Vaedderen* and *Triton* are part of a set of four Thetis class frigates. Built in the late 1980s and early 1990s, these vessels were given a thicker hull and had reinforced rudders and added protection to their propellers. These measures provide the vessels with an ability to enter waters that are covered by ice up to one metre thick (Sharpe 1998, 167). The South African, British, and Japanese navies/coast guards also have ships that have been given ice-strengthened capabilities. But no other navy has warships as large or numerous as the Thetis class. As a point of comparison the Thetis class are 2,600 tons and 112 metres long (Sharpe 1998, 167). This is somewhat smaller than the Canadian Halifax class frigates that were built at the same time (4,750 tons, 134 metres long). The Canadian vessels are robust enough to sail anywhere, except for northern Canadian waters in winter! Unlike their Danish counterparts

A view of Hans Island.

they were not given the capability to operate in ice-covered waters. The Danish government also stations three small patrol craft in Grønnedal, Greenland that are also ice-strengthened (Sharpe 1998, 170).

The existence of the Thetis class and smaller patrol vessels demonstrates that the Danish government takes seriously its commitment to providing a northern capability for Greenland. All of these vessels are used for fishery inspections and pollution enforcement. They are also used for hydrographic services in the maritime region around Greenland. The *Thetis* has been given a specialized stern for towing seismological equipment (Sharpe 1998, 167). This is being used to determine the location of the continental shelf around Greenland in preparation for its claim to be made under the United Nations Convention on the Law of the Sea. Article 76 of the convention gives coastal states the right to determine if they have a continental shelf (United Nations Convention on the Law of the Sea 1983, Article 76(5), (6)).[9] If they are able to determine that they do, they can lay claim for an area up to 350 nautical miles from their coastline. This gives the state the right to the resources of the soil and subsurface 150 nautical miles beyond the two-hundred-nautical-mile Economic Excusive Zone. However, in order to make the claim, all states must undertake the necessary hydrographic studies and present their findings to the Commission on the Limits of the Continental Shelf (CLCS) (Commission on the Limits of the Continental Shelf 2003). This body was established to determine if the claim follows the international standards established by the convention. Thus Denmark is currently engaged in the necessary studies to determine the length and breadth of the continental shelf that extends from the north of Greenland. On the other hand, there is no corresponding Canadian commitment to the development of an equivalent northern capability. Canada has not had a naval capability to enter ice-covered waters since 1954 (Maginley and Collin 2001, 150). The Canadian Navy had taken a delivery of the icebreaker *Labrador* in 1953 to give it the capability to operate in Canadian northern waters, but quickly decided that it did not want to maintain this capability. Thus it transferred the *Labrador* to the department of transport and the Canadian Coast Guard the next year.

The Canadian Navy did send vessels into the Arctic, but only on a sporadic basis for very short periods. Since none of its vessels since 1954 are ice-strengthened, Canadian vessels can only enter far-northern Canadian waters for a short period in the open water season from late August to mid-September. Furthermore, these voyages have been restricted to the southern sections of the Canadian North. Even these limited voyages, which had occurred on an annual or bi-annual basis, were eliminated by the end of the 1980s. It was not until 2002 that they were resumed. The Canadian Navy returned to the Canadian Arctic in the summer of 2002 for the first time since 1989, but would have been incapable even at that point of sailing to Hans Island. In a combined exercise with both air and land forces, the Canadian Maritime Coastal Patrol Vessels HMCS *Summerside* and HMCS *Goose Bay* sailed as far north as Resolution Island off the southwest tip of Baffin Island (Moon 2002, 8). However, these vessels do not have ice-strengthened hulls and can enter these waters for only a short period of time. The original intention had been to send a frigate rather than a coastal patrol vessel, but demands placed on the Navy by the war on terrorism meant that no frigates were available. However, even if one was available, it would not be any more capable of sailing into ice-covered waters, as none of the Canadian naval vessels are ice-strengthened. Thus no Canadian warship could have sailed as far north as the *Vaedderen* or *Triton* on its own.

The Canadian Air Force is also losing its ability to maintain a presence in the Canadian North. It never replaced its medium-range patrol aircraft when it retired its Tracker fleet. Its fleet of long-range patrol aircraft, consisting of twenty-one Auroras (CP141) and three Arcturus, are also now facing the reduction of their total flying hours because of their age. It is expected that the three Arcturus will be retired in the next few years. Of the remaining twenty-one Auroras, fifteen are being updated. However, for the time being this means that the number of northern sovereignty patrol flights remains very low. In the 1980s and early 1990s the usual number was between eighteen and twenty-two per year. Since 1995 the number has averaged between one and three.

The two areas of improvement in terms of surveillance capabilities are the development of a satellite monitoring ability and the purchase of long endurance Arial Unmanned Aircraft (UAV) (also known as drones). Both are still in the development stage, but once either becomes operational they will provide the potential for better surveillance of the Canadian North.[10] Of course, while it is expected that both will be used in this capacity, it still remains to be seen how this will be done.

Even more troubling is the condition of the Canadian Coast Guard ice-breaking fleet. It is very professional and one of the most highly trained fleets in the world, but it is small, aging, and drastically underfunded (Canada's Coastlines 2003, 14–16). Only the Coast Guard's icebreakers can actually voyage to the region surrounding Hans Island. The current inventory of Canadian Coast Guard's heavy and medium ice-beakers consist of one heavy icebreaker – *Louis St. Laurent* built in 1969 (rebuilt 1988–93); and five medium – *Pierre Radisson* (built 1978), *Sir John Franklin* (built 1979), *Des Groseillers* (built 1982), *Henry Larsen* (built 1987), and *Terry Fox* (built 1991) (Maginley and Colin 2001, 147–58). Compounding the limited number of vessels is the fact that the operational budget of these vessels has been continually reduced in the past decade to the point where the Coast Guard cannot operate these few vessels all year round (Canada's Coastlines 2003, 14). Furthermore, a combination of age and budget cuts had forced the Coast Guard to remove the *Sir John Franklin* from service. However, a university consortium led by Laval University was able to raise the funds necessary to recommission the vessel. It has subsequently been refitted and renamed as the *Amundsen* and is now engaged as a research vessel in the Canadian North.[11] In short, the Coast Guard is now left with an effective number of four icebreakers.

There has been a growing awareness that the Coast Guard needs to rebuild its capability. In the fall of 2002, it made a presentation to Cabinet regarding its current challenges. As a result the 2002 federal budget did make an extra allocation of $94.6 million to the Coast Guard (Department of Finance Canada 2003, 17). However, this increase was for all Coast Guard activities and not specifically allocated for icebreakers. A subsequent Report on the Security of Canada's Coastline prepared by the Senate Standing Committee on National Defence has also lamented the limited capabilities that the Coast Guard has in general (Canada's Coastlines 2003, 14–17). In the chapter that focused on the Coast Guard, the Senate committee made it clear that it found the limited resources allocated to the Coast Guard deplorable and strongly urged the government to take action to change the situation (Canada's Coastlines 2003, 13–34).

The net result of the limited capabilities of the various Canadian government departments is that Canada does not have the means to match the Danish capability to visit Hans Island. It could task one of its few icebreakers to do so, but such a decision would mean that other requirements for the remaining ice-breakers would be affected. The Canadian Forces' capability of being aware of the Danish activities in the disputed waters will be improved when the satellite and UAV capabilities are made operational, but it is not known when this will occur. For the time being the Canadian Forces capability of knowing is limited, as is their ability to respond.

The problems facing Canadian capability is made worse by the lack of a coherent policy framework governing Canadian response to international challenges to its Arctic region. Canada has not been particularly aggressive in its treatment of foreign vessels entering its northern waters. In 1970 the Trudeau government passed the *Arctic Waters Pollution Prevention Act* (Beesley 1972–73, 229–31). This law was an effort by the Canadian government to respond to what it saw as a challenge to its Arctic sovereignty when the American supertanker *Manhattan* sailed through the Northwest Passage (Huebert 2001). It attempted to assert control over the passage by creating a set of regulations to protect the environment. In many regards the Act was ahead of its time (Huebert 2001). However, the Act has been limited by the manner in which it is implemented. The *AWPPA* provides for very specific requirements for vessels entering Canadian waters regarding the protection of the northern environment (Arctic Water Pollution Prevention Act). The Act seemingly gives the Canadian government the right to control shipping in all Canadian northern waters up to a distance of a hundred nautical miles from the nearest Canadian land:

> … except that in the area between the islands of the Canadian arctic and Greenland, where the line of equidistance between the islands of the Canadian arctic and Greenland is less than one hundred nautical miles from the nearest Canadian land, that line shall be substituted for the line measured seaward one hundred nautical miles from the nearest Canadian land; (Arctic Waters Pollution Prevention Act)

The intention of the Act was to give Canada a degree of control over the activity of vessels entering its Arctic waters without necessarily claiming full sovereignty (Head and Trudeau 1995, 25–64).

However, the Act has been limited by the refusal of the Canadian government to make the practical enforcement of the Act mandatory. The legislation itself is written in a manner that gives the appearance that adherence to its requirement must be followed. The problem is that ships entering Canadian Arctic waters are requested but not required to report their entry. Under the Vessel Traffic Reporting Arctic Canada Traffic Zone (NORDREG), vessels are invited to report their entry, but are not required to do so (Vessel Traffic Reporting Arctic Canada Traffic Zone (NORDREG), August 1, 2003). This creates the paradox that while the Canadian government has created an Act that requires vessels to follow Canadian law, ships are not required to report their entry! Thus if a ship wishes to disregard the Canadian effort to assert control, all it needs to do is not report its entry. There is no official reason for this contradiction. However, off the record, several Canadian officials have indicated that it is due to the fear that if NORDREG was made mandatory, it could create a challenge from the United States (confidential interviews).

The problem is compounded by Canada not having a comprehensive surveillance capability that a designated satellite system would provide. RadarSat I proves that the Canadian industry has the capability to build such a system, but the government decided some time ago that the purchase of a system only for governmental service was too expensive. Instead, the government has actively pursued a policy of encouraging the commercialization of its space assets (Godefry 2000). This situation may change if current plans to utilize RadarSat II are successful. DND is currently developing a project that will use satellite-based assets including RadarSat II for northern surveillance. However, this project is still in the planning stage and RadarSat II still needs to be launched. But if the plan is implemented, it will greatly expand current Canadian capabilities.

At the same time there has not been any effort to develop a coastal watch system that would utilize the experience of the Canadian Inuit in an manner that the Department of Defence utilizes the Rangers. It would seem reasonable that some system of reporting could be established that would combine both indigenous knowledge with high technology to provide a picture of vessels in the Canadian Arctic. However, this is not the case. Thus the Canadian government is reduced to hoping that all ships will voluntarily submit to the Canadian rules without having an independent means of knowing if ships are entering Canadian waters without reporting.

Thus, the *Vaedderen* and *Triton* were not required to inform Canadian officials that they were entering Canadian waters. Furthermore, Canada would have limited ability to independently know what a vessel is doing even if NORDREG were mandatory for all foreign vessels. The Danish government did inform Canada that the *Vaedderen* was entering Canadian waters. But once it was in disputed waters it never again contacted the relevant Canadian Coast Guard officials. The Department of National Defence had an Aurora long-range aircraft conducting fishery patrols in the region, which provided some surveillance of the vessel (confidential interview). However, this was simply a fortuitous event. The next year when the *Triton* sailed to the island, an Aurora was specifically tasked to overfly the Danish vessel. The Auroras are aging and used for fewer and fewer northern patrols for either sovereignty enforcement or fisheries inspections. It is doubtful that they will be replaced when their service life is up. In effect, the Danish effort to sail to the island and to land on it without Canadian knowledge is obviously reducing the Canadian ability to claim that it "controls" the island and the waters around it.

CONCLUSION

The Canadian government has clearly decided that the best response to the Danish actions toward Hans Island is through limited diplomatic protests. Only when media attention was focused on the issue did the government feel compelled to make a strong statement supporting the Canadian position. It is clear it was not perceived as a critical issue between Canada and Denmark.

Thus, the decision was made to protest the Danish move but not to do anything more. In part this is due to a desire to maintain good relations with Denmark. It also was made because there was little else that the Canadian government could have done.

Ultimately the problem of Hans Island presents the Canadian government with a series of difficult policy choices. The island itself is almost insignificant. However, if it is not resolved it will create a problematic precedent for the other disputes in the Canadian North. The Canadian position has been further complicated by its public statements following the coverage provided by the national media. It is now difficult for the government to take any position that would be publicly perceived as compromising Canadian interests. The other Arctic disputes are individually separate, but will be all become important as the climate change makes the North more accessible. However, the impact of climate change will be gradual and somewhat erratic. There will not be a sudden dawn of an ice-free Canadian Arctic, but a gradual easing of the harshness that has prevented southern incursions in these waters. The Canadian government will face a series of incremental challenges that it will deal with in isolation. However, over time the net result will be the slow erosion of Canadian control over this region.

What is needed is the development of a long-term policy with adequate funding to properly develop a Canadian presence in the Arctic. First, Canada should continue to press its case as to its control. It may not be able to win a claim of total and complete sovereignty over the entire region, but there is a need to ensure that it does retain the right to regulate all economic activity that enters the region. This includes international shipping. Canada needs to be aggressive in presenting its views as to the type of ships that will enter the Arctic. The Canadian government also needs to be equally strong in its demands that the training of the crews of these vessels be properly suited for operating in waters that will retain some ice, no matter to what extent warmer temperatures will reduce the permanent ice cover. The Canadian government has already established a strong regulatory system through the *Arctic Waters Pollution Prevention Act* and its regulatory system NORDREG. The problem is that the Canadian government has been reluctant to make the practical enforcement of the Act through NORDREG mandatory for foreign vessels because it fears provoking a reaction from other states such as the United States. What is required is the willingness to defend its position.

However, to ensure that the Canadian government's position is taken seriously it must ensure that it also has the means to maintain a strong presence in these waters. It needs the tools to monitor and control its Arctic region. There is a need not for ice-strengthened "gun-boats," but rather for the means to respond to standard maritime problems such as search and rescue, environmental protection, and the maintenance of law and order. By having the means to do so, Canada is then able to justify its claims of control. This will include a renewed ice-breaking fleet and surveillance capability discussed above. It will also have to include the training of personnel to have an ability to operate in the Canadian north. This will include navigational expertise, search and rescue in Arctic climates, and responses to environmental spills.

Thus, the real problem of the modern day "Vikings" is not in respect of Hans Island itself, but rather in that it demonstrates how bare the Canadian cupboard is with respect to defending the Canadian North! The Canadian ability to know what is happening in the North and to subsequently act is almost nonexistent. Given the fact that the Canadian northern boundaries are constantly being challenged by others, Canadians should be concerned. In addition, if scientific predictions are correct, the Canadian North will feel the greatest impact of global climate change. A warmer climate will undoubtedly increase access to the Canadian North which, in turn, will increase challenges. In order to meet these coming challenges, Canada needs to now begin the process of rebuilding its northern capabilities, including, but not limited to, new long-range patrol aircraft, icebreakers, and new surveillance systems. The return of the "Vikings" is only the beginning.

REFERENCES

Beesley, Alan. 1972–73. "The Arctic Pollution Prevention Act: Canada's Perspective" *Syracuse Journal of International Law* 1: 229–31.
Calgary Herald. 29 March 2004. "Editorial," A12.
Canada's Coastlines. 2003. Standing Senate Committee on National Security and Defence. October 2003. *Canada's Coastlines: The Longest Under-Defended Borders in the World.*
Churchill, Robin, and Alan Rowe. 1988. *The Law of the Sea,* 2nd ed. Manchester: Manchester University Press.
Coast Guard. 1 August 2002. E-mail correspondence Marine Safety, Prairie and Northern Region, Coast Guard to MARLANT and DFAIT JLO.
CP Wire. 2004. "Island Dispute Downplayed Canada, Denmark Claim Barren Rock," *Winnipeg Free Press* (1 April), A10.
Department of Finance Canada. 2003. *The Budget in Brief 2003: Building the Canada we Want* Ottawa: Public Works.
DFAIT. 2002. "Draft Q & A re Hans Island – Subject: 28 December 2002 *Globe and Mail* Article concerning the territorial dispute between Canada and Denmark over Hans Island." ND.
———. 30 July 2002. E-mail correspondence between DFAIT JLO and DFAIT REN.
———. 8 August 2002. "Draft Agreement – Interdepartmental Meeting on Canada-Greenland Boundary Adjustment," Ottawa.
DND. 21 July 2003. E-mail.
———. 31 July 2003. E-mail.
———. 11 July 2003. E-mail.
———. 3 August 2003. Communication "Purple 706 AMP 01."
———. 1 August 2003. E-mail.

Fisheries and Oceans Canada. 1 August 2003. *Vessel Traffic Reporting Arctic Canada Traffic Zone* (NORDREG).
http://www.ccg-gcc.gc.ca/cen-arc/mcts-sctm/mcts-services/vtrarctic_e.htm

Fogarassy, Tony. 20 April 1991. *The Alaska Boundary Dispute: History and International Law.* Vancouver: Clark, Wilson.

Franckx, Erik. 1993. *Maritime Claims in the Arctic: Canadian and Russian Perspectives.* Dordrecht: Martinus Nijhoff.

Godefry, Andrew. 2000. "Is the Sky Falling? Canada's Defence Space Programme at the Crossroads." *Canadian Military Journal* 1(2).

Gray, David. 1994. "Canada's Unresolved Maritime Boundaries." *Geomatica* 48(2) (Spring).

Harper, Kenn. 9 April 2004. "Hans Island rightfully belongs to Greenland, Denmark," *Nunastiaq News. http://www.nunatsiaq.com/opinionEditorial/opinions.html*

Head, Ivan, and Pierre Trudeau. 1995. *The Canadian Way: Shaping Canada's Foreign Policy, 1968–1984.* Toronto: McClelland & Stewart.

Huebert, Rob. 2001. "Article 234 and Marine Pollution Jurisdiction in the Arctic." In *The Law of the Sea in the Polar Oceans: Issues of Maritime Delimitation and Jurisdiction,* edited by Don Rothwell and Alex Oude Elferink. Dordrecht: Kluwer.

———. 28 December 2002. "Return of the Vikings," *Globe and Mail.*

Humphreys, Adrian. 25 March 2004. "Canada-Danish Spat Erupts over Island." *National Post,* A1.

———. 30 March 2004. "Danes Summon Envoy over Arctic Fight." *National Post,* A1.

———. 16 April 2004. "Students Protest Against Denmark's Aggression" *National Post,* A6.

——— and Chris Wattie. 26 March 2004. "Tiny Icy Island is ours, Determined Danes say." *National Post,* A1.

Macnab, Ron. 2003. "Delimiting the Juridical Continental Shelf in the Arctic Ocean: A confluence of Law, Science and Politics," *Meridian* (Fall/Winter).

Maginley, Charles, and Bernard Collin. 2001. *The Ships of Canada's Marine Service.* St. Catharines, Vanwell Publishing.

Moon, Peter. 2002. "Canada shows Flag in Unique Exercise," *Maple Leaf* 5(2) (September).

Officer, Charles, and Jake Page. 2001. *A Fabulous Kingdom: The Exploration of the Arctic.* Oxford: Oxford University Press.

Petrie, Ron. 27 March 2004. "Canada Picking a Fight with Denmark," *Regina Leader Post* B1.

Rolston, Susan, and Ted McDorman. 1988. "Maritime Boundary Making in the Arctic Region." In *Ocean Boundary Making: Regional Issues and Developments,* edited by Douglas Johnston and Phillip Saunders. London: Croom Helm.

Sharpe, Richard. 1998. *Janes Fighting Ships 1998–99.* Surrey: Janes Information Group.

Toronto Star. 5 May 2004. "Editorial." A16.

Transport Canada. 23 September 2003. *Arctic Waters Pollution Prevention Act* (AWPPA). *http://www.tc.gc.ca/acts-regulations/GENERAL/A/awppa/act/awppa.htm*

United Nations, Oceans and Law of the Sea, Division for Ocean Affairs and the Law of the Sea. 2003. *Commission on the Limits of the Continental Shelf. http://www.un.org/Depts/los/clcs_new/clcs_home.htm*

United Nations. 1983. *The Law of the Sea, United Nations Convention on the Law of the Sea with final Index and Final Act of the Third United Nations Conference on the Law of the Sea.* New York: United Nations.

U.S. Office of the Geographer. 1976. Office of the Geographer, U.S. Department of State. 4 August, 1976. *Limits in the Seas: Continental Shelf Boundary: Canada Greenland no. 72.*

Wattie, Chris, and Adrian Humphreys. 31 March 2004. "Government will not Surrender." *National Post*, A6.

Winnipeg Free Press. 30 March 2004. "Editorial." A12.

Worthington, Peter. 29 March 2004. "Who Cares." *Ottawa Sun*, 14.

NOTES

1 The Canadian Government also issued an *Aide Memoir* to the Danish Government on 11 August, 2000 inquiring about the alleged hunt. The Danish response is unknown.

2 DFAIT, "Draft Agreement – Interdepartmental Meeting on Canada-Greenland Boundary Adjustment," 8 August, 2002, Ottawa. While most of the document was censored, the page listing the issue areas of concern was not. This includes the voyage of the MV *Professor Polshkov* for seismic surveys of Greenland waters. The document does not provide the location of the vessel.

3 The most recent statement of the American position was made by American Ambassador Paul Cellucci when he was visiting Yellowknife in May 2003. When asked about the Passage he stated "[a]s you know, we have a disagreement, Canada and the United States on the Northwest Passage." Transcript of press conference Yellowknife, 6 June, 2003. On file with author.

4 The first European expedition to see the island was led by Sir George Nares in 1875–76.

5 The Gap is between points 122 and 123 of those used to determine the boundary.

6 No information on the two other requests was released at the time of the freedom of information request on Hans Island.

7 One of the most frustrating elements of attempting to determine the sequence of events is the refusal of Danish officials to officially comment. Off the record, several officials have indicated that not only did Danish sailors raise the flag, but they have been doing so for an extended period. However, as one can understand, these same officials will not speak on the record to a non-Dane about such a politically sensitive issue.

8 Peter Worthington's editorial suggested that this entire issue was nothing more than a plot on the part of the Liberal government to divert attention from other issues. This of course assumes that the *National Post* takes its direction from the Liberals!

9 These two articles state the limits of the continental shelf. Paragraph 5 states that the continental shelf "... either shall not exceed 350 nautical miles from the baselines from which the breadth of the territorial sea is measured or shall not exceed 100 nautical miles from the 2,500 metre isobath, which is a line connecting the depth of 2,500 metres."

10 The development of a new satellite capability was publicly acknowledged by the Commander of Canadian Northern Area. "Arctic Coast to be Spied on from Above," Webposted 21 November, 2003.

11 It is interesting to note that the vessel was renamed after a Norwegian explorer. Several commentators have lamented the fact that it did not have a Canadian name.

CHAPTER 16

ISSUES, PRIORITIES, & RESEARCH DIRECTIONS FOR OCEANS MANAGEMENT IN CANADA'S NORTH

Derek Armitage (Wilfrid Laurier University)
Douglas Clark (Wilfrid Laurier University)

INTRODUCTION

> *We, as aboriginal people, are part of the land and water. We recognize and respect the delicate balance of nature for the total existence of all living things including those we see physically and those we don't.... Once these natural processes are disturbed and denied their natural flow, the aboriginal people of this country are adversely affected. Taking away the land and water takes away our pride, dignity and ability to survive.*
> — Donald Saunders in *Voices from the Bay.* (McDonald *et al.* 1997, 5)

Canada's North is a special place. Special because of the forces of nature that create a unique and ecologically vulnerable landscape. Special as well because of the Inuit and First Nations people who have shaped and moulded lives rich with culture, knowledge, and understanding of the patterns and currents of life. Yet, those patterns and currents of life no longer serve as a source of certainty or comfort. They represent, instead, the local manifestations of political decisions, economic development activities, and socio-cultural changes that are crossing geographic boundaries and temporal scales. The ebb and flow of northern life is now connected with the uncertain and rapidly evolving dynamics of political negotiation and land claims agreements, diamond mining, oil and gas exploration, tourism, issues of sovereignty and security, climate change, and industrial pollutants from the South. Such externally driven agents of change, furthermore, catalyze and exacerbate processes of change embedded in local livelihood strategies, social dynamics, and political relationships.

Change is indeed a powerful concept with which to understand the North. As the title to a recent volume on climate change and adaptation in the circumpolar

North aptly suggests, the world is moving faster (Krupnik and Jolly 2002), and the ability of northern people and ecosystems to respond to socio-ecological change – both the type and speed of change – is a point of increasing uncertainty. Understanding, accommodating, and responding to socio-ecological change is a complex endeavour – interactions between people and ecosystems, including the management of human activities, encompass multi-faceted issues and varied scales (Folke *et al.* 2002; Gunderson and Holling 2002; Berkes *et al.* 2003). Management of ocean and coastal resources in Canada's North in the context of uncertainty and change, therefore, represents both a challenge and an opportunity for innovative research and practice. For example, what are the catalysts and implications of change in the North? How are linked northern socio-ecological systems responding and adapting to change? And how can northern individuals and societies create new, innovative opportunities for sustainability in the context of change and uncertainty?

As a focal area of applied research within Canada's Oceans Management Research Network (OMRN), the Integrated Management (IM) Node has been exploring the challenges of change, complexity, and uncertainty associated with oceans resource and coastal management. As this volume attests, many issues must be addressed, including the role of institutions and organizations, governance structures and processes, knowledge systems and the monitoring of human-ecological connections. Yet, in the midst of uncertain and rapid change in the North, many chapters in this volume highlight how individuals and communities are already seeking to maintain and reinvigorate historical socio-cultural and human-ecological connections. The questions they are asking are pragmatic and fundamental. How, for example, can ocean and marine resources in the North be developed in ways that benefit current and future generations? What are the implications of resource development for societies and cultures that have profound connections with ocean and marine resources? How can First Nations and Inuit participate in decision-making processes and ensure that the interests of northerners are articulated and defended in the context of cross-scale economic, political, and cultural influences?

These questions are of central importance to the ongoing development of an oceans management and policy framework as articulated in Canada's *Oceans Act* (1997) and associated 'Oceans Strategy.' As a key component of *Oceans Act* implementation, Canada's Oceans Strategy seeks to accomplish several strategic objectives, including: greater collaboration and consensus among participants in oceans management; the expansion of partnership arrangements and the sharing of responsibilities; enhanced governance of ocean resources; the development of integrative management strategies; and further efforts to link traditional and Western scientific knowledge. Efforts to meet the objectives articulated in the Oceans Strategy, however, will require the identification of priority issues and a commitment to research directions and management strategies that respond to change and uncertainty.

Achieving oceans sustainability, or meeting the policy and management commitments set forth in Canada's Oceans Strategy, will not be easily accomplished, but a path forward is being articulated in community halls and hamlet offices, and at a multitude of workshops, conferences, and community meetings across the North. As a region characterized by the complex interplay of historically embedded socio-cultural, biophysical, political, economic, and institutional dynamics, northerners recognize that answers and solutions to the complexities of change are most likely to emerge in the context of interdisciplinary understanding and meaningful collaboration. This and the preceding chapters in this volume represent, therefore, an effort to foster a collaborative and interdisciplinary approach to research and practice. As well, the volume is further recognition of the value of diverse voices in framing research priorities, analyzing information, and offering options and solutions to the challenges of sustainable ocean resource management in Canada's North.

Before commencing with this task, it is important to highlight the origins and evolution of this chapter and, in the process, the work of the IM Node. An important goal of the IM Node was to involve a variety of northern partner organizations in order to build and implement a collaborative oceans research agenda. Correspondingly, a number of individuals representing northern research and resource management organizations were involved in the establishment of the IM Node. A key concern of IM Node researchers and practitioners centred on how the outcomes of Node research could benefit northern communities, practitioners and policy makers in ways that respond to the issues they consider most critical. This chapter represents an effort of the IM Node to address that issue by synthesizing and articulating the priority issues, concerns, and objectives of northern individuals and groups. In particular, this chapter provides an opportunity to reflect on how, and to what extent, current research activities nest within the broad issues, concerns, and priorities articulated by northern individuals, communities, and organizations. This chapter also provides an opportunity to reflect on the procedural challenges of northern research and the importance of research collaboration.

The objectives of this chapter, therefore, are as follows: (1) to identify and elaborate key themes, priority issues, and concerns relevant to ocean resource management as articulated by northern communities and other key stakeholder groups; (2) to explore how those priority issues and concerns are differentially highlighted by different groups in the North, identifying in the process areas of commonality and difference; and (3) to illustrate potential directions for collaborative research in support of oceans management in Canada's North. As such, this analysis provides a reference point for the work of the IM Node, as well as a 'touchstone' for other researchers and practitioners concerned with oceans management who may wish to connect their work to the identified needs and aspirations of northern stakeholder groups.

The identification of a broad-based set of issues, concerns, and priorities is an inductive process, drawing necessarily upon existing materials and already completed studies that have sought to capture the issues and concerns of northern stakeholders. As such, this chapter draws primarily upon two sources of information. The first source includes reviews of the literature documenting outcomes of multi-stakeholder consultation processes associated with a range of environment and resource management initiatives in northern Canada in the past two decades, and particularly those related to ocean and marine resources (see Table 16.1). In identifying relevant initiatives, effort has been made to encompass different perspectives, including those which emerge from local, regional, and international scales. The following criteria were utilized in an effort to identify appropriate initiatives and activities:

- The geographical scope of the initiative should involve northern regions of Canada;
- The thematic focus of the initiative should emphasize, or have a significant concern with, ocean and marine resource issues;
- The initiative should involve broad-based, multi-stakeholder consultation activities; and
- The outcome(s) of the initiative should include the identification of priority ocean resource management issues and concerns.

The second main source of information results from literature reviews, as well as a review of the priorities and objectives articulated in the mandates of selected management and advocacy agencies in the North, including the recently created Institutions of Public Government, co-management boards, and research institutes. Based on these two sources of information, a list of priority issues and concerns associated with each initiative was identified, and then subsequently grouped into four categories: (1) livelihoods, jobs and economic development; (2) environment and human health; (3) knowledge systems; and (4) governance and decision making.

Finally, a caveat to this chapter is required. The identification of priority concerns and issues at the scale of Canada's North is fraught with intellectual pitfalls. Issues and concerns inevitably vary across and within regions, while those concerns and issues that are shared are likely to manifest themselves differently depending on context. As well, the range of resource and environmental management processes and initiatives reviewed has been opportunistic, guided in part by accessibility and knowledge of existing initiatives, and does not fully represent all geographic regions of the North. We are, for example, aware that significant geographic regions, such as Nunavik and Labrador, have not been adequately covered. In addition, we have not attempted to survey the many health-oriented projects or related literature. The initiatives reviewed are also thematically and temporally diverse, and include 'expert'-driven sessions and regionally specific management planning processes, as well as strategies and

Table 16.1 SUMMARY OF INITIATIVES

International and National Initiatives	Description	Sources
Protection of the Arctic Marine Environment (Arctic Environmental Protection Strategy)	The Protection of the Arctic Marine Environment (PAME) program was established in 1993 and is one of four programs of the Arctic Environmental Protection Strategy (AEPS). The AEPS operates under the auspices of the Arctic Council. PAME functions as an international inter-governmental program linking all countries of the circumpolar north.	www.pame.is/sidur; see also CAFF/IUCN/PAME 1999; PAME 1996
National Programme of Action for the Protection of the Marine Environment from Land-Based Activities	The National Programme of Action (NPA) is a response to the United Nations Environment Program (UNEP) led Global Programme of Action (GPA) to protect the marine environment. The NPA is a partnership among federal, provincial and territorial governments emphasizing the prevention of marine-pollution from land-based activities, as well as protecting nearshore and coastal zone habitats in Canada.	Environment Canada 2001; see also Arctic Council (www.arctic-council.org)
Arctic Environmental Strategy	An initiative undertaken by the Government of Canada focused on developing a comprehensive approach to the maintenance of the integrity of the Arctic environment. The priority objectives of the strategy include: ensuring the health and well-being of northern ecosystems; protecting and enhancing the environmental quality and sustainable use of resources; incorporating the perspectives of indigenous people in the planning and management process; improved decision making; and developing international agreements to better conserve, use and manage resources and protect the circumpolar environment.	Department of Indian Affairs and Northern Development (DIAND) 1996
Task Force on Northern Conservation	A consultative body, the Task Force was established in 1983 under the auspices of the federal government to develop a framework for the creation of a comprehensive conservation policy and implementation strategy for the Yukon and Northwest territories. The Task Force considered both terrestrial and marine environments.	Task Force on Northern Conservation 1984
Arctic Marine Conservation Strategy	The development of the Arctic Marine Conservation Strategy (AMCS) emerged from the 1984 Task Force on Northern Conservation. The purpose of the AMCS was to ensure the future health and well-being of Arctic marine ecosystems and provide for the sustainable use of Arctic marine resources, particularly for Arctic peoples. Preparation of the AMCS was based on consultation with a broad range of northern groups and other interested parties.	Department of Fisheries and Oceans (DFO), 1987
House Standing Committee Report – Canada and the Circumpolar World	The House Standing Committee report was the outcome of a comprehensive, government-led initiative to support the development of a post-Cold War northern foreign policy framework. A key component of the Standing Committee process was the consultation process which was undertaken with northern individuals and communities, government, non-government and industry representatives. The Standing Committee process led to the development of forty-nine recommendations covering a range of environmental, economic, political, institutional, transboundary and security issues, many of which are of direct relevance to ocean and marine systems in Canada's North. Importantly, the Standing Committee process led to the development of the "Northern Dimensions of Canada's Foreign Policy."	Government of Canada (1997) (7th Report of Standing Committee on Foreign Affairs and International Trade)
Calgary Working Group	An expert-based working group of northern practitioners and academics assembled to report on the priority issues and concerns influencing Canada's foreign policy in the circumpolar Arctic, and constituted in response to the Standing Committee report described above. One outcome of the Calgary Working Group was the identification of selected research needs that could enhance the policy development process in the circumpolar Arctic.	Calgary Working Group 1997

Table 16.1 SUMMARY OF INITIATIVES

	Description	Sources
Local and Regional Initiatives		
Baffin Marine Issues Scan	The DFO-supported Baffin Marine Issues Scan involved a comprehensive consultation process to identify stakeholder priorities, opportunities and constraints associated with the development of an integrated oceans resource management approach. Key activities included community meetings in seven of the eight Baffin Island communities, meetings with Hunter and Trapper Organizations, and consulting with a range of government, non-government and industry groups. The Baffin Marine Issues Scan culminated in a multi-stakeholder workshop in Iqaluit to chart a way forward in the context of *Oceans Act* implementation.	Terriplan/IER 2002
Beaufort Sea Conference on Renewable Marine Resources	The Beaufort Sea Conference, held in Inuvik in 1999, brought together representatives of a broad range of groups and organizations concerned with the management and use of renewable resources in the Beaufort Sea region. The conference was guided by three objectives: to review the current understanding of renewable resources in the region; to review the factors that affect those resources; and to develop a vision for the management and use of renewable resources in the Beaufort Sea.	Beaufort Sea 2000, 1999; Ayles *et al.* 2002; Day 2002
Inuvialuit Renewable Resource Conservation and Management Plan	The Plan outlines directions and rationale for all renewable resource management activities within the Inuvialuit Settlement Region. The Plan was prepared by the Wildlife Management Advisory Council and the Fisheries Joint Management Committee, and establishes a strategy for the long-term use and management of fish and wildlife.	WMAC/FJMC 1988
Hudson Bay TEKMS ("Voices from the Bay")	The Inuit and Cree communities around Hudson Bay initiated a project from 1992 to 1995 to document traditional ecological knowledge and observations of environmental change. A major motivation for them was to understand the cumulative environmental effects of hydroelectric developments on rivers draining into the bay. The study's results were published as the book "Voices from the Bay."	McDonald *et al.* 1997
Hudson Bay Working Group Issues Scan	In 2000–2001 the Department of Fisheries and Oceans hosted a series of meetings in Manitoba and Nunavut to develop an integrated management plan for Hudson Bay, initially focusing on the Kivalliq/Manitoba region. The impetus for this initiative came from Canada's Oceans Strategy, the implementation plan for Canada's 1997 *Oceans Act*, designed to consolidate existing federal marine legislation and promote an integrated approach to ocean management. The group identified issues of both common and specific concern to the different communities in the region, including climate change, local pollution sources, and a lack of communication between government scientists and communities.	DFO 2001a; 2001b
Nunavut Regional Land Use Planning Processes (Keewatin, Kitikmeot and North Baffin)	The Nunavut Planning Commission (NPC) undertook land use planning processes in three regions in Nunavut (Kitikmeot, Keewatin and North Baffin). The basis for two of the land use plans (Keewatin and North Baffin) was the former Lancaster Sound Regional Land Use Planning Commission. The plans provide a preliminary foundation for development decision making and were based on consultation with communities, Inuit organizations, as well as government, industry and other stakeholder groups.	Nunavut Planning Commission 1997a; 1997b; 1997c

Local and Regional Initiatives	Description	Sources
Yukon North Slope Long-Term Research and Monitoring Plan (and the Yukon North Slope Conference, 2001)	The Yukon North Slope Plan is the result of extensive consultations with the Inuvialuit Game Council, co-management bodies established under the Inuvialuit Final Agreement, federal, territorial and Alaskan government agencies, universities, NGOs and interested residents. The Plan identifies priority research and monitoring issues related to both marine and terrestrial environments, as well as the proposed actions required to address them.	Wildlife Management Advisory Council 2000; see also IFA Implementation Secretariat 2001.
Paulatuk Community Conservation Plan[1]	The purpose of the community conservation plan is to respond to the conservation principles mandated in the Inuvialuit Final Agreement. The plan is the result of discussions and analyses led by the community of Paulatuk in collaboration with the WMAC, and the Joint Secretariat. An initial plan was drafted in 1990 and subsequently revised in 2000.	Secretariat 2000
Sachs Harbour Climate Change Study	From 1999 to 2001 the Inuvialuit of Sachs Harbour, NWT and the International Institute for Sustainable Development (IISD) worked together to document the impacts of climate change in the Arctic and communicate the issue to national and international audiences. Besides a number of publications illustrating the collaboration of traditional and scientific perspectives towards understanding climate change, the project produced a video which was broadcast worldwide.	Ashford and Castleden 2001; Berkes and Jolly 2001; Riedlinger and Berkes 2001; Jolly et al. 2002
Wapusk National Park Management Planning	Established in 1996 on the western coast of Hudson Bay, in Manitoba, Wapusk (which means "white bear" in Swampy Cree) is one of the new cohort of co-managed northern national parks. The Wapusk National Park Management Board is a Ministerially appointed co-management body which makes recommendations on park management issues, and is very influential. The board is responsible for the preparation of the park's first management plan, a project which has been underway since 1997. The plan is nearly complete, but the board and Parks Canada staff have encountered numerous challenges during this process, including the need for the board to discover and define its role, developing trust and constructive working relationships between board members and park staff, the changing roles and expectations of park planners in this evolving process, integrating different knowledge systems (including scientific, traditional ecological knowledge, and "corporate memory"), and ensuring continuity when staff and board members turn over.	Wapusk National Park Management Board 1998; Weitzner 2000; Weitzner and Manseau 2001

1 The inclusion of the Paulatuk Community Conservation Plan is intended to provide one example of a local process that involves the identification of key issues. The Community Conservation Plans created by other communities under the Inuvialuit Final Agreement are similarly valuable sources of information on priority issues and concerns.

initiatives that are national in scope. Some initiatives were undertaken almost twenty years ago, while others have been completed much more recently. There are, in addition, many other specific management initiatives that are focused on discrete resources or wildlife species, including those dealing with bowhead whale (see NWMB 2000), caribou, grizzly bears, and beluga that could not be incorporated into the review. Nevertheless, this chapter is premised on the assumption that the initiatives surveyed adequately represent the voices of northern communities and other stakeholder groups. It is worth stressing, finally, that the intent of this review has never been to speak on behalf of northerners, or articulate a northern 'vision.' Rather, the intention has been to synthesize, principally, the results of those initiatives in which multiple stakeholder groups were assembled to identify and discuss their priority issues and concerns, and subsequently, to reflect on how, where and the extent to which the work of the IM Node has responded to those issues.

SCANNING THE NORTHERN LANDSCAPE:
ISSUES, CONCERNS AND PRIORITIES

Undertaking a process of issue prioritization is much like reading a map. It is easy to wander among place names and geographical features unless you pick a point from which to start your journey. The results of a recent survey by the Aurora Research Institute (ARI) in the Northwest Territories, as well as the Nunavut Research Agenda published by the Nunavut Research Institute (NRI 1997), offer that starting point. The results of the ARI (2002) survey identify ten research priority areas or themes, including 'health and well-being,' 'land resources' and 'social issues.' Other priority areas identified in the survey of particular relevance to the work of the IM Node include 'wildlife management' and 'fisheries and marine resources.' Likewise, the Nunavut Research Agenda (NRI 1997) highlights a number of key themes and related sub-themes, including social, medical and health-related research needs, 'other' social science research needs (e.g., governance and linguistics), environment and ecosystems studies, education and training, physical sciences and engineering, and technology development and transfer. The broad themes and issues identified in both the ARI (2002) survey and NRI (1997) research agenda find further resonance in the objectives and mandates specific to the institutions and organizations governing the use and management of oceans resources in the North. Created largely in the context of land claims agreements and a renegotiation of federal, territorial, and Aboriginal relationships, emergent Institutions of Public Government and various boards are changing the manner in which ocean resources are being managed.

There is much agreement among southerners and northerners, nevertheless, about the need for greater research in Canada's North. Yet, it is important to offer additional context for the general research themes and the identification of baseline ecological information needs articulated in the ARI (2002) survey and NRI (1997) research agenda. Reviews of several national and regional-scale initiatives relevant to integrated oceans resource management (see Table 16.1) provide this additional context and reveal a long list of priorities and concerns. To make sense of the range of issues and concerns identified, and in order to facilitate analysis, each of the identified issues was grouped into one of four themes: (1) livelihoods, jobs, and economic development; (2) environment and human health; (3) knowledge systems; and (4) governance and decision making. This was done to encourage the identification of a set of concerns that can and should serve as a basis for reflection on current work and a guide for future oceans management research and practice. A summary of priority issues and concerns for each category is provided below.

LIVELIHOODS, JOBS AND ECONOMIC DEVELOPMENT

Canada's North is characterized by a great deal of duality. One area of duality involves the livelihood and economic development strategies pursued by First Nations and Inuit. The survey of initiatives (Table 16.2) illustrates, for example, that individuals and communities are fundamentally concerned with their long-term ability to secure country food and maintain their subsistence livelihoods – an issue that is closely connected to concerns about cultural integrity and resilience (see Freeman 1997). Hunting, going out on the 'land,' and maintaining a reliance on country food are central values and objectives of northern groups. As Myers (2001) has noted, for example, recent surveys of Inuit families in Nunavut illustrate that the majority of households continue to hunt and secure country food for family consumption. Further evidence of the importance of subsistence activities to local economic development is provided by the responses of individuals and communities that participated in the Baffin Marine Issues Scan (Terriplan/IER 2002), as well as by the five-year harvest study being completed by the Nunavut Wildlife Management Board. The five-year harvest study illustrates the close connection between people, wildlife, and their basic needs.

While seeking to achieve subsistence goals and maintain cultural connections to the land, however, First Nations and Inuit have also identified a desire to expand on the opportunities provided by wage-based employment and resource development. Oil and gas exploration and development, tourism, and mining have the potential to support growing populations and can contribute to the desire of Aboriginal communities to participate more actively in the wage economy. However, as noted in several initiatives, job training and capacity building are central to any effort to engage more First Nations individuals in productive wage-based employment opportunities (see NRTEE 2001). As well, the issue and priority for economic development is not simply related to increasing employment in the wage economy. Rather, there is a desire and expectation that Aboriginal and First Nations groups will play a partnership (*i.e.*, joint venture) and/or lead role in resource development activities. Recent agreements signed by the Aboriginal Pipeline Group and a consortium of oil and gas multinationals interested in building the Mackenzie Valley Pipeline offers an excellent example of the role Aboriginal groups are taking in resource development.

Table 16.2 LIVELIHOODS, JOBS AND ECONOMIC DEVELOPMENT[1]

	Benefiting from oil, gas and mining developments	Tourism / Ecotourism Development opportunities	Traditional economies (maintaining mixed economy)	Influence of animal rights groups on harvesting	Jobs training and education	Opportunities for economic participation (e.g., joint ventures)	Country food as economic priority (ability to harvest)	Importance of transportation
International and National Initiatives								
Protection of the Arctic Marine Environment								
National Programme of Action for the Protection of the Marine Environment from Land-Based Activities								
Task Force on Northern Conservation			•				•	
Arctic Marine Conservation Strategy			•				•	
Arctic Environmental Strategy			•				•	
House Standing Committee Report	•		•			•		
Calgary Working Group			•					
Local and Regional Initiatives								
Baffin Issues Scan			•				•	•
Beaufort Sea Conference on Renewable Marine Resources			•	•				
Inuvialuit Renewable Resource Conservation and Management Plan			•					
Hudson Bay TEKMS: "Voices from the Bay"			•	•	•	•	•	
Hudson Bay Working Group			•		•	•		
Yukon North Slope Research and Monitoring Plan								
Paulatuk Community Conservation Plan		•	•			•		
Sachs Harbour Climate Change Study			•		•	•	•	
NPC Keewatin LUP Process	•	•						•
NPC Kitikmeot LUP Process	•						•	•
NPC North Baffin LUP Process	•	•	•				•	

1 Table summaries (Tables 16.2–16.5) provide a 'snapshot' of key themes/ priorities associated with each initiative, rather than a comprehensive list of concerns.

Wage-based economic development and subsistence resource activities, it would appear, are not, and will not become, mutually exclusive. Settled land claims have given many northern Aboriginal peoples a measure of control over development. The desire to maintain mixed economies remains strong (Usher 2002), although people are still coping with the impacts of past developments, particularly hydroelectric projects around Hudson Bay (McDonald *et al.* 1997). Rather than outsiders asking how northern people are to adapt to imposed changes, the questions are now being asked by northern peoples of themselves: how can we develop our resources responsibly while maintaining our health, well-being, and culture? And how can we sustain our waters, lands, and the traditional activities that depend on them? These questions indicate a firm grasp of the challenges of sustainability and are of universal relevance.

Northerners increasingly see themselves as active, central participants in economic development in all its forms, rather than marginalized, passive recipients of its effects. Northerners are effectively dealing with the global process of modernization (Nuttall 2000), and perhaps more than many of us, they have a wider perspective on it and are acutely aware of both its benefits and costs. Because of the Arctic's geopolitical location and history, globalization – expressed as increasing interconnection of societies, institutions, and markets – is not a new phenomenon to northern peoples either (Nuttall 2000; Chaturvedi 2000). There is potentially a great deal that could be learned about the sustainability of social and ecological systems from these ongoing northern experiments with modernity and globalization. Such information may allow northerners to become active participants in the global knowledge economy as well.

Finally, the heterogeneity of the Canadian North is becoming more visible as different economic development priorities are asserted in different areas. For example, the Hudson Bay Working Group (DFO 2001a, 22) documented a revealing (but unattributed) quote in Iqaluit, Nunavut:

> Nunavut is not a homogenous territory. When you are planning you
> need to consider the differences in the people and their communities.
> The Kivalliq communities are focused on traditional lifestyles and
> health concerns. Kitikmeot is focused on mining and technology.

ENVIRONMENT AND HUMAN HEALTH

This inclusive grouping (see Table 16.3) contains a number of interrelated issues occurring at a range of scales from global (climate change, transboundary contaminants) and regional (heavy metal contamination; fish and wildlife populations, distribution, and habitat; impacts and benefits of economic development), to local (sewage and waste management, water quality, melting permafrost, and coastal erosion). Concern about effects on human health are interwoven with these issues at all scales. There are no surprises in this list – these issues and concerns are well established, and recent studies and initiatives provide extensive documentation of their specific manifestations and

Table 16.3 ENVIRONMENT AND HUMAN HEALTH

	POPs / Transboundary Contaminants	Climate change	Sewage and waste management	Heavy metals (local, long-range)	Habitat alteration (wetlands, saltmarshes)	Impacts on human health	Water quality / quantity	Melting permafrost	Fish and wildlife population / distribution changes	Oil, gas and mining developments	Impacts from tourism activities	Transportation impact	Hydroelectric impacts
International and National Initiatives													
Protection of the Arctic Marine Environment	•		•							•			
National Programme of Action for the Protection of the Marine Environment from Land-Based Activities	•	•	•	•	•					•			
Task Force on Northern Conservation										•			
Arctic Marine Conservation Strategy													
Arctic Environmental Strategy	•		•			•	•			•			
House Standing Committee Report	•			•		•							
Calgary Working Group												•	
Local and Regional Initiatives													
Baffin Issues Scan	•	•	•							•	•	•	
Beaufort Sea Conference on Renewable Marine Resources	•	•	•							•	•		
Inuvialuit Renewable Resource Conservation and Management Plan	•				•		•			•	•		
Hudson Bay TEKMS: "Voices from the Bay"	•	•		•		•	•		•	•			•
Hudson Bay Working Group	•	•	•				•			•			•
Yukon North Slope Research and Monitoring Plan	•	•	•		•	•	•			•			
Paulatuk Community Conservation Plan						•						•	
Sachs Harbour Climate Change Study		•				•		•	•			•	
NPC Keewatin LUP Process			•	•		•	•			•		•	
NPC Kitikmeot LUP Process			•	•					•	•		•	
NPC North Baffin LUP Process	•	•	•	•						•	•	•	

resulting impacts on communities (see Table 16.1, Table 16.3). Environment and health issues are most often expressed in terms of their local impacts, and it is not surprising that the regional initiatives – more often started by northerners themselves – define issues of concern more precisely and in greater detail than the national ones.

The issues identified in individual projects depended to some degree on their scope. Studies with broad aims documented a wide range of impacts and issues. The Hudson Bay TEKMS project synthesized and mapped observations to provide a comprehensive picture of impacts (hydroelectric and otherwise) and recent environmental changes throughout Hudson and James Bay and their coastal regions. In the process, a number of common interests and cross-cutting issues emerged, including respect and environmental responsibility, the significance of environmental change, indigenous perspectives on development, and co-operation in the future (McDonald et al.1997). While such broader projects were the norm in the 1990s, there has been a recent increase in the number of more focused, management-oriented studies. The Sachs Harbour/IISD climate change study (Ashford and Castleden 2001) is typical of these efforts. The community worked in collaboration with a multidisciplinary team of scientists and specialists to document significant environmental changes impacting the traditional lifestyles and livelihoods of the community, including melting permafrost, reduced sea ice cover and seal abundance, unpredictable weather, and new animal species. Other current examples include the Beaufort Sea Beluga Management Planning Process, additional indigenous climate change studies, the Tuktut Nogak project, the Nunavut Bowhead traditional knowledge study, and the Baker Lake Inuit quaujimajatuqangit grizzly bear study (see *www.nwmb.com*; *www.fjmc.ca*). Because of the intimate connection between people and place, much northern-initiated research is characterized by a high degree of problem orientation (Clark 2002). For example, the Sachs Harbour climate change video provided graphic illustrations of what those people are coping with and sent a powerful message to a large, international audience (Ashford and Castleden 2001; Jolly *et al.* 2002). Given the magnitude of expected environmental changes in the Arctic, such powerful communication will be necessary to keep Arctic issues on the global agenda, and similar efforts will probably increase.

Development in the North generally evokes images of large-scale industrial exploitation of natural resources, such as the proposals that prompted the Mackenzie Valley Pipeline Inquiry (Berger 1977). Here we consider development activities to include those, but we also extend the term to include less consumptive economic endeavours (*e.g.*, tourism) and the traditional activities which contribute to peoples' well-being in mixed economies (Usher 2002). Concerns about oil, gas, and mining developments were the most commonly identified issues, followed by impacts from tourism, and hydroelectric impacts. From a regional perspective, environmental impacts from specific local/regional infrastructure and activity development appear to be the primary concerns, as opposed to the more generic concerns about impacts from distant or global industrial activity, even though the latter are also considered in terms of their local environmental effects. Unanticipated concerns have also surfaced, such as the impacts from tourism on wildlife, important cultural sites, and traditional harvesting patterns.

At first glance, issues appear little changed since the Berger Commission, with the impacts from oil, gas, and mining developments being the most common concern expressed. However, unease about transboundary pollution and the health impacts – direct and indirect via the consumption of country food – of persistent organic pollutants (POPs) has emerged as a priority over the past ten to fifteen years (Downie and Fenge 2003). In particular, more recent local and regional initiatives highlight POPs and transboundary pollution as a major issue with far-reaching consequences for the livelihood strategies of First Nations communities. The implementation of comprehensive programs to analyze POPs issues – Canada's Northern Contaminants Program (NCP) and the Arctic Monitoring and Assessment Program (AMAP) – illustrate the scope and extent of the problem (Shearer and Han 2003; Reiersen *et al.* 2003). However, the implications for human health suggested by volumes of research are still being analyzed (see Kuhnlein *et al.* 2003).

KNOWLEDGE SYSTEMS

Knowledge and information form the cornerstone of integrated oceans resource management. Specifically, the manner in which knowledge and information are obtained, utilized and shared among researchers, practitioners, and communities is central to interdisciplinary understanding. The importance of information and knowledge systems also stems from a recognition of the value of ocean resources to the livelihoods of northerners and the manner in which northern ecosystems and wildlife may serve as indicators of regional and global ecological change (see, for example, the Arctic Monitoring and Assessment Program – *www.amap.no/amap.htm* – and the Northern Contaminants Program – *www.ainc-inac.gc.ca/ncp/index_e.html#*). However, as articulated by northern communities and other stakeholder groups involved in the regional and national-scale initiatives surveyed for this chapter, the process by which different forms of knowledge are constructed and shared creates a number of challenges and opportunities. In particular, two dimensions are typically highlighted: (1) the integration of traditional knowledge and Western-based science; and (2) ongoing science and baseline information needs (see Table 16.4).

The need for, and value of, integrating traditional knowledge and Western science in order to improve resource management in Canada's North is well established (GNWT 1993; Stevenson 1996; Wenzel 1999; Usher 2000). However, despite the advances being made to integrate traditional and Western-based knowledge systems (McDonald *et al.* 1997; Ashford and Castleden 2001), there does remain a degree of resistance to this process (Howard and Widdowson 1996). Several significant challenges must still be overcome, including the

Table 16.4 KNOWLEDGE SYSTEMS

	Integrating TEK and Western science	Baseline science and information needs (marine and terrestrial)	Understanding and monitoring cumulative effects	Data sharing, standardization and management	Impacts of research activities on wildlife (human-animal interactions)	Determining carrying capacity of human activities	Identifying, documenting cultural resources and heritage
International and National Initiatives							
Protection of the Arctic Marine Environment	•	•	•	•			
National Programme of Action for the Protection of the Marine Environment from Land-Based Activities							
Task Force on Northern Conservation		•					
Arctic Marine Conservation Strategy	•	•					
Arctic Environmental Strategy		•					
House Standing Committee Report	•	•					
Calgary Working Group	•					•	
Local and Regional Initiatives							
Baffin Issues Scan	•	•					
Beaufort Sea Conference on Renewable Marine Resources	•			•	•		
Inuvialuit Renewable Resource Conservation and Management Plan		•					
Hudson Bay TEKMS: "Voices from the Bay"	•	•			•		
Hudson Bay Working Group	•	•			•		
Yukon North Slope Research and Monitoring Plan	•	•	•	•		•	•
Paulatuk Community Conservation Plan	•	•	•	•			
Sachs Harbour Climate Change Study	•						
NPC Keewatin LUP Process•							•
NPC Kitikmeot LUP Process	•		•				
NPC North Baffin LUP Process	•	•	•				•

appropriate use and control of traditional knowledge, the impacts of documentation and codification (*e.g.,* the abstraction of traditional knowledge), as well as methodological issues that stem from efforts to compare different types of knowledge to inform decision makers about the impacts of specific activities (see also Kuhn and Duerden 1996; Sillitoe 1998; Sherry and Myers 2002). As outcomes of the initiatives reveal, ongoing efforts foster planning, management, and decision-making processes that build on the strengths and insights offered by different knowledge systems (see Krupnik and Jolly 2001; Jolly *et al.* 2002). One participant in the Baffin Marine Issue Scan suggested, for example, that while one *hears* about the importance of incorporating traditional knowledge in studies, you actually rarely *see* it [*emphasis added*].[1] Likewise, the Yukon North Slope Long Term Research and Monitoring Plan (WMAC 2000) identified a variety of concerns associated with the use of traditional knowledge, including the potential for its misuse by different interests, issues of accessibility, and the continuing loss of the oral history that provides an essential mechanism for the transmission of traditional knowledge.

In addition to concerns about the integration of traditional knowledge and Western science, disparate stakeholder groups identified a number of ocean resource science and baseline information needs as a priority. For industry groups working in the North, one concern of note involves the lack of adequate and up-to-date hydrographic data (*e.g.,* for marine navigation and exploration purposes). More broadly, however, a number of northern communities and resource industry interests identify similar concerns about adequate baseline data on key ocean resource stocks. As identified in the Baffin Marine Issue Scan, for example, limitations on fisheries research and/or stock assessment science (in large part because of its high cost) creates a number of challenges, including: (1) the inability to establish an adequate baseline understanding of ecosystem conditions; (2) the uncertainty with which to set appropriate and sustainable quotas for a variety of marine and ocean species; and (3) the subsequent challenge associated with developing community or commercially based experimental fisheries and management processes.[2]

Specific marine science and information needs for sustainable oceans resource management vary by region, and outcomes of the initiatives surveyed for the chapter reveal a concern with different species, ecosystem dynamics, and development pressures (*e.g.,* oil and gas, tourism, mining). In general terms, however, northern regions such as Nunavut are behind other Canadian jurisdictions in terms of understanding key stocks and the baseline ecosystem science required to inform integrated ocean resource management processes (see also Conference Board of Canada 2001). The scale and scope of oceans resource management in the North makes this a somewhat intractable problem. Efforts to address this information deficit will necessarily involve different levels of government, as well as industry and community input. A lack of baseline information points as well to a further priority issue identified in several of the initiatives – the importance of data sharing, standardization, and management (regionally and internationally). Identifying and implementing tools and mechanisms that foster data sharing, and which draw on different knowledge systems, is necessary to better understand and monitor the cumulative effects of development activities on wildlife (*e.g.*, caribou), habitats and ecosystems, and human health (AXYS 2000; Waehdoo Naowo Ko 2000; Lutsel K'e Dene First Nation 2001; see also *www.ceamf.ca*).

GOVERNANCE AND DECISION MAKING

Effective governance involves creating conditions in which the political and economic power of diverse stakeholder groups (formal and informal) is exercised in an appropriate manner, and several specific governance and decision-making-related priorities have been articulated by northern communities and other stakeholders in the North (Table 16.5). Central among issues of governance are ongoing concerns about the need for better communication, collaboration, and partnership building. According to the outcomes of the initiatives, opportunities for integrated oceans resource management across Canada's North require greater transparency, commitment, and coordination among community-based resource management organizations (*e.g.*, Hunters' and Trappers' Organizations, Renewable Resource Councils), regional bodies and institutions of public government, and industry. However, the barriers to the collaboration, coordination, and communication identified are neither recent nor unique. Existing barriers include: (1) the presence of numerous federal and territorial government agencies with mandates and programs relevant to marine and ocean resources; (2) a still evolving framework for oceans resource management (Chudczak 2002; DFO 2002); and (3) the lack of a shared understanding of relatively new and still evolving institutional, organizational, and community roles, responsibilities, and relationships created by northern claims agreements.

Across much of the North, comprehensive claims agreements and evolving legislation (*e.g.*, the *Mackenzie Valley Resource Management Act*) have created new governance and decision-making structures and processes that are still being internalized by Aboriginal, government, non-government, and industry groups (see also Rodon 1998; White 2001). There is, arguably, a still evolving recognition among some Inuit and First Nations groups about the increased authority and new roles they have in ocean resource governance and decision making.[3] Embedded in this re-framing of governance and decision-making structures and processes, however, are the capacity issues management organizations (at local and regional scales) and communities confront as a result of funding, infrastructure, and technical limitations, and education, training, and human resource constraints. As explicitly highlighted in a number of regional-scale initiatives in particular, education and awareness are a focal point in ongoing efforts to foster greater management collaboration, communication, and partnership. A key component of relationship-building exercises must certainly include education and awareness about marine issues and management

Table 16.5 GOVERNANCE AND DECISION MAKING

	Marine Conservation Areas / Protected Areas	Capacity constraints (local organizations)	System of licensing and quotas for resources	Security and sovereignty issues[1]	Communication, collaboration and partnerships	Education and awareness	Enforcement of regulations	International relations and cooperation
International and National Initiatives								
Protection of the Arctic Marine Environment	•	•	•		•			•
National Programme of Action for the Protection of the Marine Environment from Land-Based Activities								
Task Force on Northern Conservation	•	·			•			
Arctic Marine Conservation Strategy	•				•	•		
Arctic Environmental Strategy		•			•	•		•
House Standing Committee Report				•	•			•
Calgary Working Group				•				
Local and Regional Initiatives								
Baffin Issues Scan	•	•	•	•	•	•	•	•
Beaufort Sea Conference on Renewable Marine Resources	•	•			•	•		
Inuvialuit Renewable Resource Conservation and Management Plan	•					•	•	
Hudson Bay TEKMS: "Voices from the Bay"	•	•			•	•		
Hudson Bay Working Group	•				•	•		
Yukon North Slope Research and Monitoring Plan	•							•
Paulatuk Community Conservation Plan	•					•		
Sachs Harbour Climate Change Study					•	•		•
NPC Keewatin LUP Process	•				•			
NPC Kitikmeot LUP Process	•							
NPC North Baffin LUP Process	•			•	•			

1 A traditional interpretation of security and sovereignty is utilized (*e.g.*, border integrity, geopolitical threats), and corresponds to the way in which security and sovereignty were defined in the various initiatives surveyed. More recent interpretations of this terminology include many of the issues highlighted in the tables, such as transboundary environmental issues that may threaten local communities and livelihood systems.

strategies directed at resource users and harvesters. However, raising the awareness of government, non-government, industry, and co-management organizations about local resource use strategies and the complex socio-cultural, economic
and political interests that drive resource use is essential as well.

Finally, the importance and role of marine protected areas and related conservation measures has been highlighted in many of the initiatives surveyed for
this chapter. Although the focus is on species and habitats, the fundamental
challenge associated with the creation of marine conservation areas and protected
areas is one of governance: creating and supporting through regulation and
enforcement the necessary legislative framework, developing and implementing multi-stakeholder management planning processes, and establishing and
sustaining collaborative, formal and informal management and decision-making
frameworks (Weitzner and Manseau 2001).

REFLECTING ON CHANGE: DIVERSE PRIORITIES, COMMON
PERSPECTIVES AND DIRECTIONS FOR RESEARCH

> *The Southerners usually just go ahead with whatever they are planning be-*
> *cause they have the resources and the money.... But, they cannot go ahead*
> *as planned if a group of people do not agree with their intentions.*
> — Peter Matte in *Voices from the Bay*. (MacDonald *et al.* 1997, 67)

A key objective of this chapter has been to identify and elaborate
on the priority issues and concerns relevant to oceans resource management
as articulated by northern communities and other stakeholder groups. To accomplish this task, a range of initiatives, programs, and relevant literature has
been assessed and categorized according to four broad themes of inquiry and
action: livelihoods, jobs and economic development; environment and health;
knowledge systems; and governance and decision making. There is much overlap and connectivity among these themes, and the speed at which these issues,
concerns, and priorities coalesce to create socio-ecological change and uncertainty represents a fundamental challenge for oceans sustainability in Canada's
North. For researchers, managers and policy makers, therefore, this chapter
can offer a valuable frame of reference with which to identify the dynamics of
change, explore how linked social-ecological systems in Canada's North are
adapting to change, and examine how northern individuals and societies may
be able to create innovative strategies that foster sustainability. The concept of
change, moreover, can help connect the questions and inquiries of researchers
and practitioners with the complexities of oceans resource management in
Canada's North, and the intimate knowledge of northern people. When seeking to understand the impact and influence of change, for example, it is useful
to explore how different groups in the North identify, highlight, and prioritize
different catalysts of change. Highlighting the different priorities and concerns
among different groups is not easy when based on a broad-based review of local,

regional, and national initiatives, all of which were undertaken differently in disparate places and at different times. Nevertheless, a few patterns do emerge that are worth highlighting.

First Nations and Inuit communities consistently highlight and prioritize a number of issues, including the economic and cultural importance of country food, the desire to develop and maintain a well balanced mixed economy (*i.e.*, one that provides subsistence and wage-based employment opportunities), and the need to better integrate traditional environmental knowledge and Western science as a basis for decision making and management. Highlighted as well are expectations that non-renewable (and renewable) resource development activities (*e.g.*, oil and gas development, mining and tourism) will directly benefit First Nations communities, and that the negative implications of resource development must be addressed and mitigated. Of particular concern as well are the problems of sewage and waste management and the potentially negative impacts of transboundary pollution (*e.g.*, POPs) on human health and wildlife.

Although those initiatives that had a more involved government planning orientation (*e.g.*, the Arctic Environmental Strategy, Task Force on Northern Conservation) also identified some of these issues and concerns, there were some differences in focus and orientation. Higher-level national scale initiatives, for example, are more likely to raise sovereignty and international security as an issue, along with the need for more baseline information (*e.g.*, ecological, hydrographic). The different emphasis placed on some issues is also worth noting – conservation and marine protected areas offering one example. Although the need for protection and conservation was a widely cited priority, Aboriginal groups are concerned with sustainability of stocks for harvesting, while other groups whose voices come out in these multi-stakeholder initiatives raise the idea of protection as a goal in and of itself. Although perhaps more prevalent in previous decades, ideology and the 'clash of cultures' (see Freeman 1997) regarding northern livelihoods continues to create a subtext for the socio-ecological priorities identified by different groups. Finally, it is worth noting that the higher-level initiatives surveyed tended not to emphasize those issues related to livelihoods, jobs, and economic development.

Despite these patterns and different areas of emphasis, there is significant overlap among many of the priority issues and concerns identified across northern regions, the types of initiatives surveyed, and the groups who participated in those initiatives (see Figure 16.1). There are, for example, broad-based concerns with POPs, climate change, the cumulative effects of development, and the importance of fostering education, awareness, and the capacity of northern individuals and communities to engage rapid change. Underlying these issues, as well, are more fundamental concerns not explicitly identified as priorities: specifically, the speed and pace of change, the consequent uncertainty created for individuals and communities, and the well-recognized need to achieve a degree of balance in the context of this uncertainty. And finally, for First

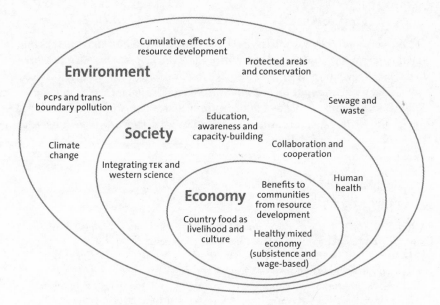

Figure 16.1 Northern perspectives, priorities, and directions for oceans research.

Nations and Inuit communities in particular, there is a strong commitment to greater autonomy, self-reliance, and self-government as a means of achieving that balance.

It would be erroneous to suggest that Figure 16.1 represents a vision or comprehensively articulated set of northern goals and objectives. Yet, the figure does capture those priorities – priorities that cross artificial social, economic, and environmental boundaries – identified by northern individuals, communities, and organizations as fundamental. Finally, the figure also raises some questions pertinent to this analysis: To what extent does this volume address these identified issues and help to forge lines of inquiry that benefit northerners? And to what extent are northerners themselves (individuals and organizations) directly engaged in identifying those lines of inquiry and carrying out priority research? In asking these questions, the purpose is neither to justify this volume nor to highlight its limitations. Rather, the purpose is to continue to articulate northern perspectives so that the issues and priorities they identify remain central to the type of research undertaken (regardless of disciplinary orientation) and the collaborative process by which that research should take place.

ACKNOWLEDGMENTS

The authors gratefully acknowledge the direction provided by Fikret Berkes during the development of this chapter, as well as the helpful comments of three anonymous reviewers. We would also like to acknowledge the participants (too numerous to mention) at the Kananaskis meeting of the IM Node, and Sarah Kalhok and Andrew Applejohn of the Aurora Research Institute, for their comments and feedback on an earlier draft. Finally, we acknowledge the support of the Social Science and Humanities Research Council through the IM Node of the Oceans Management Research Network.

REFERENCES

Ashford, G. and J. Castleden. 2001. *Inuit Observations on Climate Change*. Winnipeg, MB: International Institute for Sustainable Development.

ARI (Aurora Research Institute). 2002. NWT *Research Agenda Survey 2001*. Aurora College, Aurora Research Institute, Inuvik, NT. 37 pp.

AXYS (AXYS Environmental Consultants). 2000. *Regional Approaches to Managing Cumulative Effects in Canada's North*. Report submitted to Department of Environment, Government of Canada. Yellowknife, NWT.

Ayles, B., R. Bell, and H. Fast. 2002. Beaufort Sea Conference 2000 on the Renewable Marine Resources of the Canadian Beaufort Sea. *Arctic*. 55(1): iii–v.

Beaufort Sea. 2000. 1999. *Beaufort Sea 2000: Renewable Resources for Our Children*. Conference Summary Report. September 15–18, Fisheries Joint Management Committee, Inuvik, NT.

Berger, T.R. 1977. *Northern Frontier, Northern Homeland. The Report of the Mackenzie Valley Pipeline Inquiry*. Vol. 1. Toronto: James Lorimer & Co.

Berkes, F. and D. Jolly. 2001. "Adapting to climate change: social-ecological resilience in a Canadian western Arctic community." *Conservation Ecology* 5(2): 18. *http://www.consecol.org/vol5/iss2/art18*

Berkes, F., C. Folke, and J. Colding. 2003. *Navigating Social-Ecological Systems: Building Resilience for Complexity and Change*. Cambridge: Cambridge University Press.

CAFF/WCPA/PAME. 1999. *Circumpolar Marine Workshop: Report and Recommendations*. Conservation of Arctic Flora and Fauna, IUCN World Conservation Union and Protection of the Arctic Marine Environment. November 28–December 2, 1999. Montreal.

Calgary Working Group. 1997. *A Report and Recommendations for Canadian Foreign Policy in the Circumpolar Arctic*. Calgary Working Group. *http://www.carc.org/calgary/r9.htm*

Chaturvedi, S. 2000. Arctic geopolitics then and now. In *The Arctic: Environment, People, Policy*, edited by M. Nuttall and T.V. Callaghan, 441–58. Amsterdam: Harwood Academic Press.

Chudczak, C. 2002. *Canada's Oceans Strategy: Working Towards Tomorrow*. Presentation to the Oceans Management Research Network. October 26, 2002, University of Ottawa, Ottawa, Ontario.

Clark, T.W. 2002. *The Policy Process: A Practical Guide for Natural Resource Professionals*. New Haven, CT: Yale University Press.

Community of Paulatuk, WMAC and Joint Secretariat. 2000. *Paulatuk Community Conservation Plan*.

Conference Board of Canada. 2001. *Nunavut Economic Outlook*. Ottawa: Conference Board of Canada (Economic Services).

Day, B. 2002. "Renewable Resources of the Beaufort Sea for Our Children: Perspectives from an Inuvialuit Elder." *Arctic*. 55(1): 1–3.

DFO (Department of Fisheries and Oceans). 1987. *Canadian Arctic Marine Conservation Strategy*. Ottawa: Department of Fisheries and Oceans.

———. 2001a. Trip report: Churchill, Kivalliq and Iqaluit community tour, March 7–19, 2001. Ottawa: Department of Fisheries and Oceans.

———. 2001b. Draft minutes, Hudson Bay Oceans Working Group Planning Meeting, Siniktarvik Hotel, Rankin Inlet, NU, October 3–4, 2001. Ottawa: Department of Fisheries and Oceans.

———. 2002. *Policy and Operational Framework for Integrated Management of Estuarine, Coastal and Marine Environments in Canada*. Working Document for Canada's Oceans Community. Fisheries and Oceans Canada, Oceans Sector, Winnipeg. 35 pp.

DIAND (Department of Indian Affairs and Northern Development). 1996. *The Arctic Environmental Strategy: Five Years of Progress*. Ottawa: Department of Indian Affairs and Northern Development.

Downie, D.L., and T. Fenge. 2003. *Northern Lights Against POPs: Combatting Toxic Threats in the Arctic*. Montreal and Kingston: McGill-Queens University Press.

Environment Canada. 2001. *Implementing Canada's National Programme of Action for the Protection of the Marine Environment from Land-Based Activities: National Report to the 2001 Intergovernmental Review Meeting on Implementation of the Global Programme of Action*. Federal/Provincial/Territorial Advisory Committee on Canada's National Programme of Action for the Protection of the Marine Environment from Land-Based Activities, Ottawa.

Folke, C., S. Carpenter, T. Elmqvist, L. Gunderson, C.S. Holling, B. Walker, J. Bengtsson, F. Berkes, J. Colding, K. Danell, M. Falkenmark, L. Gordon, B. Kasperson, N. Kautsky, A. Kinzig, S. Levin, K. Goran-Mäler, F. Moberg, L. Ohlsson, O. Olsson, E. Ostrom, W. Reid, J. Rockström, H. Savenjie, and U. Svedin. 2002. *Resilience and Sustainable Development: Building Adaptive Capacity in a World of Transformations*. Science Background Paper for World Summit on Sustainable Development. Environmental Advisory Committee to the Swedish Government. 33 pp.

Freeman, M. 1984. "Contemporary Inuit exploitation of the sea-ice environment." In CARC. Sikumiut: 'The people who use the sea,' 77–96. Ottawa: Canadian Arctic Resources Committee.

———. 1997. "Issues affecting subsistence security in Arctic societies." *Arctic Anthropology*. 34(1): 7–17.

GNWT. 1993. *Traditional Knowledge Policy*. Government of Northwest Territories, Yellowknife.

Government of Canada. 1997. *Canada and the Circumpolar World: Meeting the Challenges of Cooperation into the Twenty-First Century*. 7th Report of the Standing Committee on Foreign Affairs and International Trade. Ottawa.

Gunderson, L. and C.S. Holling. 2002. *Panarchy: Understanding Transformations in Human and Natural Systems*. Washington, DC: Island Press.

Howard, A., and F. Widdowson. 1996. "Traditional knowledge threatens environmental assessment." *Policy Options* 17(9): 34–36.

IFA Implementation Secretariat. 2001. *Yukon North Slope Conference: The Challenge of Change*. IFA Implementation Secretariat. Department of Renewable Resources, Whitehorse: Government of Yukon.

Jolly, D., F. Berkes, J. Castleden, T. Nichols, and the community of Sachs Harbour. 2002. "We can't predict the weather like we used to: Inuvialuit observations of climate change, Sachs Harbour, western Canadian Arctic." In *The Earth is Faster Now: Indigenous Observations of Arctic Environmental Change*, edited by I. Krupnik and D. Jolly, 92–125. Fairbanks, AK: Arctic Research Consortium of the United States.

Krupnik, I., and D. Jolly, eds. 2002. *The Earth is Faster Now: Indigenous Observations of Arctic Environmental Change*. Fairbanks, AK: Arctic Research Consortium of the United States. 356 pp.

Kuhn, R.G., and F. Duerden. 1996. "A Review of Traditional Environmental Knowledge: An Interdisciplinary Canadian Perspective." *Culture* 16(1): 71–84.

Kuhnlein, H., L. Chan, G. Egeland, and O. Receveur. 2003. "Canadian Arctic Indigenous Peoples, Traditional Food Systems, and POPs." In *Northern Lights Against POPs: Combatting Toxic Threats in the Arctic*, edited by D. L. Downie and T. Fenge, 22–40. Montreal and Kingston: McGill-Queens University Press.

Lutsel K'e Dene First Nation. 2001. *Watching the Land: Cumulative Effects Assessment and Management in the Denesoline Territory*. Final Report submitted to the NWT CEAM Steering Committee and the Canadian Arctic Resources Committee.

McDonald, M., L. Arragutainaq, and Z. Novalinga. 1997. *Voices from the Bay: Traditional Ecological Knowledge of Inuit and Cree in the Hudson Bay Bioregion*. Canadian Arctic Resources Committee and the Environmental Community of the Municipality of Sanikiluaq, Ottawa, ON and Sanikiluaq, NT.

Myers, H. 2001. "Changing Environment, Changing Times: Environmental Issues and Political Action in the Canadian North." *Environment* 43(6): 34–44.

Nunavut Planning Commission. 1997a. *Keewatin Regional Land Use Plan*. Nunavut Planning Commission, Rankin Inlet.

———. 1997b. *West Kitikmeot Regional Land Use Plan*. Nunavut Planning Commission.

———. 1997c. *North Baffin Regional Land Use Plan*. Nunavut Planning Commission.

NRI (Nunavut Research Institute). 1997. *Nunavut Research Agenda: Research Policy and Needs for Nunavut*. Iqaluit, NU: Nunavut Research Institute.

NRTEE (National Round Table on the Environment and the Economy). 2001. Aboriginal Communities and Non-Renewable Resource Development. Ottawa, Canada: National Round Table on the Environment and the Economy.

NWMB (Nunavut Wildlife Management Board). 2000. Final Report of the Inuit Bowhead Knowledge Study. Iqaluit, Nunavut: Nunavut Wildlife Management Board.

Nuttall, M. 2000. "Indigenous peoples, self-determination and the Arctic environment." In *The Arctic: Environment, People, Policy*, edited by M. Nuttall and T. V. Callaghan, 377–410. Amsterdam: Harwood Academic Press.

PAME. 1996. *PAME Report to the Third Ministerial Conference on the Protection of the Arctic Marine Environment*. March 20–21, 1996. Inuvik, NWT.

Reiersen, L.O., S. Wilson, and V. Kimstach. 2003. "Circumpolar Perspectives on Persistent Organic Pollutants: The Arctic Monitoring and Assessment Program." In *Northern Lights Against POPs: Combatting Toxic Threats in the Arctic*, edited by D. L. Downie and T. Fenge, 60–86. Montreal and Kingston: McGill-Queens University Press.

Riedlinger, D., and F. Berkes. 2001. "Contributions of traditional knowledge to understanding climate change in the Canadian Arctic." *Polar Record* 37(203): 315–28.

Rodon, T. 1998. "Co-management and self-determination in Nunavut." *Polar Geography* 22(2): 119–35.

Shearer, R., and S.L. Han. 2003. "Canadian Research and POPs: The Northern Contaminants Program." In *Northern Lights Against POPs: Combatting Toxic Threats in the Arctic*, edited by D. L. Downie and T. Fenge, 41–59. Montreal: McGill-Queen's University Press.

Sherry, E., and H. Myers. 2002. "Traditional Environmental Knowledge in Practice." *Society and Natural Resources* 15: 345–58.

Sillitoe, P. 1998. "The development of indigenous knowledge: a new applied anthropology." *Current Anthropology* 39(2): 223–52.

Stevenson, M. 1996. "Indigenous knowledge in environmental assessment." *Arctic* 49(3): 278–91.

Task Force on Northern Conservation. 1984. *Report of the Task Force on Northern Conservation*. Ottawa: Department of Indian and Northern Affairs.

Terriplan/IER. 2002. *Toward Integrated Oceans Resource Management in the Baffin Island Region, Nunavut: Workshop Summary Report*. Terriplan Consulting Ltd./IER Planning, Research and Management report prepared for Fisheries and Oceans Canada, Iqaluit and Winnipeg.

Usher, P. 2000. "Traditional ecological knowledge in environmental assessment and management." *Arctic* 53(2): 183–93.

———. 2002. "Inuvialuit Use of the Beaufort Sea and its Resources, 1960–2000." *Arctic*. 55(Supp. 1): 18–28.

Waehdoo Naowo Ko. 2000. *Developing a Plan to Include Indigenous Knowledge in the NWT Cumulative Effect Assessment and Management Framework*. Dogrib Treaty 11 Council Report submitted to the NWT Cumulative Effects Assessment and Management Program.

Wapusk National Park Management Board. 1998. *Wapusk National Park Interim Management Guidelines*. Churchill, MB: Parks Canada. 32 pp.

Weitzner, V. 2000. *Taking the Pulse of Collaborative Management in Canada's National Parks and National Park Reserves: Voices from the Field*. Final Report for an Independent Research Project. Natural Resources Institute, University of Manitoba, Winnipeg. 61 pp.

——— and M. Manseau. 2001. "Taking the pulse of collaborative management in Canada's national parks and national park reserves: Voices from the Field." In *Crossing Boundaries in Park Management: Proceedings of the 11th Conference on Research and Resource Management in Parks and on Public Lands*, edited by D. Harmon, 253–59. Hancock, MI: The George Wright Society.

Wenzel, G. 1999. "Traditional ecological knowledge and Inuit: Reflections on TEK research and ethics." *Arctic* 52(2): 113–24.

White, G. 2001. "And Now for Something Completely Northern: Institutions of Governance in the Territorial North." *Journal of Canadian Studies* 35(4): 80–99.

WMAC/FJMC. 1988. *Inuvialuit Renewable Resource Conservation and Management Plan*. Yellowknife, NWT: Wildlife Management Advisory Council and Fisheries Joint Management Committee.

WMAC. 2000. *Yukon North Slope Long-Term Research and Monitoring Plan*. Prepared by Wildlife Management Advisory Council. *http://www.taiga.net/wmac/researchplan*

NOTES

1 See also letter to the Editor, *Nunatsiaq News*, titled, "Qallunaaq knowledge rears its ugly head" (14 December 2001), for an example of the increasing frustration with the lack of effective integration of different knowledge systems.

2 A cautionary note is required – recognizing the inadequacies of current scientific data sets on ocean resources in the North does not imply a focus on 'positivist' or scientific resource management processes.

3 See also Weitzner (2000) and Weitzner and Manseau (2001) for a discussion of role discovery, identification and evolution in the context of parks management in the North.

CHAPTER 17

CONCLUSION:

INTEGRATION, INNOVATION, AND PARTICIPATION

Rob Huebert (University of Calgary)
Micheline Manseau (Parks Canada and
University of Manitoba)
Alan Diduck (University of Winnipeg)

The main objectives of the book were to learn from the work of diverse local and regional resource management institutions in northern Canada, analyze the underpinning structure, and explore policy options to build capacity to adapt to change. A major focus was on how northern communities can be resilient and sustainable in the face of behavioural and historical conflict, dynamic socio-ecological conditions, and high levels of uncertainty about the future. The North has significant experience to offer the theory and practice of resource and environmental management, particularly with respect to joint problem solving and new institutional arrangements. It is a very dynamic region, presenting a diversity of responses to rapid social, cultural, economic, and environmental changes. Northern Canada has been the home of Aboriginal peoples for a period greatly exceeding the time since its "discovery" by European explorers. Inuit and First Nations have shown an understanding of and appreciation for the delicate ecological balances of northern ecosystems that southern Canadians are only now beginning to fully comprehend. Thus, any effort to understand and address present and future challenges in the Canadian North should strive to ensure full participation of northern peoples. State-centred methods of resource management and public policy making are becoming increasingly obsolete in the North, and thus innovative and bold thought is required.

In addition, the volume addresses an important objective of Canada's new *Oceans Act*, namely developing and implementing integrated management for ocean use in Canada's Arctic Ocean region. All the chapters are relevant to this issue, providing insights from interdisciplinary or multidisciplinary perspectives into stimuli and guidance provided by the Act, and problems and successes related to its implementation. In doing so, however, the papers largely kept the Act in the background as a broad framework encompassing ecosystem management, sustainable development, and the precautionary principle. Consequently, as much consideration (or more) was given to land claims agreements, including the emergence of Nunavut, with its committees for wildlife and its concern for sustainable development. Further, the volume includes analyses of two international concerns (persistent organic pollutants in the Arctic and ownership of Hans Island) typical of the new complex cross-scale challenges facing the Arctic.

The book examines important challenges facing the Canadian North, largely pertaining to how people, organizations, and communities are responding and adapting to profound and diverse changes. It has done so in a manner that brings together different perspectives from a range of professional backgrounds and disciplines, including researchers, managers, policy makers, and resource users. As a complement to the different chapters, the views of three northern communities on community-based monitoring are conveyed through a thirty-minute DVD, "Watching, listening and understanding changes in the environment." The main objectives of the DVD were to (1) learn from experiences with community-based monitoring efforts, (2) apply critical thinking to the phenomena of change and the way in which northern communities understand, respond, and adapt to change, and (3) examine the dynamics of change and explore options to build capacity to adapt to change.

As noted in Chapter 1, the book and DVD emerged from the Ocean Management Research Network, a unique collaborative project focusing on new directions in ocean use management. Although many of the book's topics have been dealt with elsewhere, this volume is unique for specifically addressing Arctic development and resource use in the context of the *Oceans Act*, and for doing so within a framework of adaptive management and resilience. From a networking perspective, of note is that people who had not previously worked together undertook many of the joint efforts reported in this volume. As well, the book includes several examples where local communities were brought into the research in a meaningful way. Further, the book includes reports not only from well-recognized researchers, but also from policy makers, practitioners, and new (and young) scholars. This is a real strength that is demonstrated, for example, in Chapter 6, which dealt with sustainability through the lens of a younger person, and in the various chapters on the experiences of co-management boards.

Given its origins and interdisciplinarity, and the diversity of its authors, the book will be of interest to both academic readers (including undergraduate and graduate audiences) and policy makers. In particular, it should appeal to people in territorial, provincial, and federal governments and to non-governmental and community-based groups. In addition, given the growing interest in comparative experiences of integrated management of marine and coastal regions, the book is relevant internationally because it presents current and careful analysis of Canada's experiments and new mechanisms for co-management as an approach to northern governance. Further, the book could appeal to at least part

Watching, listening and understanding changes in the environment
Community-based monitoring in northern Canada

Aboriginal people have traditionally been keen observers of their environment. Their observations are based on many generations of living and working on the land, on detailed observations of all living and non-living components of the environment, under different conditions and under different seasons. These observations have enabled them to understand and communicate about change. They capture key aspects of the environment, they provide insights into the condition of wildlife species, and they reflect on the connections between themselves, their community and the "Land." It is a way of life.

The DVD takes you across Canada's Arctic regions to three northern communities. In the Yukon, the Arctic Borderlands Ecological Knowledge Co-op is a collaborative alliance of indigenous communities, First Nations, Inuvialuit organizations, co-management boards, government agencies, and university researchers. The US-Canada Arctic borderlands are defined by the range of the internationally known Porcupine Caribou Herd and its near-shore environment. The region is known for its history of gas and oil development proposals. It is also considered a hot spot of recent and projected future climate change scenarios. Randall Tetlichi monitors this environment and shares his findings.

Into the Northwest Territories, Lutsel K'e is a community of three hundred people located on the east arm of Great Slave Lake. It is the most northerly and isolated of the *Denesoline* communities and depends strongly on hunting, fishing, and trapping. In recent years, diamond exploration and development in the traditional territory of the Lutsel K'e Dene has raised concerns about the health of the land, water, and wildlife. As a result, the Lutsel K'e Dene Band began an environmental and socio-cultural research and monitoring program guided by the Wildlife, Lands and Environment Committee, a group of eight local harvesters and elders.

And finally from Nunavut, Qikiqtarjuaq is a community of about five hundred people located on the east coast of Baffin Island. The Inuit have a traditional lifestyle and strongly depend on hunting and fishing for subsistence. Davidee Kooneeliusie, Park Warden in Auyuittuq National Park, has been conducting a polar bear monitoring program for the past twenty-five years, one of the longest time-series of observations in the eastern Arctic. The monitoring protocol is based on very detailed knowledge of the species, its behaviour, and its ecology.

This DVD is the output of many discussions and workshops on community-based monitoring and the generous contributions of many individuals and communities involved in this type of work.

of the growing general audience attracted to stories of current and past Arctic explorers and adventurers, and to media accounts of dramatic environmental change in the North such as ice melt impacts.

A major lesson of the book is that integrated management (IM) can provide both an analytical framework and a means of developing and implementing policy. Furthermore, the use of IM carries certain normative elements when applied to the North. In terms of analysis, it provides a means of identifying and understanding the competing demands on the environment at different spatial and temporal scales, from both a social and biophysical perspective. At the same time, it provides a means of synthesizing different forms of knowledge to define the nature of the problem. In terms of policy, IM allows (or perhaps requires) decision makers to go beyond state-centred policy responses. The involvement of communities is integral to identifying and responding to problems. In addition, the forces of change that are now accelerating many of these problems require management and policy capacities that must respond in a timely fashion. And, of course, there is the requirement that any management and policy response be carried out in the face of severe resource limitations.

From a normative perspective, IM engages resource users, managers and policy makers. Rather than accepting the state-centred approach that characterizes conventional management and policy actions, IM requires government officials and scientists to work in partnership with local communities or key publics in those communities. Further, it postulates that this is the best means of proceeding in a complex, conflictual, and uncertain management environment. A second key normative element of IM is that it is based on the assumption that humans do not "own" the environment, but rather have custodial or stewardship responsibilities to maintain natural resources and environmental services at levels that are adequate for successive generations. Thus, IM assumes a long-term perspective that is based on core sustainability principles.

The contributors of this book have helped us better understand the significance of integrated management. They have reinforced awareness that IM differs greatly from conventional resource management, which too often focuses on single activities and single species at one spatial scale. IM, which in effect strives for a practical holism, emphasizes a systems approach, diverse components of an ecosystem, different resource uses, various impacts, and multiple scales. It is a sophisticated and integrative way of understanding and framing environmental problems and corresponding policy responses, and does so in a manner that incorporates different values and needs. In many ways, IM is highly compatible with northern indigenous perspectives in which land, water and sea are an indivisible, coherent whole – "The Land."

This book was organized into four main sections: (i) Understanding the issues: learning from experience; (ii) Responding and adapting to new challenges; (iii) Resilience and institutions; and, (iv) Governance, policy, and future directions. However, it should be evident to the reader that many of the chapters defy easy classification. Rather, many deal with all four of the subheadings and/or focus on issues that link the main themes. This is due to the complex means by which the authors dealt with their specific topics and to the emergent nature of IM theory and practice. The book's coherence is rooted in its heavy reliance on ideas from resilience thinking and adaptive management, its commitment to systems perspectives (especially regarding the health of people, communities, and ecosystems), and its focus on new institutions established under land claims agreements in the western Arctic and Nunavut. Some of the main cross-cutting issues and themes deal with: (1) the utilization of indigenous knowledge in resource management; (2) new approaches to resource management in the Canadian north; (3) participatory governance and policy.

The Utilization of Indigenous Knowledge in the Canadian North: One of the most significant findings of this book is the importance of indigenous knowledge (including traditional ecological knowledge and Inuit Qaujimajatuqangit) in responding to change (broadly defined – see Chapter 1) and in generating positive change in social-ecological systems. Most of the contributors to this volume agreed that indigenous knowledge often reflects a complex and sophisticated understanding of social-ecological linkages. Moreover, all would likely agree that indigenous knowledge has long included stewardship ideas that respect and protect the resource needs of future generations. In this respect, indigenous northerners have utilized sustainability practices long in advance of the southern discovery of the sustainability concept.

The key issue that many of the chapters raised is how can managers and policy makers best utilize indigenous knowledge? There are no simple answers to this question, but many authors conveyed that in order for knowledge – any knowledge – to be used in decision making, it needs to be communicated, linked to other information, and interpreted validly within any given context. When the information originates from different knowledge systems, the challenge is greater, requiring increased emphasis on communicating, listening, and learning. It is clear from Chapters 6, 8, and 13 that appreciation and comprehension of different types of knowledge can only be achieved by working closely together, fostering collaborative interactions among knowledge holders, managers, and policy makers. Although there has been reluctance in conventional resource management to accord indigenous knowledge a proper place in decision making, several chapters in this book demonstrate how this is changing and how there is an increased understanding, respect, and trust of indigenous knowledge systems. This is the result of significant efforts made by indigenous communities at documenting their knowledge, formalizing local data collection processes, developing internal capacity, and communicating results in a form and format that is readily accessible. Scientists have made similar efforts, spending more time presenting and discussing research methods and results beyond academic and scientific communities. The process of acceptance is by no means complete, but as the chapters noted above have shown, southern-based scientists, managers, and policy makers are increasingly accepting the legitimacy of indigenous knowledge.

Chapters 6, 12, and 13 describe Aboriginal-led resource management cases where both indigenous and Western knowledge were used in research activities, monitoring, and management decisions. Furthermore, the cases showed that community-based approaches go beyond simply being manageable to being desirable. Given the long-standing sensitivity that Inuit and First Nations have to ecosystem health and integrity, it is clear that indigenous perspectives on and applications of some of the main elements of sustainability are very mature relative to those that exist among southern Canadian scientists, managers, and policy makers.

Thus it is readily apparent from this book that the need to incorporate indigenous knowledge into resource management practices exists on several levels. First, it provides an important way of understanding the northern environment that complements or exceeds the capabilities of Western science. Second, the use of traditional knowledge results in additional community capacity, enhanced confidence, and increased control over management decisions. Third, it is clear that indigenous knowledge provides an important means of ensuring that the principles of sustainability are incorporated into decisions pertaining to future activities in the northern environment.

That being said, it is important to remember that there is still much to learn about how to use traditional and conventional ecological knowledge and related management practices. It is more than a matter of merging databases or amalgamating information sources. It requires a true synthesis in a dialectic involving fundamentally different epistemologies and worldviews. Moreover, it entails overcoming an often tragic history rife with profound behavioural, interest, and value conflict. To complicate matters, accomplishing a legitimate and equitable integration is not within the neat domain of instrumental rationality. It is essentially a communicative task in the messy realms of politics and social learning (see Chapters 6, 7, 9, 13 and 15). It is dependent on the collective construction (or negotiation) of meaning in conditions that are free of coercion, and on the capacity for innovative (or double-learning) societal learning.

New Approaches to Resource Management in the Canadian North: Almost every chapter touched on the issue of co-management, shared responsibilities, or increased community participation in resource management in the Canadian North. Overall, an important unifying theme is the strong need to give greater consideration to the interests, knowledge, and values of those with the experience and resilience honed over thousands of years in the harsh environment of the Arctic. The imperative is a collaborative, highly inclusive process for addressing future action and research. For example, Chapter 16 indicates that some of the key issues identified by northerners in relation to oceans sustainability revolve around the long-term ability to secure country food and maintain traditional livelihoods, and this has led to a strong desire for meaningful local participation in resource management and/or greater local authority over natural resources. Various chapters describe management initiatives pertaining to a wide range of resources (including terrestrial, freshwater and marine species), ecosystems (*e.g.*, Beaufort Sea, Quttinirpaaq National Park), and non-renewable resource extraction activities (*e.g.*, mining, gas and oil exploration) that reflect deep local stewardship values and responsibilities. In the context of traditional livelihoods, the level of commitment is particularly strong, since northern Canadians depend on a healthy environment, healthy fish, and healthy animals for much of their diet. As Chapters 2 and 3 conveyed, when these resources become depleted or contaminated, the diet and hence the health of northern peoples immediately suffers.

Many of the chapters clearly illustrate that the use of IM is integral to ensuring that the resources of the North are managed in a manner that will ensure their availability for future generations. In addition, it was made clear that it was necessary to think in terms of systems and not a single type of resource. In effect, what is needed is an understanding of the key social and ecological processes that support the system, and how these processes and relevant structures respond to social, cultural, economic, and environmental changes. One of the key themes that emerges from many of the authors is the recognition that the term "management" in integrated management may in fact be misleading. The idea that resources can be "managed" is one with which some of the authors took issue. It was argued that while there is a need to ensure that resources are utilized in a manner that will ensure their availability for future generations, northern social-ecological systems are so complex and interconnected that it is impossible to "manage" them. The practices and polices that communities and larger political entities develop need to be thought of as means of adapting to both the existing environment and the changes that it faces. Thus, what becomes increasingly important is the need to understand and manage for resilience. Connected to this is the recognition that social and natural systems interact in complex, dynamic, and adaptive ways, and that static, linear, and reductionist views of human-environmental interactions do not provide adequate understanding for responding to uncertainty. Instead, what is required is an understanding of the capacities of systems to self-organize, cope with change, and adapt and learn (Chapters 1, 11, 13, 14). Once these capacities are understood it then becomes possible to develop practices that will ensure that the resources and functions in the system are maintained in a healthy balance.

There are several key implications that flow from these new IM-related management approaches. First, if it is necessary to understand the ability of a system to adapt to change rather than trying to increase the productive capability of the system, policy makers must be more willing to accept their limitations. In effect, it requires that the role of governance in resource management be understood as providing a much more limited capability to shape the system. Thus, policy makers need to be more humble in deciding what they can and cannot do regarding the management of resources in Canadian northern coastal regions. And they must be more prepared to accept and learn from policy and management errors (*i.e.*, when the outcomes of policy or management actions do not match the intended consequences).

A second important point (one that was made earlier but bears repeating) is the absolute need to involve communities and stakeholders in key resource management functions, including planning, decision-making, research, and education. Unfortunately, while most managers and policy makers accept this in principle, several of the authors demonstrate the difficulty in applying this in practice. However, as shown in Chapters 5, 7, 8, 12, and 13, not only is it desirable to ensure community involvement in the management of their resources, it is feasible and highly beneficial. Chapter 5 shows how effectively the Inuvialuit participated in the establishment of a marine protected area. Chapter 13 describes how community-based management can represent (and facilitate) double-loop organizational learning (manifested by fundamental changes in an organization's values and goals).

Still, with regard to involving communities in basic resource management functions, Chapters 2, 3, and 8, as well as the DVD, establish clearly the desire of communities to contribute to monitoring environmental changes; "watching, learning and understanding changes in the environment." Alone or in collaboration with government organizations, communities have participated in the development of meaningful environmental indicators. They have developed their own methodology to gather and communicate the observations, often building on observations of many generations. Chapters 2 and 3 focus on the issue of food security, and both chapters make it clear that there is confusion as to what is safe and what is not safe to eat in terms of country food, *i.e.,* the nature and effects of outside contaminants are not always fully understood. This has two potential impacts. First, it can cause northern peoples to turn away from country food, which often means adopting a diet that is high in fat and low in nutrition. On the other hand, a lack of understanding can also lead to a minimization or denial of real health risks associated with eating some country foods. Thus it is necessary for those on the land to be able to identify and understand the key indicators of environmental stress on country food. In some cases, this requires a close collaboration between scientists and traditional knowledge holders.

A third point that must be acknowledged illustrates a limitation of this book. While the case studies that were examined provide important new insights into the successful application of IM in the North, most of the evidence revolves around small-scale resource utilization. There is a need to examine the impact of a large-scale development that is driven by southern needs rather than by northerners. For example, is the dramatic expansion of the diamond industry in northern Canada being conducted with the principles of integrated management in mind? What of the coming energy extraction developments that are now being prepared for northern oil and gas production? What types of cross-scale linkages can help reconcile community-oriented IM imperatives with drivers rooted in economic globalization? Similarly, will IM be important in the search for the right balance between efficiency and profit making on the one hand and conservation of natural capital on the other? Is harmonization feasible (or even possible) in the face of high-level negotiations led by transnational corporate interests?

Another limitation of the book is that it does not examine in detail the impacts of federal institutional interplay on local situations (although some chapters did encompass aspects of federal-international interactions, *e.g.,* Chapters 11, 13, and

15). Implementation of the *Oceans Act* has been a struggle internally within the Department of Fisheries and Oceans, and interactions among various federal departments in the North have been tricky to say the least. These dynamics compound uncertainty, complexity, and conflict in the management environment, and can have important adverse effects on the implementation and outcomes of community-based models. As such, they deserve dedicated research attention, but were unfortunately beyond the scope of the original research design and are not discussed in this volume.

Governance and Policy: One of the greatest challenges that now face those who live in the North, resource managers, and policy makers centres around governance issues. A major lesson from the book is that Canada and the people of the North have developed new and robust approaches to law, knowledge sharing, and institutions. This is tempered, however, by a lack of capacity of the overall governance system to adapt to rapid social and technological change and to anticipate and prevent foreseeable problems, such as climate change, persistent organic pollutants, and a youth culture that lacks direct connections to the land. Most of the chapters of this book argue that the North is facing increasingly intractable problems requiring difficult and challenging management decisions and policies. It is clear that there will be a need to ensure that the state-centred mode of governing is replaced by newer understandings of governance. Several of the most important points have already been touched on in the preceding sections, but still warrant consideration.

First, it is clear that members of northern communities need to be included in all decisions concerning the North. It is important to stress that this inclusion must be based on meaningful participation (*e.g.,* shared decision making over normative planning issues). All of the authors who have examined this issue have found that when there has been proper involvement, the net result has been better resource management. This will mean that, at times, decisions may not be taken as quickly as possible. But this is a small price to pay.

A second theme that emerges from the new governance is the need to better understand cross-scale linkages and their potential as determinants of sustainable development. Chapters 11 and 14 repeatedly returned to the challenge of managing resources at difference scales – the local, the regional, the national, and the international. Each scale has a different set of institutions, actors, and requirements. In many instances there are inherent difficulties in coordinating action. For example, some of the most significant changes in the Canadian North are being caused by climate change. The impact on the Canadian North requires actions at all levels of governance. However, it is immediately apparent that any effort to coordinate policy is going to be difficult. There is a need for those involved to come to a shared understanding of the problem and the action needed to respond. As Chapter 14 shows, even when addressing a relatively localized resource such as the northern fisheries in Nunavut, this is difficult enough. Thus, when facing a challenge as broad as climate change, shared understandings will be difficult to reach.

The next step is the development of a policy response that will utilize the experience provided by IM practices. That is, we need truly adaptive policy development. But of course this implies a need for sound baseline information and an understanding of how a particular system is responding to different forces and types of change. Thus there is a need to identify the ability of the system to respond to the stresses that are requiring policy action. Monitoring, tight feedback loops, and the capacity for social learning at organizational and community levels are essential in this regard.

In practice this will be difficult. Since the end of the 1980s, fiscal realities have limited the ability of all levels of government to provide resources to support policies. As a result, there are significant limitations to what can be done. This has been particularly clear in terms of Canada's actions in the international context. It is clear that a response to the challenges created by climate change will be expensive and will require significant actions at the international level. However, as Chapter 15 argues, the Canadian government has not shown great willingness to devote significant resources to northern international issues. Chapter 15 focused on the ability of the Canadian government to maintain a surveillance and enforcement capability in the North, but it is easy to understand why the government will continue to employ limited resources. Thus, the need to maximize Canadian resources through cross-scale linkages will remain a critical requirement in the development of future policy initiatives.

As highlighted in the penultimate chapter, issues of resources management will remain current for many generations to come. Confronted with the complexity, uncertainty, and conflict associated with ocean resources and coastal management, solutions to questions asked by northern peoples – "How can we develop our resources responsibly while maintaining our health, well-being, and culture?" and "How can we sustain our waters, lands, and the traditional activities that depend on them?" – will likely emerge in the context of interdisciplinary understanding and collaboration, and in the context of flexibility in the approaches used and solutions adopted. Strategies will continue to evolve as new knowledge is gathered, and as new relationships between governments and stakeholders develop.

ACKNOWLEDGMENTS

The authors wish to thank two anonymous reviewers who provided insightful commentary and critique, and furnished a wealth of ideas that we have tried to incorporate into our concluding comments.

LIST OF AUTHORS

Derek Armitage
Department of Geography and
Environmental Studies
Wilfrid Laurier University
75 University Avenue West
Waterloo ON N2L 3C5
darmitag@wlu.ca

G. Burton Ayles
Canada/Inuvialuit Fisheries Joint
Management Committee
P.O. BOX 2120
Inuvik NT X0E 0T0
aylesb@escape.ca

Nigel Bankes
Faculty of Law
University of Calgary
2500 University Drive NW
Calgary AB T2N 1N4
ndbankes@ucalgary.ca

Fikret Berkes
Natural Resources Institute
University of Manitoba
70 Dysart Road
Winnipeg MB R3T 2N2
berkes@cc.umanitoba.ca

Douglas B. Chiperzak
Fisheries and Oceans Canada
200 Kent Street
Ottawa ON K1A 0E6
Chiperzakd@DFO-mpo.gc.ca

Douglas A. Clark
Department of Geography and
Environmental Studies
Wilfrid Laurier University
75 University Avenue West
Waterloo ON N2L 3C5
clar2207@wlu.ca

Donald Cobb
Fisheries and Oceans Canada
501 University Crescent
Winnipeg MB R3T 2N6
cobbd@DFO-mpo.gc.ca

Kelly J. Cott
Fish Habitat Management
Western Arctic Area
Central and Arctic Region
Fisheries and Oceans Canada
101 – 5204 50th Avenue
Yellowknife NT X1A 1E2
CottK@DFO-mpo.gc.ca

Alan Diduck
Environmental Studies Program
University of Winnipeg
515 Portage Avenue
Winnipeg MB R3B 2E9
a.diduck@uwinnipeg.ca

Gina Elliott
Fisheries and Oceans Canada
Central and Arctic Region
BOX 1871
Inuvik NT X0E 0T0
elliottg@DFO-mpo.gc.ca

Helen Fast
Fisheries and Oceans Canada
Central and Arctic Region
501 University Crescent
Winnipeg MB R3T 2N6
fasth@DFO-mpo.gc.ca

Rob Huebert
Department of Political Science and
Centre for Military and Strategic
Studies
University of Calgary
2500 University Drive NW
Calgary AB T2N 1N4
rhuebert@ucalgary.ca

Brock Junkin
Department of Sustainable
Development
Government of Nunavut
P.O. BAG 002
Rankin Inlet NU X0C 0G0
bjunkin@gov.nu.ca

Mina Kislalioglu Berkes
    Natural Resources Institute
    University of Manitoba
    70 Dysart Road
    Winnipeg MB R3T 2N2
    mberkes@mts.net

Allan H. Kristofferson
    Fisheries and Oceans
    Central and Arctic Region
    501 University Crescent
    Winnipeg MB R3T 2N6
    Kristofa@DFO-mpo.gc.ca

R. Harvey Lemelin
    Department of Recreation and
        Leisure Studies
    University of Waterloo
    Waterloo ON N2L 3G1

Present address:
School of Outdoor Recreation,
    Parks and Tourism
    Lakehead University
    955 Oliver Rd.
    Thunder Bay ON P7B 5E1
    harvey.lemelin@lakeheadu.ca

Melissa Marschke
    Natural Resources Institute
    University of Manitoba
    70 Dysart Road
    Winnipeg MB R3T 2N2
    mjmarschkeca@yahoo.com

Micheline Manseau
    Parks Canada
    145 McDermot Avenue
    Winnipeg MB R3B 0R9
        and
    Natural Resources Institute
    University of Manitoba
    70 Dysart Road
    Winnipeg MB R3T 2N2
    Micheline.Manseau@pc.gc.ca

Heather Myers
    International Studies Program
    University of Northern
        British Columbia
    3333 University Way
    Prince George BC V2N 4Z9
    myers@unbc.ca

Brenda Parlee
    Natural Resources Institute
    University of Manitoba
    70 Dysart Road
    Winnipeg MB R3T 2N2

Present address:
Department of Native Studies /
    Rural Economy
    517 Education Building
    University of Alberta
    Edmonton AB T6G 2H1
    brenda.parlee@ualberta.ca

Michelle Schlag
    Natural Resources Institute
    University of Manitoba
    70 Dysart Road
    Winnipeg MB R3T 2N2

Present address:
Resource Manager 1
    Woodland Caribou Provincial Park
    P.O. 5003
    Red Lake ON P0V 2M0
    michelle.schlag@mnr.gov.on.ca

Shirley Thompson
    Natural Resources Institute
    University of Manitoba
    70 Dysart Road
    Winnipeg MB R3T 2N2
    s_thompson@umanitoba.ca

INDEX

A

Aboriginal knowledge. *See* indigenous knowledge; traditional ecological knowledge; traditional knowledge

Aboriginal land claims. *See* land claims agreements

Aboriginal peoples, 9, 15, 55, 76, 78, 238, 244. *See also* subsistence harvesting
  Athapaskan peoples, 166
  creativity and adaptation, 158
  Cree, 24, 78, 82, 87–88, 165–66, 175
  definition, 3
  Denesǫline, 15, 30, 149, 166–72, 174–77, 179–80
  Dogrib, 78
  government relocation, 47, 60
  holistic approaches, 34, 74–75, 85, 366
  indigenous peoples-to-indigenous peoples agreements, 236
  Inuit, 27, 29, 34, 47–49, 53–54, 58, 83, 85, 87–88, 143, 154, 158, 165, 175, 239, 249, 252, 305, 310, 356–57, 367–68
  Inuvialuit, 12–13, 95, 99, 111, 131
  Lutsel K'e Dene First Nation, 14, 86–87, 143, 147–50, 154, 158, 166–80
  Metis Nation-NWT, 30
  notions of respect, 87, 171, 349
  sensitivity to ecosystem health, 159, 363, 367–68
  social and cultural dislocation, 52
  stewardship, 130–31, 367
  sustainable lifestyles, 48–49 (*See also* subsistence harvesting)
  views of conservation, 155, 159
  wage-based employment and resource development, 25, 28, 36, 51, 123, 249, 253, 345, 356

Aboriginal People's Survey, 56

Aboriginal Pipeline Group, 345

Aboriginal rights
  Calder case, 63
  Under Canadian Constitution, 148, 160, 203
  conflicts with federal fishery and wildlife rules, 305
  repositioned in hierarchy, 300
  resource conflicts, 12

Aboriginal worldviews, 82, 85

adaptation, 226, 233, 239
  capacity for, 42, 45, 311
  creativity and, 158
  indigenous coping mechanisms, 55
  sharing hunting opportunities, 236

adaptive co-management, 6, 16, 228–29, 250, 265

adaptive management, 6–9, 226, 228, 244, 366
  dynamic learning, 250
  phases of, 265
  rediscovery of traditional systems of knowledge, 263
  and resilience, 364

advocacy, 120
AEPS. *See Arctic Environmental Protection Strategy*
Aiviit HTO, 206–7, 210
Aklavik, 85, 99, 105, 126–27
  population, 122
  wage employment, 123
Aklavik Elders Committee, 229
Aklavik HTC, 229
Alagalak, David, 35
Alaska Beluga Whale Committee, 143, 153, 156
Alaska/Yukon border, 321
Alert, 151
Allan, Dwight, 194
AMAP. *See* Arctic Monitoring and Assessment Programme
anthropogenic activities, 175–76
  in Wapusk National Park (WNP), 194
anti-whaling, anti-sealing, and anti-trapping movements, 62, 123
aquaculture, 8
Arctic Bay, 39–40, 273
Arctic Borderlands Ecological Co-op, 89
Arctic Borderlands Knowledge Coop, 175
Arctic char, 39, 51, 59, 144, 249–66
  biological complexity, 250–51
  circumpolar distribution, 251
  conventional *vs.* traditional fishery management, 251
  mixed-stock, 258
  tagging program, 260–61
Arctic Climate Impact Assessment, 2
Arctic Council, 2, 13, 237–38, 241
Arctic Environmental Protection Strategy (AEPS), 237–38
Arctic Foods Limited (AFL), 209
Arctic Islands, 95
Arctic Monitoring and Assessment Programme (AMAP), 237, 350
Arctic Waters Pollution Prevention Act, 324, 331, 333
Arctic wide POPS case, 240
"areas of open water," 172
Arial Unmanned Aircraft, 329
arsenic, 57
Artillery Lake area, 167, 172, 175
Artillery Lake caribou crossing
  as community indicator, 178
arts and crafts economy, 36, 38
  carving and printmaking, 64–65
Athapaskan peoples, 166
Atlantic Fishery Regulations, 304
Attawapiskat, 24

Aurora College, 14, 123
Aurora Research Institute, 123, 344
Auyuttuq National Park, 151

B

Baffin Bay narwhal population, 276
Baffin communities, 32
  mercury, 58
  PCBs, 58
Baffin Island region, 233
Baffin Marine Issues Scan, 345, 352
Baffin Region fisheries, 38
Baker Lake Inuit ququjimajatuqangit grizzly bear study, 349
balance of power, 141–42, 155–56, 235
Banks Island, 121
baseline information, 350, 352, 356
Bathurst and Beverly caribou movement, 168
Bathurst Mandate, 36
bearded seals, 95
Bears (documentary), 195
Beaufort Sea, 13, 51, 59, 95, 113, 121, 146
  oil and gas, 96, 99
Beaufort Sea 2000 Conference, 2
  exchange between scientific and traditional knowledge, 243
Beaufort Sea Beluga Management Plan, 77, 101, 105, 109, 145, 349
  adaptive co-management approach, 228
  aerial surveys, 146
  consultation strategies, 115
  management zones, 98–99
  voluntary compliance, 98, 101
Beaufort Sea Integrated Management Planning Initiative (BSIMPI), 77, 102, 108, 110–14, 133
  youth and, 135
Beaufort Sea Integrated Management Planning Initiative (BSIMPI) Working Group, 102, 105, 109–10, 115, 119, 123
Beaver Dam (sacred site), 172
beluga, 32, 48, 58, 95, 144, 301. *See also* Beaufort Sea Beluga Management Plan
  aerial surveys, 146
  in area of oil and gas development, 101
  mercury levels, 59
  muktuk, 51
  stocks, 153, 308
beluga harvest, 13, 99
  quotas, 146, 303
  traditional values related to, 98, 145
beluga monitoring program, 145, 156
  conflicts, 146

diabetes, 28, 51
diachronic indicators, 173–74
diamond mining, 148–49
    effects of, 172
    environmental assessment, 154
DIAND (Department of Indian and Northern Affairs), 97
Diavik Diamond Mine, 148
dieldrins, 32
documentation of traditional knowledge, 158, 167
dog teams, 26, 205
Dogrib, 78
Dolly Varden char, 101
double-loop learning, 282, 284–85, 287, 370
    dominant management worldview and, 286
    key processes, 271–72

E
"The earth is faster now," 2, 48
E.coli bacteria, 54
ecological footprint, 48, 64
ecological indicators, 165–66, 168, 173
ecology. See also environment
    placing humans inside, 142, 159
economic development. See commercial/economic opportunities; development
ecosystem, 3, 8, 173
    definition, 2
ecosystem-based management, 3, 71–73, 77, 85
    broadened definition, 74
    Innu forest management plan, 154
ecosystem health, 11
ecosystem management, 363–64
ecosystem monitoring studies, 80
'ecosystem people,' 49
ecotourism, 13, 15
Ecotourism and Development Working Group, 13
eda cho "big caribou crossing," 172
edo aja "something has happened to it," 176
education, 65
    benefits of staying in school, 130, 132, 136
    contaminants information in school curricula, 30, 65
    levels, 54
    mandatory schooling, 55
    Oceans 11 Arctic marine science curriculum, 127–28, 134–35
    post-secondary education, 136
    of resource users and harvesters, 355

school/traditional knowledge conflicts, 129–30, 132
    standards, 130, 134
eider ducks, 51, 59, 89
Ekalluk River, 255
elders, 124–25, 131, 149, 156, 172, 176
    elder-scientist retreats, 30
    transmission of TEK, 135
elders' committees, 108, 148, 167
Ellesmere Island, 319–20, 322, 324
employment, 53, 340, 345–47. See also wage economy
    CHCDC, 217
    employment, 36, 51, 249
    job creation, 209, 213
    opportunities, 100
    unemployment, 56
endosulf, 32
environment, 13, 48. See also ecology; ecosystem
    competing demands on, 1
    quality of land and water, 40, 90, 168, 171
    source of livelihood and basis of culture, 75, 78
environmental and human health links, 165, 177, 340, 345, 347–48
    Denesoline conception of, 173
environmental change, 23–24, 48–49, 75, 81, 85
    effects on country food, 28
    effects on food security, 47
    significance of, 349
environmental indicators, 74, 90, 370
Environmental Monitoring Advisory Board, 148
enzyme electrophoresis, 256
epistemic communities, 240–42, 284
Eureka, 151
Eurocentric Canadian policies. See government policies
European Union (EU), 206, 214, 239
Exclusive Economic Zone (EEZ), 11, 321, 328
experts and expertise, 31, 108, 242. See also scientific knowledge
    expert knowledge, 243
    expert management, 9

F
Federal-Provincial Polar Bear Administrative Committee (PBAC), 234, 277
Federal-Provincial Technical Committee (PBTC), 234, 277

support for TEK, 154
Inuvialuit Game Council (IGC), 97, 101, 143, 235
Inuvialuit-Inupiat Polar Bear Management Agreement, 235–36, 242
Inuvialuit Regional Corporation (IRC), 97, 101, 127, 133, 136
Inuvialuit Renewable Resource Conservation and Management Plan (IRRCMP), 97
Inuvialuit Settlement Region, 15–16, 32, 35, 78, 95, 97, 102, 142
  marine stewardship, 120–36
  use of TEK, 144, 146
Inuvik, 99, 124, 126–27
  population, 122
  wage employment, 123
invasive species, 11
IPEN (International POPS Elimination Network), 239
Iqaluit, 53
iron, 52
ivory, 272, 301–2

J

James Bay, 28, 82
James Bay and Northern Quebec Agreement, 10, 259
James Bay Cree fishery, 252–53
Jayco River, 255
Jensen, Alex, 326
Jolly, D., 2

K

kahdele (areas of open water), 172
Keewatin Meat and Fish Plant, 211, 213
Kendall Island, 99
Kennady Lake (*Gahcho Kue*), 149
"key players"
  cross-scale linkages, 238, 241–42
kinship principles, 49
Kitikmeot Foods Ltd., 255
Kitikmeot region, 32
Kivalliq Arctic Foods, 207, 214
Kivalliq communities, 32, 35, 54
  mercury, 58
knowledge
  local, 10, 14, 225, 228, 231
knowledge networking
  Indigenous peoples, 110, 175, 229, 239
knowledge systems, 89, 141, 156–57, 340, 345, 350–52, 367. *See also* indigenous knowledge; scientific knowledge; traditional ecological knowledge (TEK);

traditional knowledge
  combining, 7, 14, 17, 78, 89, 227, 232
  inherent differences, 159
  knowing limitations of, 158
  people conversant in both, 157, 160
  respect for, 156–57, 243
Knuvik, 105
Kugaaruk, 273
Kugmallit Bay, 99–100

L

La Grande hydro development, 88
La Grande River, 82
Labrador, 33, 41, 340
  Innu Nation forest management, 143, 154, 158
Labrador tea, 168
Lac de Gras, 148
Lake Hazen, 151
lake herring, 168
lake trout, 59, 168
lake whitefish, 168
Lancaster Sound, 279
land, 366
  cultural land use data, 154
  going out on the land, 127, 148, 151, 345
  meaning in Indigenous languages, 75, 173
  protecting "for use rather than from use," 203
Land Administration of the Kivalliq Association, 14
land-based economy, 15, 24, 229. *See also* traditional economy
land-based knowledge
  erosion of, 26, 166
land-based peoples, 173
land claims agreements, 16, 31, 63, 72, 77, 122, 244, 273, 344, 364
  adaptability, 293
  co-management under, 227, 250, 259, 261–62
  comprehensive claims agreements, 353
  context of, 10–11
  cross-scale management through, 227
  facilitating social learning, 308
  Inuvialuit Final Agreement (IFA), 10–11, 63, 95, 97, 111, 122, 143, 146, 154, 225, 227–28, 243, 259
  James Bay and Northern Quebec Agreement, 10, 259
  Nunavut Land Claims Agreement (NLCA), 11, 16, 36, 63, 87, 143, 151, 154, 204, 231, 241, 243, 250, 259, 261–62, 269, 294, 303, 305–11

William Barr, general editor · Copublished with The Arctic Institute of North America ISSN 1701-0004

NORTHERN LIGHTS SERIES

University of Calgary Press and the Arctic Institute of North America are pleased to be the publishers of the Northern Lights series. This series takes up the geographical region of the North (circumpolar regions within the zone of discontinuous permafrost) and publishes works from all areas of northern scholarship, including natural sciences, social sciences, earth sciences, and the humanities.